Mineral nutrition of tropical plants

Renato de Mello Prado

Mineral nutrition of tropical plants

Renato de Mello Prado
São Paulo State University
Jaboticabal, São Paulo, Brazil

Translated by Priscila Porchat with the support of the grant 20/05310-1 from the São Paulo Research Foundation (Fundação de Amparo à Pesquisa do Estado de São Paulo – FAPESP). The opinions, hypotheses and conclusions expressed in this material are the authors' responsibility and don't necessarily reflect FAPESP's vision.

ISBN 978-3-030-71264-8   ISBN 978-3-030-71262-4   (eBook)
https://doi.org/10.1007/978-3-030-71262-4

This Springer imprint is published by the registered company Springer Nature Switzerland AG
The registered company address is: Gewerbestrasse 11, 6330 Cham, Switzerland

Foreword

Humans have been living with plants since they first appeared on earth. Plants enable animal life on the planet by providing most necessities from the oxygen we breathe to the food we eat. We have learned to grow vegetables for food; therefore, through agricultural science and technology, we have become even more dependent on plants.

Plant nutrition is still a subject little known to the majority of the population living in cities, which in Brazil is over 80% today. The following example illustrates this: when asked what my profession is and I answer that my expertise in the agronomic area is plant nutrition, most interlocutors look at me astonished and seem to think: what is plant nutrition? When I explain that human nutrition and plant nutrition are closely linked, that is, plants need to feed themselves so that humans and other animals can nourish themselves, it seems that people get a little more interested. This relationship needs to be better known to give due importance to the foods and plant nutrition techniques that are necessary to produce them.

The study of plant nutrition is interesting because plants need to be fed to manifest their genetic potential and produce the food that the growing population needs. Until the sixteenth century, when the human population was less than 1 billion, little was known about the composition of vegetables. With the scientific revolution, scholars began to discover the composition of plants. At the beginning of the nineteenth century, Earth's population reached 1 billion, and most of the chemical elements that make up plants were already known, as well as the fact that removing them from the soil caused its impoverishment. It was then proposed to return them from where they were being extracted to maintain the soil fertility (i.e., the production capacity), and then there was a great advance in food production. The sources of nutrients, mainly nitrogen and phosphorus, were limited and insufficient to supply the soil with what the crops extracted from it. The great leap in supplying crops with enough nitrogen for them to manifest their full genetic potential occurred in the first quarter of the twentieth century with the industrial fixation of nitrogen in the atmosphere, when the world population was just under 2 billion inhabitants. A virtuous circle was established as the most properly fed plantations produced more food, leading to a surge in population, which in turn created a demand for more food, requiring an increase in agricultural production. Thus, with the advance in agronomic techniques, it became possible to produce food in abundance to satisfy the growing population and allowing it to cross the 6 billion mark at the end of the twentieth century and more than 7 billion today. My conviction is that the projected population for 2050, of almost 10 billion inhabitants, will only be attained if there is an increase in the availability of food from an increase in the productivity of crops supplied with adequate nutrition.

The book *Mineral Nutrition of Tropical Plants* joins other literature on the subject, but it has the specificity of dealing mainly with knowledge related to the nutrition of tropical crops. This book will provide readers with knowledge and, above all, value and allow them to marvel at plant life and this field of human knowledge, giving plants and what they represent the deserved recognition.

The book is didactically divided into chapters that facilitate reading and learning for those interested in the subject. In Chap. 1, as an introduction, basic concepts about plant mineral nutrition are given, such as the essentiality criteria, the relative composition of nutrients and their accumulation during plant life, and other elements of interest in plant nutrition, and the hydroponic cultivation technique is explained. Chapters 2 and 3 are the foundations on which the subsequent chapters are based and deal with root and leaf absorption, the transport and distribution of nutrients, and the factors that influence these processes as well as some aspects of foliar fertilization. Thus, following a common scheme, Chaps. 4 to 9 address macronutrients (N, S, P, K, Ca, and Mg) and Chaps. 10 to 17 address micronutrients (B, Zn, Mn, Fe, Cu, Mo, Cl, and Ni), with each chapter beginning with an introduction, where some peculiarities and basic knowledge related to each of the essential elements are presented. Chapter 18 discusses some chemical elements, called heavy metals, which can be potentially toxic to plants. Chapter 19 discusses the various criteria that can be used to assess the nutritional status of crops, starting with the presentation of visual symptoms and concluding with the results obtained in laboratory analyses supported by the criteria for their interpretation. Chapter 20, last but not least, is dedicated to understanding the most common interactions that can occur between nutrients. Plant nutrition is of a very complex nature, because in reality there are interactions between nutrients that can reflect on the mineral composition of the crops and, therefore, on the nutritional status and consequently on the production and the final quality of the agricultural product. At the end of the book, the author adds some final remarks, followed by the vast literature consulted to write this work.

My first meeting with the author of this book, Prof. Renato de Mello Prado, together with his students, occurred when he visited the Laboratory for Plant Mineral Nutrition "Professor Eurípedes Malavolta" at the Nuclear Energy Center in Agriculture, University of São Paulo (USP), where I worked at that time. He had recently been hired as a professor in the School for Agricultural and Veterinary Sciences at São Paulo State University (UNESP) to teach "plant nutrition." Since then, we have met with some frequency due to common scientific and didactic interests. After reading the book, I am convinced that this work is the culmination of almost 20 years of research and mainly teaching in the area of plant nutrition as experienced by Prof. de Mello Prado. To confirm this statement, one can just go to the bibliographic references of this work and see that much of what is included in the book is the result of research by the author and his collaborators, in addition to the nearly 400 works published in indexed journals. Another aspect that I am pleased to comment on Prof. de Mello Prado is about his teaching activities, because besides teaching plant nutrition, he has already supervised more than a hundred students from different university levels and he leads a group of studies on plant nutrition at UNESP (Genplant) that periodically holds symposia, which are attended by many professionals and students to which national and international collaborators who update their knowledge of plant mineral nutrition are invited. The author is also proficient in literary production in the agronomic field.

The book *Mineral Nutrition of Tropical Plants* is recommended for everyone who is interested in knowing how plants feed so that food can reach our tables, thus marveling and revering these sessile green beings on whom animal life on Earth depends. Anyone who reflects on the importance of plants for human beings with regard to food production realizes that this fact is a great miracle that has been happening every day since the beginning of human civilization. Miracle, in the sense that I wish to express, is not that extraordinary event which cannot be explained by natural laws, but in the sense that it is something so great that it stirs our admiration and should to be seen with attentive eyes and admired with the heart. The fact is so great that a favorable environment for peace between peoples is only possible through the availability of food in quantity and quality compatible with its demand.

Thus, making a reflection, I conclude this foreword, which was honorably requested of me by my friend Prof. de Mello Prado, stating that anyone who works in agriculture is a "producer of life, contained in food." Convinced that the food produced is a gift from nature but also dependent on the effort, knowledge, and tiring work of those who dedicate themselves to producing it – a feeling that I am sure is the same as the author's – we wish for everyone who performs such an important service to humanity:

- Strength to take care of the soil formed on Mother Earth and respect it as a living entity
- Consciousness to conserve soil life and improve it through the respectful work of man
- Wisdom to sow in the soil and produce healthy food
- Joy in sharing the food produced so that everyone eats what is good and tasty and that no one dies of hunger
- Certainty that by producing food and preserving the living nature, everyone collaborates with the production of life contained in food
- Courage to render the honorable service to humanity by producing food for all, thus promoting peace between peoples
- Hope that when producing food, it will arrive on all tables, making it possible to make a better world, where the feeling of brotherly love for all forms of life will be watered by the sweat of nature's collaborators

Finally, I hope that this book will inspire many young people to work with great commitment to this field of knowledge so important for humanity, that is, plant nutrition, and face challenges with joy and a sense of accomplishment.

Piracicaba, São Paulo, Brazil   Antônio Enedi Boaretto

Acknowledgment

We would like to thank the support of the São Paulo State Research Support Foundation (FAPESP) for helping with the publication and research carried out. FAPESP (fapesp.br/en) is a public foundation, funded by the taxpayer in the State of São Paulo, with the mission to support research projects in higher education and research institutions in all fields of knowledge.

Contents

Chapter 1
Introduction to Plant Nutrition

Keywords Criteria essentiality · Nutrient accumulation · Nutritional requirement · Nutritional efficiency · Hydroponic cultivation · Silicon

The introduction to plant nutrition addresses basic and general topics on the importance of this area to meet nutritional requirements and promote crop growth, development, and yield. We will address important topics, such as (1) concepts of plant nutrition and its relationship with related disciplines; (2) the concept of nutrient and criteria of essentiality; (3) relative composition of nutrients in plants; (4) nutrient accumulation by crops and crop formation; (5) other chemical elements of interest in plant nutrition, such as potentially toxic and beneficial elements, with emphasis on silicon; and (6) hydroponic cultivation, preparation, and use of nutritional solutions.

1.1 Concepts of Plant Nutrition and Its Relationship with Related Disciplines

Current knowledge of the concept of plant nutrition is historically recent. We will briefly present some occurrences throughout the history of plant nutrition. The first inference on some aspects of plant mineral nutrition started in antiquity, when Aristotle (384–322 BC), the Greek philosopher and biologist, already made statements about how plants fed. At that time, he indicated that plants were like inverted animals, with their mouths on the ground. For him, food would be previously digested by the soil, as vegetables did not show visible excretions, as animals do. Investigations on the ways through which plants fed continued, and new discoveries were found at each stage of human history.

In the nineteenth century, the Swiss researcher De Saussure (1804) made an important publication establishing that the plant obtained C from CO_2 of the atmosphere, with hydrogen and oxygen being absorbed along with carbon. This publication also established that plant dry matter increased mainly due to the absorbed C, H, and O, and that the soil supplied minerals indispensable to plant life. In that same century, the chemist Justus von Liebig (1803–1873), "father of plant mineral nutrition," established in Germany that foods of all green plants are inorganic or mineral substances. This study was presented at an event of the British Association for the Advancement of Science and resulted, in 1840, in the publication of the book Organic Chemistry in its Application to Agriculture and Physiology. Liebig, with his dominant vigor, convinced the scientific community of the time with his theory, although being a compilation of studies by other authors (De Saussure, Sprengel, and others) (Browne 1942). Sprengel (Professor of Agronomy) published a study in 1826 recognizing 20 elements as nutrients, with macronutrients among them.

According to Epstein (1975), Liebig's main contribution to plant nutrition was to end the humus theory that believed soil organic matter was the plant carbon source. According to Liebig's theory, plants lived on carbonic acid, ammonia (azotic acid), water, phosphoric acid, sulfuric acid, silicic acid, magnesian lime, caustic potash, and iron. Thus, throughout the end of the nineteenth century, the classic list of plant nutrients consisted basically of N, P, S, K, Ca, Mg, and Fe, defining the requirement of plants, especially regarding macronutrients and iron. At that time, Liebig contributed to the rise of fertilizer industries. The twentieth century saw the establishment of the concept of micronutrients, nutrients equally essential, but required in smaller quantities by plants. A new era for plant nutrition began in the 1930s–1950s with Hoagland, who determined the ideal nutrient solution for plant growth. Afterward, modern scientists wrote classic books on plant nutrition, such as Epstein (1972), Mengel and Kirkby (1987), Marschner (1986), and Malavolta (1980) in Brazil. The establishment of the first teaching and research institutions in Brazil took place in the late 1800s (Federal University of Bahia (UFBA) in 1877, the Agronomic Institute of Campinas (IAC) in 1887, and the Luiz de Queiroz College of Agriculture (ESALQ) in 1901), establishing the basis for plant nutrition studies, which began in the 1950s.

Although plant nutrition is a new science, with only 180 years old, it advanced extraordinarily from the demystification of the humus theory, in 1840, to recent discoveries regarding nutrient absorption through identification of genes that encode proteins (carriers).

Thus, the study of plant nutrition establishes what are the essential elements for the plant life cycle, how they are absorbed, transported, and accumulated, and how the plant redistributes its functions, requirements, and disturbances when nutrients are in deficient or excessive quantities.

It is clear that plant nutrition has links from nutrient acquisition by the roots, related to soil science, to the functions nutrients perform in plants, and to aspects studied by plant biochemistry and physiology. More broadly, plant nutrition and agronomy are closely related, as it is known that the main objectives of agronomic science are the production of food, fibers, and energy. For this, there are more than 50 factors of production that must be considered for maximum efficiency in agricultural production systems. These factors of production are arranged in three major systems, namely: soil, plant, and environment. The plant nutrition area centers on the plant system, like other areas (plant physiology, molecular biology, plant breeding, and phytotechnics). The areas of soil fertility, fertilizers/correctives, and fertilization, among others, are centered on the soil, while irrigation, drainage, and climatology are centered on the environment. Most factors of production can be controlled in the field by the producer. However, some are difficult to control, such as light and temperature.

Environmental factors are highlighted with climate change, especially due to increasing air temperature, CO_2, and water irregularity (flood and drought), affecting plant nutrition and production (Viciedo et al. 2019a, b; Barreto et al. 2020; Carvalho et al. 2020a), and forage quality (Habermann et al. 2019). However, increased CO_2 and temperature can increase nutritional efficiency and plant growth in certain species, such as Stylosanthes capitata Vogel (Carvalho et al. 2020b). In this species, warming combined with well-watered conditions increased leaf biomass production by 38%, presumably due to a higher level of stoichiometric homeostasis (Viciedo et al. 2021).

Plant nutrition is closely related with agronomy, specifically with the disciplines of soil fertility, fertilizers/correctives, and crop fertilization. Fertilization = $(QP - QS)$. factor f; where QP = nutrient amount required by the plant (nutritional requirement); QS = nutrient amount in the soil; and f = fertilizer efficiency factor; which can be reduced by losses in the soil (volatilization, adsorption, leaching, erosion, etc.). In a cultivation system with soil tillage, efficiency factors of 50%, 30%, and 70% are admitted for N, P, and K, respectively, corresponding to the f values equal to 0.50, 0.30, and 0.70, respectively. The use of twice as much N; 3.3 times more P; and 1.4 times more K in fertilization ensures adequate plant nutrition.

In addition, nutrient cycling increases only in a consolidated no-tillage system with live roots throughout the crop year. This improves efficiency factors for all nutrients and, in the case of phosphorus, may double its efficiency.

There are other areas related to plant nutrition, such as microbiology, plant breeding, and molecular biology, among others.

1.2   Concept of Nutrient and Criteria of Essentiality

There are many chemical elements in nature without considering isotopes, as stated in the periodic table, with more than a hundred chemical elements. This number may increase with new discoveries by science, which can occur even by synthesis in laboratory.

However, when plant tissue is chemically analyzed, it is common to find approximately 50 chemical elements, and not all of them are considered plant nutrients. Plants have the ability to absorb chemical elements in the soil or nutrient solution with little restriction, which could be a nutrient, nonmineral nutrients from the atmosphere, or a beneficial and/or toxic element (Kathpalia and Bhatla, 2018, Fig. 1.1).

Considerations regarding the beneficial and/or toxic element are addressed in the next item.

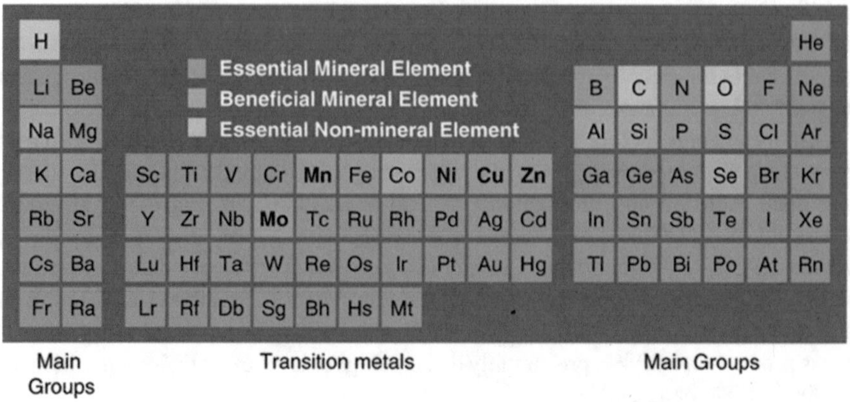

H																	He
Li	Be	■ Essential Mineral Element									B	C	N	O	F	Ne
Na	Mg	■ Beneficial Mineral Element									Al	Si	P	S	Cl	Ar
		■ Essential Non-mineral Element														
K	Ca	Sc	Ti	V	Cr	Mn	Fe	Co	Ni	Cu	Zn	Ga	Ge	As	Se	Br	Kr
Rb	Sr	Y	Zr	Nb	Mo	Tc	Ru	Rh	Pd	Ag	Cd	In	Sn	Sb	Te	I	Xe
Cs	Ba	Lu	Hf	Ta	W	Re	Os	Ir	Pt	Au	Hg	Tl	Pb	Bi	Po	At	Rn
Fr	Ra	Lr	Rf	Db	Sg	Bh	Hs	Mt									

Main Groups   Transition metals   Main Groups

Fig. 1.1 Distribution of essential and beneficial mineral elements. All, except molybdenum, are among the 30 lightest elements. Elements in bold letters are hyperaccumulators in plants

The nutrient is defined as a chemical element essential to plants, that is, without it the plant cannot complete its life cycle. For a chemical element to be considered a nutrient, it is necessary to meet the two criteria of essentiality, direct and indirect, or both. The criteria were proposed by Arnon and Stout (1939), physiologists at the University of California, thanks to the advancement of science regarding analytical chemistry, allowing the determination of trace elements, and to the advancement of cultivation techniques in nutritive solution. The essentiality criteria are described below:

Direct:

- The element participates in some compound or reaction, without which the plant does not live.

Indirect:

- In the absence of the element, the plant cannot complete its life cycle.
- The element cannot be replaced by any other element.
- The element, with its presence in the medium, must directly affect plant life and not only play the role of neutralizing physical, chemical, or biological effects unfavorable to the plant.

Epstein and Bloom (2006) proposed an adaptation of the criteria of essentiality, that is, an element is essential if it meets one or both of the following criteria:

- The element is part of a molecule that is an intrinsic component of the plant structure or metabolism.
- The plant is so severely deprived of the element that it causes abnormalities in growth, development, or reproduction – that is, its performance – compared to plants without its deprivation.

The world literature considers 17 chemical elements as plant nutrients, namely: C, H, O, N, P, K, Ca, Mg, S, Fe, Mn, Zn, Cu, B, Cl, Mo, and Ni. Nutrients are important for life as they play significant roles in metabolism, either as substrate (organic compound) or as enzymatic systems. Briefly, such functions can be classified as follows (Malavolta et al. 1997):

- Structural (part of the structure of any vital organic compound for the plant).
- Constituents of enzymes (part of a specific structure, prosthetic group, or active site of enzymes).
- Enzyme activator (not part of the structure). It should be noted that the nutrient not only activates but also inhibits enzymatic systems, affecting the speed of many reactions in plant metabolism.

Epstein and Bloom (2006) proposed another classification of nutrients with more detail, organized by their functions in plants, as follows:

- Nutrients that are integral elements of organic compounds, for example, N and S.
- Nutrient for acquisition and use of energy and for the genome, for example, P.
- Nutrients structurally associated with the cell wall, for example, Ca and B (Si).
- Nutrients that are integral compounds of enzymes and other essential entities of metabolism, for example, Mg, Fe, Mn, Zn, Cu, Mo, and Ni.
- Nutrients that activate or control enzyme activity, for example, K, Cl, Mg, Ca, Mn, Fe, Zn, and Cu (Na).
- Nonspecific functions: nutrients that serve as counterions, for positive or negative charges, for example, K^+, NO_3^-, Cl^-, SO_4^{-2}, Ca^{+2}, Mg^{+2}, and (Na$^+$).
- Nonspecific functions: nutrients that serve as osmotic agents of cells, for example, K^+, NO_3^-, Cl^-, and (Na$^+$).

The next chapters discuss the role of each nutrient after reaching the location where they exercise their various functions, the cell (cell walls, cytoplasm, and vacuole).

When a nutrient performs its function in the plant, that is, the integration of biochemical functions, it affects one or several physiological processes (photosynthesis, respiration, etc.), affecting crop growth and production.

Photosynthesis is the most important physicochemical reaction on the planet, as all life forms depend on it. Organic compounds are synthesized with light (visible from 400 to 740 nm) by photosynthetic pigments in plants (chlorophylls, carotenoids, and phycobilins). However, a small fraction of solar radiation (~5%) reaching the Earth is converted into organic compounds by leaf photosynthesis.

In summary, the physicochemical reaction of photosynthesis occurs in two phases. During the photochemical or light-dependent phase, sunlight is used to break down the water molecule (H_2O) into oxygen (O_2) – conversion of light energy into electrical energy – which generates chemical energy, with adenosine triphosphate (ATP) and nicotinamide adenine dinucleotide phosphate (NADPH) as primary products. Thus, light energy is captured to allow electron transfer by a series of compounds that act as electron donors and receivers. Photolysis of the water molecule and electron transport allow the creation of a proton gradient between the thylakoid lumen and the chloroplast stroma. Ultimately, most electrons reduce NADP⁺ to NADPH. Light energy also generates proton motive force across the thylakoid membrane, which synthesizes ATP through the ATP synthase complex. The light-independent phase or photosynthetic carbon reduction cycle is basically an enzymatic phase, where light is not required, and the primary products from the previous phase are used to obtain carbon hydrates $(Cn(H_2O)n)$, such as glucose, from carbon dioxide (CO_2). The free energy to reduce one mole of CO_2 to glucose level is 478 kJ mol⁻¹.

The photosynthetic process occurs inside chloroplasts (Fig. 1.2), plastids located in cells of palisade and spongy mesophylls. The number of chloroplasts per cell ranges from one to more than one hundred, depending on the plant type and growing conditions. Chloroplasts are discoid shaped with diameter from 5 to 10 microns, limited by a double membrane (outer and inner). The inner membrane acts as a barrier controlling the flow of organic molecules and ions inside and outside the chloroplast. Small molecules, such as CO_2, O_2, and H_2O, pass freely through chloroplast membranes. Internally, the chloroplast consists of a complex system of thylakoid membranes containing most of the proteins required for the photochemical phase of photosynthesis. The proteins required for CO_2 fixation and reduction are located in a colorless matrix called stroma. Thylakoid membranes form thylakoids, flat vesicles with an aqueous inner space called lumen. Thylakoids, in certain regions, are arranged in stacks called granum. Thus, the first phase of photosynthesis occurs in the chloroplast inner membranes, the thylakoids, while the second phase occurs in the chloroplast stroma, the aqueous region surrounding the thylakoids. Therefore, the products formed in photosynthesis, carbon sources, are accumulated as sucrose

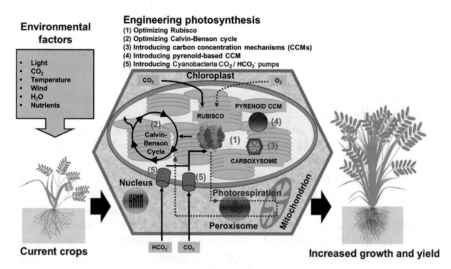

Fig. 1.2 Scheme of the photosynthetic process responsible for crop growth, development, and production

in vacuoles and starch in chloroplasts, in order to be subsequently used in photosynthesis, such as in respiration and synthesis of reserves and structural materials.

The photosynthetic process depends on some nutrients acting with structural or enzymatic function. Moreover, the products formed by photosynthesis also depend on nutrients to produce other vital organic compounds for plant development and production, which are detailed in the next chapters.

Thus, we perceive the importance of nutrients for plant life. According to the history of plant nutrition, these chemical elements, which are currently considered plant nutrients, were discovered and had their essentiality demonstrated from 1804 until recently, with the last inclusion in 1987 (Table 1.1).

There are disagreements on the author who demonstrated the essentiality of some nutrients, often due to problems in the scientific rigor of the research, for example boron, whose authorship of essentiality is attributed to Warington (1923).

Although these nutrients are equally important for plant production, they are classified based on the proportion in which they are found in the dry matter of vegetables. Therefore, there are two major groups of plant nutrients (not considering C, H, and O), namely:

- Macronutrients – Nutrients absorbed or required by plants in larger quantities: N, P, K, Ca, Mg, and S (expressed in g kg^{-1} of dry matter). Macronutrients can be further divided into primary macronutrients (N, P, and K) and secondary macronutrients (Ca, Mg, and S).
- Micronutrients – Nutrients absorbed or required by plants in smaller quantities: Fe, Mn, Zn, Cu, B, Cl, Mo, and Ni (expressed in mg kg^{-1} of dry matter).

Table 1.1 Discovery and demonstration of essentiality of plant nutrients

Nutrients	Atomic mass	Discoverer	Year	Demonstration	Year
C	12.01	–	–	De Saussure	1804
H	1.01	Cavendish	1774	De Saussure	1804
O	16.00	Priestley	1774	De Saussure	1804
N	14.01	Rutherford	1772	De Saussure	1804
P	30.98	Brand	1772	Ville	1860
S	32.07	–	–	Von Sachs, Knop	1865
K	39.10	Davy	1807	Von Sachs, Knop	1860
Ca	40.08	Davy	1807	Von Sachs, Knop	1860
Mg	24.32	Davy	1808	Von Sachs, Knop	1860
Fe	55.85	–	–	Von Sachs, Knop	1860
Mn	54.94	Scheele	1744	McHargue	1922
Cu	63.54	–	–	Sommer	1931
Zn	65.38	–	–	Sommer & Lipman	1926
B	10.82	Gay Lussac and Thenard	1808	Sommer & Lipman[a]	1939
Mo	95.95	Hzelm	1782	Arnon & Stout	1939
Cl	35.46	Schell	1774	Broyer et al.	1954
Ni	58.69	–	–	Brown et al.	1987

[a]Researchers defined the essentiality of B for nonleguminous plants

In some cases, crops that accumulate certain micronutrients may have higher contents of this nutrient than of a macronutrient. Prado (2003) found for *Averrhoa carambola* that the leaf content of Mn (1.7 g kg^{-1}) exceeded that of the macronutrient S (1.4 g kg^{-1}). Thus, other classification systems for nutrients emerged based not on the amount accumulated by the plant, but on the (biochemical) role they play in the plant life. Mengel and Kirkby (1987) classified nutrients into four groups. The first group consists of C, H, O, N, and S, structural nutrients in the constitution of organic matter, which also participate in the enzymatic system and in assimilation in redox reactions. The second group consists of P, B, and Si in some crops, nutrients that easily form connections of ester type (energy transferors). The third group consists of K, Mg, Ca, Mn, Cl, and (Na), nutrients responsible for enzymatic activity, maintenance of the osmotic potential, ion balance, and electrical potential, especially K and Mg. Finally, the last group consists of Fe, Cu, Zn, and Mo, which act as prosthetic groups of enzymatic systems and also participate in electron transport (Fe and Cu) to different biochemical systems.

The author highlights that the list of 17 chemical elements considered essential may increase with the progress of research. There are isolated studies whose authors indicate certain elements as essential to plants, such as silicon in tomato plants (Miyake and Takahashi 1978), sodium in Atriplex vesicaria (Brownell and Wood 1957), cobalt in lettuce (Delwiche et al. 1961) and alfalfa (Loué 1993), and Se (Wen et al. 1988). However, for a chemical element to be included in this list (if it occurs, it is more likely to be a micronutrient), additional studies are necessary in order to satisfy the criteria of essentiality for a considerable number of plant species and to convince the international scientific community. There are strong candidates for inclusion in the list of nutrients, such as Si, Na (Malavolta et al. 1997), Se, and Co (Malavolta 2006).

1.3   Relative Composition of Nutrients in Plants

In a freshly harvested plant, depending on the species, we observe that the largest proportion of its mass, from 70% to 95%, consists of water (H$_2$O). After drying this plant in a forced air circulation oven (±70 °C for 24–48 h), the water evaporates and we obtain the dry matter or dry mass; and when the plant is subjected to mineralization, either in a muffle furnace (300 °C) or in strong acid, the organic and mineral components (nutrients) are separated. After analyzing this dry plant material, we generally observe predominance of C, H, and O, composing 92% of plant dry matter (Table 1.2).

The results of chemical analysis of plant material are expressed based on the dry matter for being more stable than fresh matter, which varies according to the medium, that is, the time of day, soil water availability, and temperature, among others.

It is noteworthy that C comes from atmospheric air in the form of carbon dioxide (CO$_2$); H and O come from water (H$_2$O); and minerals (macronutrients and

Table 1.2 Relative composition of nutrients in plant dry matter

Classification	Nutrient (element)[a]	Participation %	Total
Organic macronutrients	C	42	
	O	44	
	H	6	
			92
Macronutrients	N	2.0	
	P	0.4	
	K	2.5	
	Ca	1.3	
	Mg	0.4	
	S	0.4	7
Micronutrients	Cl, Fe, Mn, Zn, B, Cu, Mo, Ni		1
Total			100

[a]The element form of nutrients is not always the chemical form that plants absorb

micronutrients) come from the soil, directly or indirectly. Therefore, we notice that plant nutrients come from the three following systems: air, water, and soil, and approximately 92% of plant dry matter comes from air and water systems, with only 8% coming from the soil. However, although the latter is less important quantitatively in relation to the others, it is the most discussed topic in plant nutrition studies. In addition, it is the most expensive material in agricultural production systems, especially if we consider that air and rainwater have no cost (in nonirrigated production systems).

1.4   Nutrient Accumulation by Crops and Crop Formation

Nutrient accumulation in plants reflects their nutritional requirement, ranging according to several factors, such as the level of production, species, cultivar, soil fertility and/or fertilization, climate, and crop treatments.

In general, crops have nutritional needs representing the amount of macronutrients and micronutrients that plants remove from the soil during cultivation to meet all development stages, resulting in adequate harvests.

Crops in general, such as sugarcane, soybean, and wheat, have high demand of nitrogen and/or potassium and low demand of copper and molybdenum (Table 1.3). However, the order of requirements for other nutrients may vary between crops and even between cultivar/hybrid.

The standard, decreasing order of extraction of crops is generally as follows:

Macronutrients: N > K > Ca > Mg > P $\leftrightarrow$ S
Micronutrients: Cl > Fe > Mn > Zn > B > Cu > Mo > Ni

Table 1.3 Total extraction (shoot) and export (stems/grains) of nutrients by commercial crops.

Nutrient		Sugarcane (100 t ha^{-1})			Soybean (5.6 t ha^{-1})			Wheat (3.0 t ha^{-1})		
		Stems	Leaves	Total	Grain	Crop residue	Total	Grain	Crop residue	Total
		kg ha^{-1}								
Macronutrient	N	90	60	150	152	29	181	75	50	125
	P	10	10	20	11	2	13	15	7	22
	K	65	90	155	43	34	77	12	80	92
	Ca	60	40	100	8	43	51	3	13	16
	Mg	35	17	52	6	20	26	9	5	14
	S	25	20	45	4	2	6	5	9	14
		g ha^{-1}								
Micronutrient	B	200	100	300	58	131	189	100	200	300
	Cu	180	90	270	34	30	64	17	14	31
	Fe	2500	6400	8900	275	840	1115	190	500	690
	Mn	1200	4500	5700	102	210	312	140	320	460
	Mo	–	–	–	11	2	13	–	–	–
	Zn	500	220	720	102	43	145	120	80	200

Considering the crops shown in Table 1.3, we note that the order of total extraction of nutrients changed. For macronutrients, sugarcane requires more K compared to N, while S is the third most required nutrient in wheat. For micronutrients, we note that Cl was the most extracted (not mentioned). The same change in the standard order occurs especially between Zn and B, for example, sugarcane demands more Zn, while soybeans and wheat demand more B.

Regarding the export of nutrients from the agricultural area, a significant amount of elements are mobilized to the harvest product (stem or grain) (Table 1.3). We note that a significant part of N, S, P, and Zn, among others, is mobilized in the grains. Thus, nutrients are stored in seeds in the form of specific organic compounds, for example, N and S, accumulated in specific storage proteins (Müntz 1998), and P and various cations, accumulated in the form of phytates (Raboy 2001). Each phytate molecule contains six phosphate groups that form complexes with cations, and most of K, Mg, Mn, Ca, Fe, and Zn in seeds are associated to the phytate (Epstein and Bloom 2006).

Consequently, for living beings (humans and animals), seeds are more nutritious than the rest of the plant. Thus, higher nutrient levels in the seeds benefit food quality, reducing human malnutrition, which is high in several regions of the world (Stein 2010). In addition, this quality would influence the initial growth of the new crop in seed production fields. Many plants can live on the P in the seed for approximately two weeks (Grant et al. 2001).

In practice, crops that export a large part of the absorbed nutrients in harvest, or those in which the harvested product is the whole shoot (sugarcane, maize silage, pasture), leave little crop residue. These crops require more attention regarding the need to replace these nutrients through maintenance fertilization.

Studies on nutrient extraction can identify the requirement for a given nutrient in crops, enabling to meet its demand and increase crop production.

In Brazilian agriculture, fertilizer application often may not meet the nutritional requirements of crops, consequently limiting agricultural production. This fact is verified by comparing plant nutritional requirements with the average use of fertilizers in the respective crops (Table 1.4).

In addition to fertilizers, there are other nutrient sources, such as steel slag (Ca, Mg, and micronutrients) (Prado et al. 2002a, b), calcium silicates (Ca) (Prado and Natale 2005), and biomass ash, among others, which were not considered in Table 1.4.

We obtained crop requirements for yield levels close to the Brazilian average. Based on these average results, we infer that soil depletion (in a soil already poorly fertile) may be occurring. However, there are many areas in Brazil with adequate fertilizer application, reaching the highest yields in the world, such as for soybean.

Researchers at the International Fertilizer Development Center (IFDC) stated that most of the agricultural soils in the world are undergoing depletion of some nutrients, except for North America, Western Europe, and Australia/New Zealand. The authors conclude that we need to maintain this agricultural technology; otherwise, we will not achieve the food production required for the future. The growth of the world economy and the use of agricultural raw materials for fuel production increase the world demand for food, which will only be ensured with the increase of agricultural production by meeting crop nutritional requirements.

Still regarding nutritional requirements, nutrient extraction from the soil does not occur constantly throughout the crop production cycle. The curve of nutrient extraction or accumulation over the period of cultivation (uptake rate) follows that of plant growth, explained by a sigmoid curve. It is characterized by an initial phase of decreased growth and nutrient uptake and, in the next phase, fast (almost linear)

Table 1.4 Nutritional requirement and apparent consumption of fertilizers ($N + P_2O_5 + K_2O$) of some crops in Brazil

Crop	Nutritional requirement[b]		Fertilizer consumption
	$N + P + K$	$N + P_2O_5 + K_2O$[c]	$N + P_2O_5 + K_2O$
Soybean[a] (2.8 t ha^{-1})	90(54) + 7 + 38	152 (97)	145
Sugarcane (73.0 t ha^{-1})	73 + 9.7 + 76	186	206
Citrus (26 t ha^{-1}) (fresh fruit)	66.5 + 8.3 + 52	192	122
Maize (3.7 t ha^{-1})	176 + 32 + 149	430	110
Rice (3.2 t ha^{-1})	82 + 8 + 47	157	77
Bean (1 t ha^{-1})	102 + 9 + 93	235	31
Cassava (16,600 plants)	187 + 15 + 98	339	8

[a]For soybean, the author estimates that 60% of N requirements come from biological fixation, and the rest from the soil (54 kg ha^{-1} of N)
[b]Fertilization needs are higher than nutritional requirements due to nutrient loss in the soil of 50, 70, and 30%, on average, for N, P, and K, respectively
[c]$P \times 2.29136 = P_2O_5$; $K \times 1.20458 = K_2O$

plant growth and increased nutrient uptake/accumulation. In the last phase, plant growth/development and nutrient uptake stabilize until completing the production cycle. However, at the end of the latter phase, accumulation of certain nutrients (K, N) can stabilize or even decrease due to the loss of senescent leaves and rain washing out the nutrient (K) from the leaf. The sigmoid curve explains nutrient uptake during the crop life cycle (perennial or annual). P accumulation for the formation of sugarcane seedlings is high during a period from 65 to 110 days after transplanting, requiring fertilization (Fig. 1.3).

In soybeans, Bataglia and Mascarenhas (1977) verified a period of maximum demand for each nutrient, which would correspond to the maximum speed of nutrient accumulation by the plant. The authors found that most nutrients reach the period of maximum uptake along with the period of maximum accumulation of dry mass (60–90 days), except for K, P, Cl, Mn, Mo, and Zn, which reached this period at 30–60 days. In maize, the maximum speed of accumulation would be in the period of 60–90 days for most nutrients, also corresponding to the maximum accumulation of dry matter. For P and K, the maximum speed of accumulation occurred earlier, at 30–60 days. However, nutrient application, for example, K, should preferably occur until 30 days, corresponding to the beginning of the period of maximum requirement by the crop.

Although most studies in the literature established the nutrient uptake rate using chronological data (in days), plants develop as thermal units and accumulate above a base temperature, while growth stops below this temperature. Through thermal accumulation, also known as degree-days, we obtain excellent correlations with the duration of the crop cycle or with the stages of phenological development of a given cultivar. Therefore, new studies on nutrient uptake rates or accumulation curves as a function of degree-days accumulated during the crop cycle are relevant.

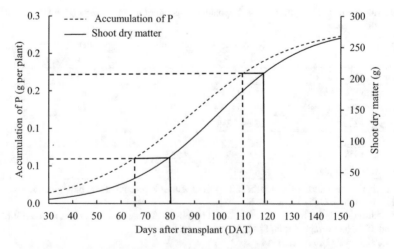

Fig. 1.3 Phosphorus accumulation by sugarcane seedlings as a function of plant age

Studies on uptake rates are important to detect at which development stage the crop demands more of a given nutrient, that is, at which stage the nutrient has the highest uptake rate. Through this information, it is possible to predict the moment of nutrient application to satisfy the nutritional requirement at the respective stage of crop development, avoiding nutritional deficit and decreased production.

Thus, the best period for fertilizer application depends on the evaluation of the nutrient uptake rate by the crop, besides depending on the nutrient and its release dynamics in the soil.

The author notes that crop nutritional requirements are specific to the species and to the cultivar/variety of the same species. Thus, nutritional differences between crops will be discussed for each nutrient in the following chapters. In order to ensure maximum efficiency of fertilization, that is, the most appropriate moment for nutrient/fertilizer application, it is necessary to know the plant (i.e., its nutrient uptake rate) and environmental factors, such as soil (texture), water (irrigated or nonirrigated), and cultivation system (conventional or no-till).

Modern agriculture requires maximum economic benefit regarding the environment. Thus, the Brazilian producer has a great challenge in face of the low fertility of tropical soils and the high cost of fertilizers, which are the nutrient source.

A rational solution for sustainable agriculture would be to adapt the plant to the soil by using crops/cultivars that are efficient in the process of crop formation, doing more with less. In recent decades, especially in the 1990s, agricultural production increased, although fertilizer application decreased. This could be explained by increased efficiency of nutrient use by crops (Epstein and Bloom 2006).

As nutrient uptake, transport, and use can be genetically controlled, it is possible to improve and/or select more efficient cultivars regarding nutrient use (Gabelman and Gerloff 1983). For the plant to use nutrients with increased efficiency, it is necessary to optimize several physiological and biochemical processes of crop formation. The possible mechanisms to control the nutritional needs of plants include nutrient acquisition from the environment (soil or nutrient solution), their movement through the roots and release to the xylem, their distribution in the organs, and use in metabolism and growth (Marschner 1986).

There may be a crop with the same nutritional requirement, but more efficient agronomically. For example, a hybrid wheat A can accumulate the same nitrogen amount as a hybrid B. However, hybrid B uses N for increased grain production, compared to hybrid A that prioritizes increased production of vegetative organs (crop residues). Thus, for the same dry matter production (6 t ha^{-1}), hybrid B produces 40% more grains than hybrid A, although both accumulate the same N amount in the shoot (90 kg ha^{-1}).

These results illustrate the current trend in agriculture, that is, it is better for the plant breeder to select plants with increased grain production (commercial product) and harvest index than total biomass. The harvest index is the ratio between the harvested dry matter (grains) and the total dry matter of the plant. Therefore, photoassimilate partitioning by plants is also an important production factor for modern agriculture.

In addition, modern cultivars develop shallower root systems (Rublo et al. 2003), with increased recovery rates for fertilizers or organic matter mineralization.

The term nutrient use efficiency emerged in this context, which is the ability of a species or genotype to provide high yields even in soils with deficit of the nutrient under study (Graham 1984). It is also the relationship between production and concentration of the nutrient in the tissue (Lauchli 1987). Therefore, a species or cultivar with superior nutritional efficiency develops adequately in low fertility soils due to an increased ability to absorb the necessary nutrients in less quantity and/or to distribute them more efficiently in the various plant components, ensuring adequate plant metabolism with high conversion into dry matter.

Simple ways to increase nutritional efficiency are to reduce fertilizer doses to levels that are still economical to produce or by genetic improvement through selection of plants with better nutritional indexes.

From dry matter and nutrient content in the plant, we can calculate nutritional indexes comprising the efficiency of absorption, translocation/transport, and nutrient use for conversion into dry matter. These indexes are shown below:

Uptake efficiency = (total nutrient content in the plant)/(root dry matter) (Swiader et al. 1994). This index indicates the ability of the plant to extract nutrients from the growing medium (soil). Mechanisms developed for high uptake efficiency differ between plant species. Some species grow an extensive root system and others have high uptake rates per unit of root length, that is, increased nutrient influx (Föhse et al. 1988).

Translocation efficiency = ((shoot nutrient content)/(total nutrient content in the plant)) ×100 (Li et al. 1991). This index indicates the ability of the plant to transport nutrients from root to shoot. Samonte et al. (2006) observed the correlation between N translocation index and protein content in rice grains.

Utilization efficiency (UE) (coefficient of utilization) = (total dry matter)^2/(total nutrient content in the plant) (Siddiqi and Glass 1981). This index indicates the ability of the plant to convert the absorbed nutrient into total dry matter. According to Gabelman and Gerloff (1983), the ability of a plant to redistribute and reuse minerals from an older and senescent organ characterizes utilization efficiency in the metabolism of the growth process.

Usually, the efficiencies aforementioned are more used in pot assays due to easiness to work with plant root systems compared to field conditions. This line of research in plant nutrition is very important, as adequate nutrient use is essential to increase or sustain agricultural production.

Such efficiencies will be increasingly correlated with agricultural yield if the cultivars have increased assimilate partitioning for organs of interest (such as grains).

Thus, through field experimentation, other nutritional indexes similar to the previous ones emerged, although concerned to indicate nutritional efficiencies regarding the dry matter of the commercial part (such as grains).

In this sense, Fageria (2000) observed increased correlation with rice production, agronomic efficiency (AE), agrophysiological efficiency (APE), and physiological efficiency (PE) (similar to the aforementioned utilization efficiency). In this same crop, Samonte et al. (2006), who studied N application, observed that grain yield correlated with N utilization efficiency and N content. In addition, the authors added that it is interesting to select plants that not only have increased yields, but also use the nutrient efficiently and with quality (grains with high protein content). Svecnjak and Rengel (2006) observed differences in N utilization efficiency in canola culti-vars, although they absorbed the nutrient similarly, as certain cultivars considered efficient produced more dry matter with decreased N content in different plant organs, except for the root. In maize plants, N uptake efficiency was more important than utilization efficiency (Erying et al. 2020).

Thus, the forms of calculating nutritional efficiencies (agronomic, physiological, agrophysiological, recovery, and utilization) will be presented, being used in field assays, as indicated by Fageria et al. (1997):

Agronomic efficiency (AE) = (GPwf −GPnf)/(NAa), given in mg mg^{-1}, where: GPwf = grain production with fertilizer, GPnf = grain production without fertil-izer, and NAa = nutrient amount applied. This index indicates the grain produc-tion capacity per unit of fertilizer applied in the soil.

Physiological efficiency (PE) = (TOPwf–TOPnf)/(NACwf–NACnf), given in mg mg^{-1}, where: TOPwf = total organic production (shoot and grains) with fertilizer; TOPnf = total organic production without fertilizer; NACwf = nutrient accumu-lation with fertilizer, and NACnf = nutrient accumulation without fertilizer. This index indicates the shoot production capacity per unit of nutrient accumulated in the plant.

Agrophysiological efficiency (APE) = (GPwf − GPnf)/(NACwf − NACnf), given in mg mg^{-1}, where: GPwf = grain production with fertilizer, GPnf = grain produc-tion without fertilizer, NACwf = nutrient accumulation with fertilizer, and NACnf = nutrient accumulation without fertilizer. This index is similar to the previous one, although indicating the specific capacity of grain production per unit of nutrient accumulated in the plant.

Recovery efficiency (RE) = (NACwf–NACnf)/100(NAa), given in percentage, where: NACwf = nutrient accumulation with fertilizer, NACnf = nutrient accu-mulation without fertilizer, and NAa = nutrient amount applied. This index indi-cates the amount of nutrient applied in the soil absorbed by the plant.

Utilization efficiency (UE) = physiological efficiency (PE) x recovery efficiency (RE). This index indicates the total production capacity of the shoot per unit of nutrient applied. This index differs from the index (c), as it computes the fertilizer recovery efficiency, that is, the plant capacity for nutrient uptake/acquisition from the soil.

In these experiments, plants are grown in soils with low and high content of the given nutrient. Plants with better nutritional efficiency in soils with low nutrient content are considered efficient, that is, they produce more under stress conditions, while plants with the best nutritional indexes when submitted to soils with high content of the given nutrient are considered responsive.

1.5   Other Chemical Elements of Interest for Plant Nutrition

In addition to the elements considered essential to plant life, there are elements considered beneficial, and also a group of toxic elements. The beneficial element is defined as the one which stimulates plant growth, but is not essential or is essential only for certain species or under certain conditions (Marschner 1986). Silicon, cobalt, and selenium are considered beneficial to the growth of certain plants. However, Malavolta (2006) considers only Si and Na to be beneficial. We note that even a nutrient or beneficial element, when present in high concentrations in the soil solution, can be toxic to plants. However, an element is considered toxic when it is not qualified as nutrient or beneficial element. Potentially toxic elements present high harm potential even at low concentrations in the environment as they accumulate in the trophic chain and slow growth, which may lead to plant death. We have as examples Al, Cd, Pb, and Hg, among others, and their harmful potential depends on the dose.

Aluminum has been extensively studied, considering that tropical soils have an acid reaction with high concentrations of exchangeable Al^{+3}, which is toxic. However, it can even reduce the toxicity of other elements (Cu and Mn) at low concentration (0.2 mg L^{-1}). Usually, excess Al in the soil is toxic for plants, constituting the main limiting factor for food and biomass production in the world (Vitorello et al. 2005). The presence of this element affects from germination (Marin et al. 2004) to root growth, interfering with nutrient uptake (such as P, Mg, Ca, and K) (Freitas et al. 2006). Aluminum stress increases the molecular mass of cell wall hemicellulose, making it rigid and inhibiting root elongation (Zakir Hossain et al. 2006), besides increasing membrane leakage. However, the latter may be the consequence of exposure to the element and not caused by damage of the root growth (Yamamoto et al. 2002).

Al toxicity symptoms thicken roots, making them short and brittle, sometimes developing a brown color (Furlani and Clark 1981). In the shoot, Al toxicity symptoms may not be clearly identifiable and may even be confused with nutritional imbalance (P, Ca, Fe, or Mn). According to Malavolta (1980), it is similar to P and K deficiency, that is, yellowing of the margin and drying of leaves. High Al content in the plant can block plasmodemata (preventing the transport of solutes and water via symplast), which is induced by callose production, a polysaccharide that plants produce in the phloem when subjected to pathogenic or environmental stress (temperature, Al or Ca in the cytosol).

Research on plant nutrition aims to select tolerant genotypes and the mechanisms that plants use to mitigate the toxic effect of the element. Jo et al. (1997) indicate two types of mechanisms that Al-tolerant plants have, such as (1) external mechanisms, in which tolerant plants release organic acids by the root, usually citrate and malate, which bind to aluminum forming stable complexes that prevent Al uptake by the plant; (2) internal mechanisms, in which aluminum is absorbed into the plant and, consequently, into the cell, where it is inactivated by some enzyme or isolated inside the vacuole. Menosso et al. (2001) observed that soybean cultivars considered Al-tolerant distinguish from sensitive cultivars by increased accumulation of organic acid (citric). Mendonça et al. (2005) observed in rice that the Al-tolerant cultivar could adjust its proton balance more efficiently, in order to absorb less Al and better tolerate the presence of this cation in the nutrient solution. However, Braccini et al. (2000) found that change in rhizosphere pH is not related to aluminum tolerance in coffee genotypes grown in soil.

Besides Al, other toxic elements are extensively studied to understand the mechanisms that tolerant plants use to minimize harmful effects. Research of this nature allows introducing plants in areas with high load of heavy metals. As aforementioned, plants tend to accumulate metal in vacuoles and also form chelates with these metals from two types of cysteine, phytochelatins or metallothioneins.

One of the beneficial elements most recently studied in Brazil is silicon. China, USA, and Brazil are the countries that mostly publish scientific articles on silicon. Plants normally absorb silicon in the form of acid (H_4SiO_4) with energy expenditure (active) (Rains 1976), which then interacts with pectin and deposits on the cell wall, remaining practically motionless in the plant. In the literature, Marschner (1995) suggests a division of plants regarding their accumulation capacity for this element, as follows: accumulating plants: contain 10–15% SiO_2 contents, for example, grasses, such as rice; intermediate plants: contain 1–5% SiO_2 contents, with some grasses and cereals; nonaccumulating plants: contain less than 0.5% SiO_2 contents, the majority of which are dicotyledons, such as legumes.

Ma and Takahashi (2002) propose criteria to differentiate nonaccumulating plants from accumulating plants:

"Accumulators" have a Si concentration over 1% and a[Si]/[Ca] ratio >1.
"Excluders" have a Si concentration below 0.5% and a[Si]/[Ca] ratio <0.5.
Plants that do not meet these criteria are called "intermediates."

We emphasize that all plants accumulate silicon, even those considered nonaccumulating. These plants accumulate silicon in the roots.

The evolution of the research allowed us to understand the absorption of Si by plants. Si enters the plant from the external environment in the form of $Si(OH)_4$ through specific influx channels (termed Lsi1), and efflux transporters (termed Lsi2) mediate the loading of Si into the xylem and thus facilitate root-to-shoot translocation, which, in turn, moves Si to the aerial parts of the plant, where it deposits as amorphous SiO_2 (Ma and Yamaji 2015).

Miyake and Takahashi (1985) also add that there are other differences in plants regarding Si uptake besides their content in the plant, referring to the transport rate. Si-accumulating plants have high transport rate, intermediate plants have decreased transport, and nonaccumulating plants concentrate absorbed Si in the roots. Currently, Si is not considered a universal plant nutrient, although most authors consider it beneficial, or, according to Epstein (2002), nearly essential.

The silica particles grow to a size of about 1 to 3 nm and are negatively charged such that they can interact with the local environment of the cell walls of plants. It has been suggested that the nucleation and growth of these structures are under the control of specific proteins (Perry and Tucker 2000) and that a fraction of Si form bonds with proteins, phenolic compounds (lignin, condensed polyphenols), lipids and polysaccharides (cellulose) (Kolesnikov and Gins 2001). It is possible that silicon can replace structural compounds in the cell wall. This can decrease the energy cost for the plant to form lignin or cellulose in the cell wall, favoring its growth.

In the literature, there are many reports on the benefits of Si for plants (Peixoto et al. 2020). However, the most discussed benefits are its resistance against fungal diseases in different crops, such as rice (Marschner 1986), and also decreased incidence of pests (Goussain et al. 2002) due to their increased difficulty to feed, increased jaw wear, and increased mortality rate.

Si reduces diseases not only due to a physical factor (formation of compounds that act as a physical barrier below the leaf cuticle) but also due to a chemical factor (formation of compounds that are phytotoxic to pathogens). Recent studies indicate that Si-rich plants may have holes in the epidermis without physical protection, which hampers phytopathologists in the isolation of chemical compounds produced by the plants.

The effects of Si on plants can be summarized as follows:

- Increases plant resistance to fungal diseases.
- Attenuates salinity (Calero Hurtado et al. 2019) and impacts the C:N:P stoichiometry depending on the species (Calero Hurtado et al. 2020) by increasing enzyme activity (superoxide dismutase (SOD), catalase (CAT), and glutathione reductase (GR)) and eliminating reactive oxygen species (Alves et al. 2020).
- Attenuates water deficit (Teixeira et al. 2020a), which can be seen visually in sugarcane plants (Fig. 1.4) and forage (Rocha et al. 2021).

Fig. 1.4 Silicon attenuates water deficit in sugarcane plants

- Increases cell stiffness through increased production of compounds such as lignin in some species (Alvarez et al. 2018; Deus et al. 2019). Si can play a structural role similar to lignin, reducing its synthesis, which demands high energy costs.
- Increased cell stiffness improves leaf architecture in the plant and favors photosynthesis (Flores et al. 2018).
- Reduces leaf senescence rates.
- Attenuates Fe, Mn, and Al toxicity. Si with salicylic acid attenuated B toxicity in peas (Oliveira et al. 2020a).
- Attenuates the deficiency of K, Zn, Mn, Fe, and B (Prado et al. 2018); K, Ca, and Mg in forages (Buchelt et al. 2020); Mn (Oliveira et al. 2019; Oliveira et al. 2020b), B (Souza Júnior et al. 2019), Zn (Guedes et al. 2020), and Fe (Teixeira et al. 2020b). The effect of Si can be seen visually in the attenuation of symptoms of K deficit in quinoa plants (Fig. 1.5a), and Ca deficit in cabbage plants (Fig. 1.5b) and also in tomato plants (Alonso et al. 2020) and cabbage (Silva et al. 2021).
- Increased P use by plants (Silva and Prado 2021).

Fig. 1.5 Silicon attenuates symptoms of K deficit in quinoa plants (**a**) and Ca deficit in brachiaria grass (**b**) and cabbage plant (**c**)

- Attenuates ammonium toxicity (Barreto et al. 2016, 2017; Viciedo et al. 2019b; Silva Júnior et al. 2019; Viciedo et al. 2020a).
- Increases nitrogen nutrition efficiency, as it increases nitrate reductase activity (Silva et al. 2020).
- Reduces excessive transpiration.
- Increases reproductive growth (Miyake and Takahashi 1978) and pollen grain production and viability.
- Increases C utilization efficiency (Frazão et al. 2020; Lata-Tenesaca et al. 2021).

Si-accumulating plants such as rice can accumulate 250 kg ha^{-1} of Si, being more absorbed than N or K (Körndorfer et al. 2002). Thus, there is a relationship between Si in the plant and rice production, reaching 95% of the maximum with leaf content equal to 34 g kg^{-1} in US organic soil (Korndörfer et al. 2001).

Beneficial effects of Si may occur for sugarcane with steel slag application as corrective material and Si source, which promoted a linear effect on stem production. Leaf spraying is an alternative for supplying Si with soluble sources and increasing yield (Flores et al. 2018; Deus et al. 2019; Felisberto et al. 2020) and growth of seedlings of sugarcane (Santos et al. 2020), soybeans, and common beans (Souza Júnior et al. 2020), besides biofortified vegetables (Souza et al. 2018).

Plant response to silicon is more significant in production systems with some type of stress, whether biotic (diseases/pests; poorly erect cultivar, etc.) or abiotic (water deficit; excess metals, such as Al; low pH, etc.).

Si toxicity in field conditions is not known. There are reports on orchids with leaf applications of high Si concentration for 18 months (Mantovani et al. 2018, 2020). This effect may have occurred due to decrease in leaf gas exchange.

Sodium is also considered a beneficial element, as it was a nutrient for a salt-tolerant halophyte (Atriplex vesicaria), according to Brownell and Wood (1957). It also showed beneficial effect for other plants, such as asparagus, barley, broccoli, carrots, cotton, tomatoes, wheat, peas, oats, and lettuce (Subbarao et al. 2003). One important aspect of sodium is its ability to replace part of K in nonspecific functions, such as vacuolar K when K supply is limited.

Thus, Na would replace K in its contribution to the solute potential and, consequently, in the generation of cellular turgor. This occurs significantly only in a specific group of plants, such as beets, spinach, savoy, coconut, cotton, cabbage, lupine, and oats (Lehr 1953). In these crops, it is possible to use potassium fertilizers with increased proportion of sodium (lower cost) in nonsodic soils. In addition, Na can affect photosynthesis, especially in C4 plants, although this role is not fully understood. Na would increase CO_2 concentration and chloroplast integrity in leaf sheath mesophilic cells (Brownell and Bielig 1996), as well as regenerate phosphoenolpyruvate (PEP) in the chloroplast and participate in chlorophyll synthesis. However, Na can impair the enzymatic action of K at high concentrations, dislodging it from enzyme action sites.

There are indications in the literature that Se is essential in organic compounds such as amino acids, proteins, volatile compounds, ferrodoxins, and hydrogenases

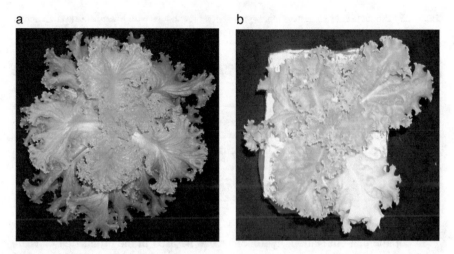

Fig. 1.6 Lettuce control plants (**a**) with symptoms of selenium toxicity (128 μM of Se in the form of selenite) (**b**)

(Malavolta 2006), transporting RNA (Wen et al. 1988) and activating some enzymes, such as superoxide dismutase, catalase, glutathione reductase, and guaiacol/ascorbate peroxidase (Djanaguiraman et al. 2005). Se decreases the leaf senescence rate by decreasing peroxidase activity, which can increase N utilization efficiency, benefiting crop production. However, it promotes oxidative stress/nutritional disturbance at high concentrations, damaging photosynthesis (Ferreira et al. 2020), inducing toxicity symptoms in the plant, reducing growth, and causing leaf chlorosis and small and brittle roots (Fig. 1.6).

Se biofortification through fertilization is feasible by increasing Se content in food. This is important because Se is a nutrient for humans and animals, although being highly toxic if ingested in excess, and this fact is rarely addressed in biofortification studies (Prado et al. 2017)

1.6   Hydroponic Cultivation: Preparation and Use of Nutritional Solutions

The term hydroponics (from Greek hydro = water and ponos = work) is relatively new, designated as soilless cultivation.

The alternative cultivation system of hydroponics can optimize production with an increased number of crops per year (Jensen and Collins 1985) and increased production compared to the conventional system (Table 1.5). However, hydroponic cultivation is restricted to crops with fast cycle and small size, such as vegetables and flowers, among others.

Table 1.5 Production of some vegetables grown in a greenhouse with hydroponic system and in the field

| Crop | Hydroponic system | | | Field |
	t/ha	Number of crops	t/ha/year	t/ha/year
Broccoli	32.5	3	97.5	10.5
Green beans	11.5	4	46.0	6.0
Savoy	57.5	3	172.5	30.0
Chinese cabbage	50.0	4	200.0	–
Cucumber	250.0	3	750.0	30.0
Aubergine	28.0	2	56.0	20.0
Lettuce	31.3	10	313.0	52.0
Bell pepper	32.0	3	96.0	16.0

In commercial cultivation, hydroponics can be used in the production of valuable, high-quality (free of microorganisms and pesticides) crops with increased added value to the product, such as various vegetables.

Thus, the hydroponic system has some advantages and disadvantages, as follows:

- Advantages

  Requires less operational work.
  Elimination of soil preparation (fuel costs and machinery purchases).
  Elimination of crop rotation.
  Reutilization of the culture medium.
  Increased production without competition for nutrients and water.
  Uniform plants regarding development.
  Better root development and products with higher quality and longer shelf life.
  Low loss of water and nutrients.
  Reduction of pests and diseases (decreased spraying).
  Better use of the agricultural area.
  Implementation without restriction to the type of area (shallow soils, low drainage, or high slope).
  Immune to climatic adversities (frost/hail).

- Disadvantages

  Increased costs and initial work.
  Higher risk of losses due to lack of electricity in automatic systems.
  Requires technical skill and knowledge of plant physiology.
  Inadequate balance of the nutrient solution can cause serious problems for plants.
  Need to support roots and shoots.
  Only inert materials should come into contact with the plants.
  Regular routines.
  Good drainage to prevent root death.
  Water contamination affects the entire system.
  In research, hydroponics can be used in several studies, such as:

Demonstration of nutrient essentiality.
Definition of symptoms of nutritional disorder, whether due to deficit or excess (toxicity).
Knowing plant nutritional requirement.
Selection of plants tolerant to nutritional stress.
Ionic uptake, transport, and redistribution mechanisms.
Disease control.
Quality of products in hydroponics (e.g., nitrate accumulation).

For hydroponic cultivation, appropriate use of the nutrient solution is fundamental for its success, either for experimental or for commercial purposes. The nutrient solution is a homogeneous system with nutrients and oxygen dispersed in appropriate proportions and quantities, in ionic or molecular form. The homogeneity of the nutrient solution is altered when it comes into contact with the plant roots, as there are organic compounds from microbial activity, especially from decomposition of root fragments or other impurities (coming with the plant or the hydroponic system), or even exudates of organic acids from the roots.

One of the classic studies in the field of nutrition employing hydroponics is the use of induced deficit or missing element technique. This technique was widely used to investigate nutrient functions through the effects of the lack of nutrient on the biochemical activity of the plant, besides being used for teaching in the discipline of plant nutrition. In this technique, a complete solution is used minus the nutrient under study, cultivating the plant until the characteristic symptoms of the missing element appear, identifying the deficit (Fig. 1.7). When the plant is subjected to stress, such as nutritional stress, it attempts to acclimatize with a series of changes in the hormonal system (Morgan 1990). If the stress continues, a series of events occur before the injury is visible, and if it occurs, this is the last biological event, that is, at tissue level (Fig. 1.8). At this stage, half the production should be compromised depending on the species. In species with long cycle, production losses can be less than 50%. The form of symptoms depends on the role that the respective nutrient plays in the plant and the place of occurrence (old or new leaf) depends on its mobility in the plant phloem.

Besides the description of symptoms, we monitor plant responses throughout the crop through growing plants with the complete nutrient solution and deficient solution (usually with total omission or with 10% of the appropriate concentration) based on growth evaluations (height, stem diameter, leaf area, and dry matter). Some plants are sensitive to nutrient deficit and excess. There are plants that indicate nutrient deficiency, such as N (maize, apple), P (lettuce, barley), K (potato), Ca (alfalfa, peanuts), Mg (cauliflower, broccoli), S (cotton, alfalfa), Zn (citrus, peach), B (beetroot, turnip, celery, cauliflower), Mn (apple, cherry, citrus), Cu (citrus, plum), Fe (cauliflower, broccoli), and Mo (tomato, lettuce, spinach) (Malavolta et al. 1997). There are several studies in the literature with the omission of nutrients and the features caused by nutritional disorder in crops such as cotton (Rosolem and Leite 2007), brachiaria (Monteiro et al. 1995), sorghum (Santi et al. 2006), rice (Alves et al. 2002), maize (Coelho et al. 2002), bean plant (Cobra Netto et al. 1971;

Fig. 1.7 Plants grown in
nutrient solution with all
nutrients (complete – with
phosphorus) (+P) and
minus phosphorus (−P)
(beans) (a), demonstrating
visual deficiency
symptoms

Dantas et al. 1979), green beans (Osório et al. 2020), sunflower (Prado and Leal
2006), sugarcane (Mccray et al. 2006), mallow (Fasabi 1996), Cyclanthera pedata
(Fernandes et al. 2005), coffee (Haag et al. 1969), acacia (Dias et al. 1994; Sarcinelli
et al. 2004), açaí palm (Viegas et al. 2004a, b), cupuassu (Salvador et al. 1994),
Myrciaria dubia (H.B.K.) McVaugh (Viegas et al. 2004a, b), stevia (Lima Filho and
Malavolta 1997), eucalyptus (Rocha Filho et al. 1978), rubber tree (Amaral 1983),
teak (Barroso et al. 2005), guava tree (Salvador et al. 1999), soursop (Avilán 1975;
Batista et al. 2003), passion fruit tree (Avilán 1974), Spondias tuberosa (Gonçalves
et al. 2006), Cinchona officinalis (Viegas et al. 1998), castor bean (Lavres Júnior
et al. 2005), black pepper (Veloso and Muraoka 1993; Veloso et al. 1998), peach
palm (Silva and Falcão 2002), beetroot (Alves et al. 2008), basil (Borges et al.
2016), aubergine (Flores et al. 2014), soybean (Malavolta et al. 1980), watermelon
(Cavalcante et al. 2019), and orchid (David et al. 2019).

Soilless cultivation techniques can be divided into several categories due to use
of different substrates (materials other than soil) (Castellane and Araújo 1995):

Water culture or hydroponics: plant roots are immersed in a solution formed by
    water and nutrients called NFT (Nutrient Film Technique) nutrient solution.
Sand culture: plants are supported by a solid substrate, with particles from 0.6 to
    3.0 mm diameter.
Gravel culture: the substrate is solid with particles larger than 3 mm in diameter.

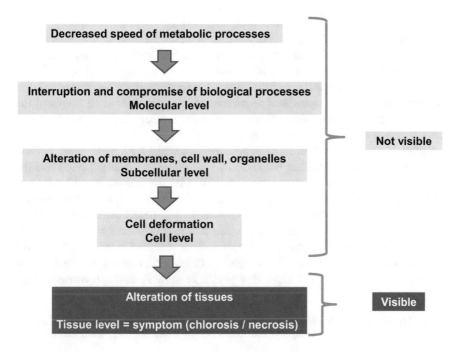

Fig. 1.8 Sequence of biological events in nutrient-deficient plants

Vermiculite culture: when the substrate is vermiculite or its mixture with other materials.

Rockwool culture: uses rockwool, glasswool, or similar material as substrate (high porosity and water uptake, with nutrients and air close to the roots). Rockwool has an inconvenience regarding the environmental aspect at disposal, since its decomposition is very slow.

We highlight that cultivation in a nutrient solution without solid components is the most used technique in plant nutrition studies.

It is difficult to consider that a nutrient solution is ideal for all crops, with a formula that ensures maximum development, and that all nutrients are supplied exactly in the proportion in which they should be absorbed. Thus, nutrient solution's composition is influenced by a number of factors, namely: plant species (nutritional requirements are genetically controlled); plant age and growth stage; time of year (length of light period); environmental factors (temperature, humidity, light); part of the plant harvested, etc. In order to calculate the chemical composition of a nutrient solution, besides crop requirements during cultivation, the environment should be considered, as it affects the luminosity and temperature, directly affecting the transpiration rate. Transpiration rate is important as conditions favoring high transpiration would increase the loss of water solution at a faster rate than the nutrient uptake, which may cause a salt effect. Thus, the higher the transpiration rate predicted for the crop, the lower nutrient concentration in the nutrient solution. For example, if a crop requires an appropriate N content equal to 50 g kg^{-1} dry matter, associated with a transpiration rate of 300 L kg^{-1} dry matter, it would have 50 g of K in 300 L of water or 166 mg L^{-1} of K. If the transpiration rate was 400 L kg^{-1} dry matter, the solution would need dilution, that is, 50 g per 400 L or 125 mg L^{-1} of K.

Therefore, the chemical composition or ideal formula of the nutrient solution is the one that meets the nutritional requirements of the species at all stages of the production cycle. The nutrient solution may contain other elements besides nutrients, reaching up to 20 elements (Jones 1998).

Santos (2000) tested four nutrient solutions using the NFT hydroponic system. From the results, the solution proposed by Castellane and Araújo (1995) had the best performance with increased yield, followed by the one proposed by Furlani (1995). Thus, the author recommends the two solutions for lettuce cultivation in a hydroponic system in the State of Rio Grande do Sul, Brazil. Although this study indicated differences between solutions, this fact may not occur frequently, as the solutions studied and more than a hundred solutions indicated in the literature derive from the Hoagland and Arnon (1950) solution (Table 1.6), one of the solutions with the highest concentration of salts. However, it has low Fe and Mn concentration,

Table 1.6 Chemical composition of some nutrient solutions, namely: Castellane and Araújo (1995), Furlani (1995), and Hoagland and Arnon (1950), water solubility (cold and hot), and salt index

Components	Castellane and Araújo (1995)	Furlani (1995)	Hoagland and Arnon (1950)[b]	Solubillity[c] (g L^{-1}) (0, 5, and 100 °C)	Salt index
	g/1000 L of solution				
Calcium nitrate	950	1000	1200	1212 and 6598	53
Monoammonium phosphate	–	150	150	224 and 1730	30
Potassium phosphate monobasic	272	–			
Potassium chloride	–	150	250	277 and 561	116
Potassium nitrate	900	600	260	134 and 2471	74
Magnesium sulfate	246	250	500	700 and 906	2
Manganese chloride	–	1.17			
Manganese sulfate	1.70	–			
Zinc sulfate	1.15	0.44			
Copper sulfate	0.19	0.10			
Boric acid	2.85	1.02		19.5 and b389	
Sodium molybdate	0.12	0.13			
Fe-EDTA[a]	1 (L)	0.5 (L)			

[a]Fe-EDTA was used as iron source in both solutions, obtained by dissolving 24.1 g of iron sulfate in 400 mL of water and 25.1 g of sodium-EDTA in 400 mL of hot water (80 °C), mixing the two solutions when cold and completing the volume to 1 L

[b]Solution of micronutrients (L/1000 L of solution) and Fe-EDTA (L/1000 L of solution)

[c]Solubility of other salts (in g L^{-1}) (0.5 and 100 °C): ammonium nitrate (1183 and 8711); diammonium phosphate (426 and 1063); ammonium sulfate (704 and 1033); magnesium sulfate (700 and 906); and potassium sulfate (67 and 239)

which can affect demanding plants, such as grasses. It should be noted that all nutritional solutions provide the essential elements for plants. Franco and Prado (2006) observed similarity in four nutritional solutions tested (Hoagland and Arnon; Sarruge; Castellane and Araújo; Furlani) in the growth of guava seedlings. In a similar study with star fruit seedlings, it was observed that the use of these nutrient solutions affected plant nutritional efficiencies (Rozane et al. 2007).

In the literature, the nutrient concentration indicated to formulate a nutrient solution is highly variable, regardless of nutrient, such as (in mg L^{-1}) N-NO_3: 70 to 250; N-NH_4^+: 0 to 33; P: 15 to 80; K: 150 to 400; Ca: 70 to 200; Mg: 15 to 80; S: 20 to 200; B: 0.1 to 0.6; Cu: 0.05 to 0.3; Fe: 0.8 to 6.0; Mn: 0.5 to 2.0; Mo: 0.01 to 0.15; Zn: 0.05 to 0.5; and Cl: 1 to 188 (Cometti et al. 2006).

The nutrient solution shown (Table 1.6) at full concentration (100%) is used in plants with some development or in very cold periods as these solutions are very concentrated, with the risk of physiological damage (salt effect) for young plants in early growth stages. We use diluted solutions (25–75%) at the beginning of growth and less diluted solutions are used as the plant develops, until reaching full concentration (100%).

Besides nutrients, the beneficial element can also be used in the nutrient solution, such as Si. Thus, the Si concentration used is 0.5 mmol L^{-1} in Na_2SiO_3 $9H_2O$ form, which must be added first, maintaining a low pH in solution at this moment (Epstein 1995). The maximum Si concentration without risk of polymerization is 2.0 mmol L^{-1}.

In basic plant nutrition studies using nutrient solution, the concentration of an element in the solution does not always explain plant growth. This is because the ion availability for plant uptake, that is, its activity in the solution can be influenced by several factors, such as the solution ionic strength, the pH value, and chelate types (Cometti et al. 2006). Ionic strength is more important when working with heavy metals, especially Al, where the ion activity is reduced by increasing the ionic strength. In a nutrient solution, unlike the soil solution, there is high pH variation throughout cultivation, which can change free and complex forms of the element. High pH (>6.0) decreases the availability of macronutrients Ca and P and micronutrients Mn, Cu, Zn, and B, due to the formation of precipitates, besides reducing nutrient transport into the cells. The use of Fe chelates in the solution can chelate Cu, Zn, and Mn. Thus, use of the Fe-EDDHA (ethylenediamine-N,N′-bis(2-hydroxyphenylacetic acid) chelating agent will only partially chelate Cu, while use of diethylenetriamine pentaacetic acid (DTPA) or ethylenediaminetetraacetic acid (EDTA) chelates also forms complexes with Zn and Mn, especially at pH >5.5 for Zn and pH >7.0 for Mn. This is important because micronutrients Cu, Mn, and Zn are absorbed in the free form, and the quality of the Fe chelate can induce deficiency of Fe and other micronutrients (Zn, Mn, and Cu). Fe-EDDHA (very stable) prevents chemical reactions with other elements in the solution, although possibly releasing less Fe^{+2} in the cell cytosol, depending on the species, causing deficiency.

Thus, in order to choose the appropriate solution, it is necessary to consider the management factors of the nutrient solution, so as to increase the productive efficiency of hydroponic systems, such as:

Sources of Fertilizers

Nutrient sources depend on the nature of cultivation, with commercial fertilizers being used for commercial cultivation and pro analyze sources, with increased degree of purity, intended especially for basic research in plant nutrition. We emphasize that pro-analysis sources must be purchased from reputable companies, with special attention to choose products free of inert materials or other elements that may compromise the results and scientific rigor of the study.

We indicate ammonium molybdate or molybdic acid as Mo source, as sodium molybdate is very alkaline, which may cause precipitation reactions with other micronutrients.

Water

In commercial hydroponics, drinking water can be used, while distilled water is generally used in scientific hydroponics. However, in some cases, deionized (up to twice) water is better, as in assays aimed to induce nutritional deficiency, especially of micronutrients.

If the water comes from the urban network, we recommend let it rest for about 24 h to eliminate the chlorine used in its treatment (Martinez and Silva Filho 2004).

Order of Nutrient Addition

In order to avoid precipitation reactions of fertilizers (poorly soluble), which would become unavailable to plants, it is necessary to follow a certain order of nutrient addition. Before adding fertilizers (nutrients), the water pH value must be adjusted to appropriate standards (discussed in the subsequent item).

1. Add calcium-free macronutrient fertilizers, such as potassium phosphate, potassium nitrate, and magnesium sulfate. This procedure is necessary due to incompatibility between calcium nitrate and salts containing phosphorus and sulfur, forming precipitated compounds with low solubility.
2. Calcium salts (calcium nitrate).
3. Iron-free micronutrients, such as fertilizer sources based on Mn, Zn, Cu, B, and Mo.
4. Iron source (Fe-EDTA, Fe-EDDHA, or other).

Then, immediately adjust the pH value and determine the EC (electrical conductivity) (discussed in the subsequent item).

Thus, if a stock solution is formulated, it is necessary to have a solution A based on calcium nitrate, another solution B with other macronutrients, and a solution C for micronutrients.

Maintenance of Adequate pH in the Nutrient Solution

As for pH, there are the following concerns:

Calibrate the instrument before each measurement.

The pH value should be adjusted daily to keep the solution pH value in the 5.5–6.5 range. In case of a pH value above this range, add drops of an acid at 0.1 M (HCl; HNO_3, H_2SO_4; H_3PO_4) or, otherwise, add a 0.1 M base (NaOH; KOH). However, the best way to adjust the pH in the solution is the management of nitrogen sources (NO_3^- or NH_4^+) by using part of N in the NH_4^+ form, in the proportion of 10–20% of the total N.

The use of solution with N in ammonium and nitrate form may initially decrease the pH value until almost all ammonia in the solution was absorbed by the plant, with subsequent increased nitrate uptake and increased pH value. Maintenance of the solution pH is important as low values ($\sim$ 3–4) affect membrane integrity (H^+ affects the permeability of cell membranes), which may cause loss of nutrients already absorbed, besides affecting cation availability and uptake. In a solution with high pH (>7), there may be problems with the electrochemical gradient and proton-anion cotransport across the membranes, besides the loss of phenolic compounds and electrolytes from cells and anion uptake reduction. In addition, high pH may cause undesired chemical reactions in the solution, leading to nutritional deficiencies in plants (Fe, Mn, B, and P).

We observed that lettuce cultivation in acid pH caused increased damage to plant growth in relation to alkaline pH, compared to the appropriate range. In the pH range from 5 to 7, plant dry matter production (relative) was 100.0%, while in other pH ranges, 4–5, 3–4, 2–3, and 7–9, the production was 94.5, 1.2, 0.6, and 60.0%, respectively.

Maintenance of Adequate Osmotic Pressure in the Nutrient Solution

The osmotic pressure in the nutrient solution must be in the range of 0.5–1.0 atm, as high values may indicate excess salts in the solution, with serious damage to the roots.

Maintenance of Adequate Temperature in the Nutrient Solution

In order to avoid heating of the nutrient solution (>25–30 °C), it is important to avoid light incidence, storing the solution in a shaded and ventilated location. We also note that light incidence in the solution allows algae to proliferate.

Maintenance of Oxygenation in the Nutrient Solution

It is a consensus that oxygenation of the nutrient solution is mandatory to maintain adequate O_2 contents for roots (~3 ppm O_2). The adequate oxygenation level ranges, with low demand (rice = 3mg L^{-1} O_2) and high demand (tomato = 16mg L^{-1} O_2) of oxygen in the nutrient solution according to the crop. As increasing temperatures in the nutrient solution decrease the dissolved O_2 content, it is important to avoid high temperatures. Aeration is applied according to each hydroponic system, since only in the fall when the nutrient solution returns to the reservoir or, in the case of isolated pots, the injection of compressed air can supply the necessary oxygen to the plants.

Maintenance of Adequate EC in the Nutrient Solution

During cultivation of plants in nutrient solution, the elements of the solution are depleted, ranging according to the nutrient. This is due to differences in the uptake rate of nutrients, which can be fast (N, P, K, and Mn), intermediate (Mg, S, Fe, Zn, Cu, and Mo), or slow (Ca and B) (Bugbee 1995). Thus, nutrients need to be replaced in the solution, which is performed using electrical conductivity data.

The conductivity of an electrolyte solution is the quantitative expression of its ability to transport electrical current. It is defined as the inverse of the electrical resistance of 1 cubic cm of liquid at a temperature of 25 °C. Electrical conductivity is the unit equivalent to 1 mhos = 1 Siemens = 10^3 mS = 10^6 μS (mS = milliSiemens; μS = microSiemens).

Normally, the nutrient solution's EC ranges from 1.5 to 4.0 mS/cm according to the solution chosen for the respective cultures. The EC obtained in a nutrient solution is the sum of the EC of all fertilizers used in the formula of that solution. Castellane and Araújo (1995) obtained an EC of 2.6 to 2.8 mS/cm. As 1 mS/cm corresponds to 640 ppm of nutrients, we note that the use of this variable during cultivation would prevent the nutrient solution from having a low nutrient concentration, which could lead to a nutritional deficiency. Normally, when the electrical conductivity reduces to a certain level of the initial solution (approximately 30–50%), it is recommended to replace it. Backes et al. (2004) suggested replacing the solution when it decreased to 50% of the initial EC. In commercial cultivation, it is possible to use EC to manage the nutrient solution as follows (Carmello and Rossi 1997):

(a) Add daily an amount of new solution equivalent to the amount of solution that reduced in the recipient. After 21 days, renew the nutrient solution if the conductivity reaches 4 mS/cm.

(b) Add only water to replace the amount of evaporated solution and monitor conductivity; when it reaches less than 1 mS/cm, add the salts to recompose it or change it. Renew the nutrient solution in 21 days.

Disease Prevention Measures

Diseases may occur in hydroponic cultivation, especially fungal, requiring preventive measures. Thus, it is important to disinfect materials for each crop. We can use sodium hypochlorite or calcium hypochlorite, based on 1.000 mg/L or 10.000 mg/L of active chlorine, respectively. It is necessary to rigorously wash the materials, as Cl residues (>0.5 mg/L) can injure plants, especially with the use of ammoniacal salts in the nutrient solution (Martinez and Silva Filho 2004). The nutrient solution can also be disinfected through pasteurization. Pasteurization occurs as follows: heat the nutrient solution from 95 to 105 °C for a period of 30 s and then cool it quickly until reaching room temperature (in 30 s) (Martinez and Silva Filho 2004).

References

Alonso TAS, Barreto RF, Prado RM, et al. Silicon spraying alleviates calcium deficiency in tomato plants, but ca-EDTA is toxic. J Plant Nutr Soil Sci. 2020;183:659–64. https://doi.org/10.1002/jpln.202000055.

Alvarez RCF, Prado RM, Felisberto G, et al. Effects of soluble silicate and nanosilica applied to oxisol on rice nutrition. Pedosphere. 2018;28:597–606. https://doi.org/10.1016/S1002-0160(18)60035-9.

Alves BJR, Santos JCF, Virgem Filho AC, et al. Avaliação da disponibilidade de macro e micronutrientes para arroz de sequeiro cultivado em um solo calcário da região de Irece, Bahia. Revista Universidade Rural. 2002;22:15–24.

Alves AU, Prado RM, Gondim ARO, et al. Effect of macronutrient omission on beet development and nutritional status. Hortic Bras. 2008;26:282–5.

Alves RC, Nicolau MCM, Checchio MV, et al. Salt stress alleviation by seed priming with silicon in lettuce seedlings: an approach based on enhancing antioxidant responses. Bragantia. 2020;79:19–29. https://doi.org/10.1590/1678-4499.20190360.

Amaral DW, Deficiências de macronutrientes e de boro em seringueira (Hevea brasiliensis L.). Dissertação, Escola Superior de Agricultura Luiz de Queiroz. 1983.

Avilán LR. Efectos de la deficiencia de macronutrientes sobre el crecimiento y la composicion quimica de la parcha granadina (Passiflora quadrangularis L.) cultivada en soluciones nutritivas. Agronomía Tropical. 1974;24:133–40.

Avilán LR. Efecto de la omisión de los macronutrientes en el desarollo y composición química de la guanábana (Annona muricata L.) cultivada en soluciones nutritivas. Agronomia Tropical. 1975;25:73–9.

Backes FAAL, Santos OSS, Pila FG, et al. Nutrients replacement in nutrient solution for lettuce hydroponic cultivation. Ciênc Rural. 2004;34:1407–14. https://doi.org/10.1590/S0103-84782004000500013.

Barreto RF, Prado RM, Leal AJF, et al. Mitigation of ammonium toxicity by silicon in tomato depends on the ammonicum concentration. Acta Agric Scand B Soil Plant Sci. 2016;66:483–8. https://doi.org/10.1080/09064710.2016.1178324.

Barreto RF, Prad RM, Schiavon Júnior PA, et al. Silicon alleviates ammonium toxicity in cauliflower and in broccoli. Sci Hortic. 2017;225:743–50. https://doi.org/10.1016/j.scienta.2017.08.014.

Barreto RF, Prado RM, Habermann E, et al. Warming change nutritional status and improve *Stylosanthes capitata* Vogel growth only under well-watered conditions. J Soil Sci Plant Nutr. 2020;20:00255. https://doi.org/10.1007/s42729-020-00255-5.

Barroso DG, Figueiredo FAMMA, Pereira RC, et al. Diagnóstico de deficiência de macronutrientes em mudas de teca. Rev Árvore. 2005;29:671–9. https://doi.org/10.1590/S0100-67622005000500002.

Bataglia OC, Mascarenhas HAA. Absorção de nutrientes pela soja. Campinas: Instituto Agronômico; 1977.

Batista MMF, Viégas IJM, Frazão DAC, et al. Effect of macronutrient omission in growth, symptoms of nutricional deficiency and mineral composition in soursop plants (*Annona muricata*). Rev Bras Frutic. 2003;25:315–8. https://doi.org/10.1590/S0100-29452003000200033.

Borges BMMN, Flores RA, Almeida HJ, et al. Macronutrient omission and the development and nutritional status of basil in nutritive solution. J Plant Nutr. 2016;39:1627–33. https://doi.org/10.1080/01904167.2016.1187742.

Braccini MCL, Martinez HEP, Braccini AL, et al. Rhizosphere pH evaluation of coffee genotypes in response to soil aluminum toxicity. Bragantia. 2000;59:83–8. https://doi.org/10.1590/S0006-87052000000100013.

Browne CA. Liebig and the law of the minimum. In: Moulton FR, editor. Liebig and after Liebig: a century of progress in agricultural chemistry. Washington, DC: American Association for the Advancement of Science; 1942. p. 71–82.

Brownell PF, Bielig LM. The role of sodium in the conversion of pyruvate to phosphoenolpyruvate in mesophyll chloroplasts of C4 plants. Aust J Plant Physiol. 1996;23:171–7. https://doi.org/10.1071/PP9960171.

Brownell PF, Wood JG. Sodium as an essential micronutrient element for Atriplex vesicaria, Heward. Nature. 1957;179:635–6. https://doi.org/10.1038/179635a0.

Buchelt AC, Teixeira GCM, Oliveira KS, et al. Silicon contribution via nutrient solution in forage plants to mitigate nitrogen, potassium, calcium, magnesium, and sulfur deficiency. J Soil Sci Plant Nutr. 2020;20:1532–48. https://doi.org/10.1007/s42729-020-00245-7.

Bugbee B. Nutrient management in recirculanting hydroponic culture. In: Abstracts of the 16rd annual conference on hydroponics. Tucson: Hydroponic Society of America; 1995. p. 15.

Calero Hurtado A, Chiconato DA, Prado RM, et al. Silicon attenuates sodium toxicity by improving nutritional efficiency in sorghum and sunflower plants. Plant Physiol Biochem. 2019;142:224–33. https://doi.org/10.1016/j.plaphy.2019.07.010.

Calero Hurtado A, Chiconato DA, Prado RM, et al. Silicon application induces changes C:N:P stoichiometry and enhances stoichiometric homeostasis of sorghum and sunflower plants under salt stress. Saudi J Biol Sci. 2020;27:3711–9. https://doi.org/10.1016/j.sjbs.2020.08.017.

Carmello QAC, Rossi F. Hidroponia: solução nutritiva – Manual. Viçosa: Centro de Produções Técnicas; 1997.

Carvalho JM, Barreto RF, Prado RM, et al. Elevated CO_2 and warming change the nutrient status and use efficiency of *Panicum maximum* Jacq. PlosOne. 2020a;15:e0223937. https://doi.org/10.1371/journal.pone.0223937.

Carvalho JM, Barreto RF, Prado RM, et al. Elevated [CO₂] and warming increase the macronutrient use efficiency and biomass of *Stylosanthes capitata* Vogel under field conditions. J Agron Crop Sci. 2020b;206:597–606.

Castellane PD, Araujo JAC. Cultivo sem solo - Hidroponia. Funep: Jaboticabal; 1995.

Cavalcante VS, Prado RM, Vasconcelos RL, et al. Growth and nutritional efficiency of watermelon plants grown under macronutrient deficiencies. HortScience. 2019;54:738–42. https://doi.org/10.21273/HORTSCI13807-18.

Cobra Netto A, Acoorsi WR, Malavolta E. Studies on the mineral nutrition of the bean plant (*Phaseolus vulgaris* L., var). An Esc Super Agric Luiz de Queiroz. 1971;28(257):274. https://doi.org/10.1590/S0071-12761971000100018.

Coelho AM, França GE, Pitta GVE, et al. Cultivo do milho: diagnose foliar do estado nutricional da planta. Sete Lagoas: Embrapa; 2002.

Cometti NN, Furlani PR, Ruiz HA, et al. Soluções nutritivas: formulação e aplicações. In: Fernandes MS, editor. Nutrição mineral de plantas. Viçosa: Sociedade Brasileira de Ciência do Solo; 2006. p. 89–114.

Dantas JP, Bergamin Filho H, Malavolta E. Studies on the mineral nutrition of Vigna Sinensis. II. Effects of deficiencies of macronutrients on growth, yield and leaf composition. An Esc Super Agric Luiz de Queiroz. 1979;36:247–57. https://doi.org/10.1590/S0071-12761979000100014.

David CHO, Paiva Neto VB, Campos CNS, et al. Nutritional disorders of macronutrients in *Bletia catenulata*. HortScience. 2019;54:1836–9. https://doi.org/10.21273/HORTSCI14284-19.

Deus ACF, Prado RM, ALVAREZ RCF, et al. Role of silicon and salicylic acid in the mitigation of nitrogen deficiency stress in rice plants. SILICON. 2019;11:1–9. https://doi.org/10.1007/s12633-019-00195-5.

Dias LE, Faria SM, Franco AA. Crescimento de mudas de *Acacia mangium* Willd em resposta à omissão de macronutrientes. Rev Árvore. 1994;18:123–31.

Djanaguiraman M, Durga Devi D, Shankler AK, et al. Selenium – na antioxidative protectant in soybean during senescence. Plant Soil. 2005;272:77–86. https://doi.org/10.1007/s11104-004-4039-1.

Epstein E. Mineral nutrition of plants: principles and perspectives. New York: Wiley; 1972.

Epstein E. Nutrição mineral das plantas e perspectivas. Portuguese edition: Malavolta E. São Paulo: Universidade de São Paulo; 1975. p. 1975.

Epstein E. Photosynthesis, inorganic plant nutrition, solutions, and problems. Photosynth Res. 1995;46:37–9. https://doi.org/10.1007/BF00020413.

Epstein E. Silicon in plant nutrition. In: abstracts of the 2rd silicon in agriculture conference. Tsuruoka: Japanese Society of Soil Science and Plant Nutrition; 2002. p. 1.

Epstein E, Bloom A. Nutrição mineral de plantas: princípios e perspectivas. In: Maria Edna Tenório Nunes, Português editors. Londrina: Planta; 2006.

Erying C, Ling Q, Yanbing Y, et al. Variability of nitrogen use efficiency by foxtail millet cultivars at the seedling stage. Pesq Agropec Bras. 2020;55:1–9. https://doi.org/10.1590/s1678-3921.pab2020.v55.00832.

Fageria NK. Potassium use efficiency of upland rice genotypes. Pesq Agropec Bras. 2000;35:2115–20. https://doi.org/10.1590/S0100-204X2000001000025.

Fageria NK, Baligar VC, Jones CA. Rice in. In: Fageria NK, Baligar VC, Jones CA, editors. Growth and mineral nutrition of field crops. M. New York: Dekker; 1997. p. 283–343.

Fasabi JAV, Carências de macro e micronutrientes em plantas de malva (Urena lobata), variedade BR-01. Dissertação, Faculdade de Ciências Agrárias do Pará. 1996.

Felisberto G, Prado RM, Oliveira RLL, et al. Are nanosilica, potassium silicate and new soluble sources of silicon effective for silicon foliar application to soybean and rice plants? SILICON. 2020;12:00668. https://doi.org/10.1007/s12633-020-00668-y.

Fernandes LA, Alves DS, Ramos SJ, et al. Mineral nutrition of *Cyclanthera pedata*. Pesq Agropec Bras. 2005;40:719–22. https://doi.org/10.1590/S0100-204X2005000700014.

Ferreira RLC, Prado RM, Souza Júnior JP, et al. Oxidative stress, nutritional disorders, and gas exchange in lettuce plants subjected to two selenium sources. J Soil Sci Plant Nutr. 2020;20 https://doi.org/10.1007/s42729-020-00206-0.

Flores RA, Borges BMMN, Almeida HJ, et al. Growth and nutritional disorders of eggplant cultivated in nutrients solutions with suppressed macronutrients. J Plant Nutr. 2014;38:1097–9. https://doi.org/10.1080/01904167.2014.963119.

Flores RA, Arruda EM, Damin V, et al. Physiological quality and dry mass production of *Sorghum bicolor* following silicon (Si) foliar application. Aust J Crop Sci. 2018;12:631–8. https://doi.org/10.21475/ajcs.18.12.04.pne967.

Föhse D, Claassen N, Jungk A. Phosphorus efficiency of plants. I. External and internal P requirement and P uptake efficiency of different plant species. Plant Soil. 1988;110:101–9. https://doi.org/10.1007/BF00010407.

Franco CF, Prado RM. Nutrition solutions in the culture of guava: effect in the development and nutricional state. Acta Scient Agron. 2006;28:199–205 https://doi.org/10.4025/actasciagron.v28i2.1042.

Frazão JJ, Prado RM, de Souza Júnior JP, et al. Silicon changes C:N:P stoichiometry of sugarcane and its consequences for photosynthesis, biomass partitioning and plant growth. Sci Rep. 2020;10:12492. https://doi.org/10.1038/s41598-020-69310-6.

Freitas FA, Koop MM, Souza RO, et al. Nutrient absorption in aluminum stressed rice plants under hydroponic culture. Ciênc Rural. 2006;36:72–9. https://doi.org/10.1590/S0103-84782006000100011.

Furlani PR, Clark RB. Screening sorghum for aluminium tolerance in nutrient solutions. Agron J. 1981;73:587–94.

Furlani PR. Cultivo de alface pela técnica de hidroponia – NFT. Campinas: Instituto Agronômico; 1995. 18p.

Gabelman WH, Gerloff GC. The search for and interpretation of genetic controls that enhance plant growth under deficiency levels of a macronutrient. Plant Soil. 1983;72:335–50.

Gonçalves FC, Neves OSC, Carvalho JG. Nutritional deficiency in "Umbuzeiro" seedlings caused by the omission of macronutrients. Pesq Agropec Bras. 2006;41:1053–7. https://doi.org/10.1590/S0100-204X2006000600023.

Goussain MM, Moraes JC, Carvalho JG, et al. Effect of silicon application on corn plants upon the biological development of the fall armyworm *Spodoptera frugiperda* (J.E. Smith) (Lepidoptera: Noctuidae). Neotrop Entomol. 2002;31:305–10. https://doi.org/10.1590/S1519-566X2002000200019.

Graham RD. Breeding for nutritional characteristics in cereals. In: Tinker PB, Lauchli A, editors. Advances in plant nutrition. New York: Praeger; 1984. p. 57–102.

Grant CA, Flaten DN, Tomasiewics DJ, et al. A importância do fósforo no desenvolvimento inicial da planta. Informações Agronômicas. 2001;95:1–5.

Guedes VHF, Prado RM, Frazão RJJ, et al. Foliar-applied silicon in sorghum (*Sorghum bicolor* L.) alleviate zinc deficiency. SILICON. 2020;13:0825–3. https://doi.org/10.1007/s12633-020-00825-3.

Haag HP, Sarruge JR, Camargo PN, et al. Studies on the mineral diet of coffee. XXVI. Effects of multiple deficiencies in mineral appearance, growth and composition. An Esc Super Agric Luiz de Queiroz. 1969;26:119–39. https://doi.org/10.1590/S0071-12761969000100011.

Habermann E, Oliveira EAD, Contin DR, et al. Warming and water deficit impact leaf photosynthesis and decrease forage quality and digestibility of a C4 tropical grass. Physiol Plantarum. 2019;165:383–402. https://doi.org/10.1111/ppl.12891.

Hoagland DR, Arnon DI. The water culture method for growing plants without soils. Berkeley: California Agricultural Experimental Station; 1950. 347p

Jensen MH, Collins WL. Hydroponic vegetable production. In: Janick J, editor. Horticultural reviews. New York: Willey Press; 1985. p. 483–557.

Jo J, Jang YS, Kim KY, et al. Isolation of ALU1-P gene encoding a protein with aluminum tolerance activity from arthrobacter viscosus. Biochem Bioph Res Co. 1997;239:835–9. https://doi.org/10.1006/bbrc.1997.7567.

Jones JB Jr. Plant nutrition manual. Boca Raton: CRC Press; 1998. 147p

Kathpalia R, Bhatla SC. Plant mineral nutrition. In: Bhatla SC, Lal MA, editors. Plant physiology, development and metabolism. Singapore: Springer; 2018. p. 37–81. https://doi.org/10.1007/978-981-13-2023-1_2.

Kolesnikov M, Gins V. Forms of silicon in medicinal plants. Appl Biochem Microbiol. 2001;37:524–7.

Körndorfer GH, Snyder GH, Ulloa M, et al. Calibration of soil and plant silicon analysis for rice production. J Plant Nutr. 2001;24:1071–84. https://doi.org/10.1081/PLN-100103804.

Körndorfer GH, Pereira HS, Camargo MS. Papel do silício na produção da cana-de-açúcar. Stab. 2002;21:6–9.

Lata-Tenesaca LF, Prado RM, Piccolo CM et al. Silicon modifies C:N:P stoichiometry, and increases nutrient use efficiency and productivity of quinoa. Sci Rep 2021;11:9893. https://doi.org/10.1038/s41598-021-89416-9.

Lauchli A. Soil science in the next twenty five years: does a biotechnology play a role? Soil Sci Soc Am J. 1987;51:1405–9. https://doi.org/10.2136/sssaj1987.03615995005100060003x.

Lavres Junior J, Boaretto RM, Silva MLS, et al. Deficiencies of macronutrients on nutritional status of castor bean cultivar Iris. Pesq Agropec Bras. 2005;40:145–51. https://doi.org/10.1590/S0100-204X2005000200007.

Lehr JJ. Sodium as a plant nutrition. J Sci Food Agric. 1953;4:460–1. https://doi.org/10.1002/jsfa.2740041002.

Li B, Mckeand SE, Allen HL. Genetic variation in nitrogen use efficiency of loblolly pine seedlings. For Sci. 1991;37:613–26.

Lima Filho OF, Malavolta E. Symptoms of nutritional disorders in stevia (Stevia rebaudiana (bert.) bertoni). Sci Agric. 1997;54:53–61. https://doi.org/10.1590/S0103-90161997000100008.

Loué A. Oligoelements en agriculture. Paris: SCPA Nathan; 1993. 577p

Ma JF, Takahashi E. Soil, fertiliser, and plant silicon research in Japan. Amsterdam: Elsevier; 2002.

Ma JF, Yamaji N. A cooperative system of silicon transport in plants. Trends Plant Sci. 2015;20:435–42. https://doi.org/10.1016/j.tplants.2015.04.007.

Malavolta E. Elementos de nutrição de plantas. São Paulo: Agronômica Ceres; 1980. 251p

Malavolta E. Manual de nutrição mineral de plantas. São Paulo: Agronômica Ceres; 2006.

Malavolta E, Vitti GC, Oliveira SA. Avaliação do estado nutricional das plantas: princípios e aplicações. Piracicaba: Associação Brasileira de Potassa e do Fósforo; 1997. 319p

Mantovani C, Prado RM, Pivetta KFL. Silicon foliar application on nutrition and growth of Phalaenopsis and Dendrobium orchids. Sci Hortic. 2018;18:83–92. https://doi.org/10.1016/j.scienta.2018.06.088.

Mantovani C, Pivetta KFL, Prado RM, et al. Silicon toxicity induced by different concentrations and sources added to in vitro culture of epiphytic orchids. Sci Hortic. 2020;265:109272. https://doi.org/10.1016/j.scienta.2020.109272.

Marin A, Santos DMM, Banzatto DA, et al. Seed germination of pigonpea (Cajanus cajan (L.) Millsp.) under water stress and aluminum sublethal doses. Bragantia. 2004;63:13–24. https://doi.org/10.1590/S0006-87052004000100002.

Marschner H. Mineral nutrition of higher plants. London: Academic Press; 1986.

Marschner H. Mineral nutrition of higher plants. London: Academic; 1995.

Martinez HEP, Silva Filho JB. Introdução ao cultivo hidropônico. Viçosa: Universidade Federal; 2004.

Mccray JM, Ezenwa IV, Rice RW et al. Sugarcane plant nutrient diagnosis. 2006. http://edis.ifas.ufl.edu/sc075. Accessed 21 Sept 2006

Mendonça RJ, Cambraia J, Oliva MA, et al. Rice cultivars ability to change nutrient solution pH in the presence of aluminum. Pesq Agropec Bras. 2005;40:447–52. https://doi.org/10.1590/S0100-204X2005000500004.

Mengel K, Kirkby EA. Principles of plant nutrition. Worblaufen-Bern: International Potash Institute; 1987.

Menosso OG, Costa JA, Anghinoni I, et al. Root growth and production of organic acids by soybean cultivars with different tolerance to aluminum. Pesq Agropec Bras. 2001;36:1339–45. https://doi.org/10.1590/S0100-204X2001001100003.

Miyake Y, Takahashi E. Silicon deficiency of tomato plant. Soil Sci Plant Nutr. 1978;24:175–89. https://doi.org/10.1080/00380768.1978.10433094.

Miyake Y, Takahashi E. Effect of silicon on the growth of soybean plants in solution culture. Soil Sci Plant Nutr. 1985;31:625–36. https://doi.org/10.1080/00380768.1985.10557470.

Monteiro FA, Ramos AKB, Carvalho DD, et al. Growth of *Brachiaria brizantha* Stapf. cv. Marandu in nutrient solution with macronutrient omissions. Sci Agric. 1995;52:135–41. https://doi.org/10.1590/S0103-90161995000100022.

Morgan PW (1990) Effects of abiotic stresses on plant hormone systems. In: Alscher RC, Cumming JR Stress responses in plants: adaptation and acclimation mechanisms. Willey-Liss, New York, pp. 113–146.

Müntz K. Deposition of storange proteins. Plant Mol Biol. 1998;38:77–99.

Oliveira KR, Souza JP Jr, Bennett SJ. Exogenous silicon and salicylic acid applications improve tolerance to boron toxicity in field pea cultivars by intensifying antioxidant defence systems. Ecot Environ Safety. 2020a;201:110778. https://doi.org/10.1016/j.ecoenv.2020.110778.

Oliveira RLL, Prado RM, Felisberto G, et al. Silicon mitigates manganese deficiency stress by regulating the physiology and activity of antioxidant enzymes in sorghum plants. J Soil Sci Plant Nutr. 2019;19:524–534. https://doi.org/10.1007/s42729-019-00051-w.

Oliveira KS, Prado RM, Guedes VHF. Leaf spraying of manganese with silicon addition is agronomically viable for corn and sorghum plants. J Soil Sci Plant Nutr. 2020b;20:00173–6. https://doi.org/10.1007/s42729-020-00173-6.

Osório CRWS, Teixeira GCM, Barreto RF, et al. Macronutrient deficiency in snap bean considering physiological, nutritional, and growth aspects. PlosOne. 2020;15:e0234512. https://doi.org/10.1371/journal.pone.0234512.

Peixoto MM, Flores RA, Couto CA, et al. Silicon application increases biomass yield in sunflower by improving the photosynthesizing leaf area. SILICON. 2020;13:0818–2. https://doi.org/10.1007/s12633-020-00818-2.

Perry C, Tucker K. Biosilicification: the role of the organic matrix in the structure control. J Biol Inorg Chem. 2000;5:537–50.

Prado RM. Effect of limestone application on development, nutritional status and fruit production of guava and star fruit during three years in orchards under implantation. Jaboticabal, São Paulo State University - Doctoral thesis, 2003.

Prado RM, Leal RM. Nutritional disorders due to deficiency in sunflower var. Catissol 01. Pesq Agropec Trop. 2006;36:173–9.

Prado RM, Natale W. Effect of application of calcium silicate on growth, nutritional status and dry matter production of passion fruit seedlings. Rev Bras Eng Agríc Ambient. 2005;9:185–190. https://doi.org/10.1590/S1415-43662005000200006.

Prado RM, Correa MCM, Cintra ACO et al. Micronutrients released from one basic slag applied a ultisol cultivated with guava plants (*Psidium guajava* L.). Rev Bras Frutic. 2002a;24:536–542. https://doi.org/10.1590/S0100-29452002000200051.

Prado RM, Coutinho ELM, Roque CG et al. Evaluation of slag and calcareous rocks as corrective of the acidity of the ground in the culture of lettuce Pesq Agrop Bras. 2002b;37:539–546. https://doi.org/10.1590/S0100-204X2002000400016.

Prado RM, Cruz FJR, Ferreira RLC. Selenium biofortification and the problem of its safety. In: Shiomi N, editor. Superfood and functional food: an overview of their processing and utilization. Rijeka: InTech; 2017. p. 221–38.

Prado RM, Felisberto G, Barreto RF. Nova abordagem do silício na mitigação de estresse por deficiência de nutrientes. In: Prado RM, Campos CNS, editors. Nutrição e adubação de grandes culturas. Jaboticabal: FCAV; 2018. p. 17–26.

Raboy V. Seeds for a better future: 'low phytate' grains help to overcome malnutrition and reduce pollution. Trends Plant Sci. 2001;6:458–62. https://doi.org/10.1016/S1360-1385(01)02104-5.

Rains DW. Mineral metabolism. In: Bonner J, Varner JE, editors. Plant biochemistry. New York: Academic; 1976. p. 561–98.

Rocha Filho JVC, Haag HP, Oliveira GD. The effects of mineral nutrient deficiencies on *Eucalyptus urophylla* growth in nutrient solutions. An Esc Super Agric Luiz de Queiroz. 1978;35:19–34. https://doi.org/10.1590/S0071-12761978000100002.

Rocha JR, Prado RM, Teixeira GCM et al. Si fertigation attenuates water stress in forages by modifying carbon stoichiometry, favouring physiological aspects. J Agron Crop Sci. 2021;207:12479. https://doi.org/10.1111/jac.12479.

Rosolem CA, Leite VM. Coffee leaf and stem anatomy under boron deficiency. Rev Bras Ciênc Solo. 2007;31:477–83. https://doi.org/10.1590/S0100-06832007000300007.

Rozane DE, Natale W, Prado RM, et al. Size of samples for nutritional status assessment of mango trees. Rev Bras Frutic. 2007;29:371–6. https://doi.org/10.1590/S0100-29452007000200035.

Rublo G, Liao H, Yan X, et al. Topsoil foraging and its role in plant competitiveness for phosphorus in common bean. Crop Sci. 2003;43:598–607. https://doi.org/10.2135/cropsci2003.5980.

Salvador JO, Muraoka T, Rossetto R, et al. Symptoms of mineral deficiencies in cupuaçu plants (*Theobroma gramdiflorum*) grown in nutrient solution. Sci Agric. 1994;51:407–14. https://doi.org/10.1590/S0103-90161994000300005.

Salvador JO, Moreira A, Muraoka T. Visual symptoms of micronutrient deficiency and of mineral content in guava young plant leaves. Pesq Agropec Bras. 1999;34:1655–62. https://doi.org/10.1590/S0100-204X1999000900016.

Samonte SOPB, Wilson LT, Medley JC, et al. Nitrogen utilization efficiency: relationships with grain yield, grain protein, and yield-related traits in rice. Agron J. 2006;98:168–76. https://doi.org/10.2134/agronj2005.0180.

Santi A, Camargos SL, Scaramuzza WLMP, et al. The macronutrients deficiency in sorghum. Ciênc Agrotec. 2006;30:228–33. https://doi.org/10.1590/S1413-70542006000200006.

Santos OS. Hidroponia da alface. Santa Maria: Imprensa Universitária; 2000.

Santos LCN, Teixeira GCM, Prado RM, et al. Response of pre-sprouted sugarcane seedlings to foliar spraying of potassium silicate, sodium and potassium silicate, nanosilica and monosilicic acid. Sugar Tech. 2020;22:00833. https://doi.org/10.1007/s12355-020-00833-y.

Sarcinelli TS, Ribeiro ES Jr, Dias LE, et al. Symptoms of nutritional deficiency in seedlings of *Acacia holosericea* submitted to absence of macronutrients. Rev Árvore. 2004;28:173–81. https://doi.org/10.1590/S0100-67622004000200003.

Siddiqi MY, Glass ADM. Utilisation index: a modified approach to the estimation and comparison of nutrient utilisation efficiency in plants. J Plant Nutr. 1981;4:289–302. https://doi.org/10.1080/01904168109362919.

Silva JRS, Falcão NPS. Characterization of symptoms of nutritional deficiencies in peach palm cultivated in nutrient solution. Acta Amazon. 2002;32:529–39. https://doi.org/10.1590/1809-43922002324539.

Silva Junior GB, Prado RM, Campos CNS, et al. Silicon mitigates ammonium toxicity in yellow passion fruit seedlings. Chil J Agric Res. 2019;79:425–34. https://doi.org/10.4067/S0718-58392019000300425.

Silva ES, Prado RM, Soares AAVL, et al. Response of corn seedlings (*Zea mays* L) to different concentrations of nitrogen in absence and presence of silicon. SILICON. 2020;12:00480–8. https://doi.org/10.1007/s12633-020-00480-8.

Silva JLF, Prado RM. Elucidating the action mechanisms of silicon in the mitigation of phosphorus deficiency and enhancement of its response in sorghum plants. J Plant Nutr. 2021;45:8155. https://doi.org/10.1080/01904167.2021.1918155.

Silva DL, Prado RM, Tenesaca LFL et al. Silicon attenuates calcium deficiency by increasing ascorbic acid content, growth and quality of cabbage leaves. Sci Rep. 2021;11:1770. https://doi.org/10.1038/s41598-020-80934-6.

Souza Júnior JP, Prado RM, Sarah MMS, et al. Silicon mitigates boron deficiency and toxicity in cotton cultivated in nutrient solution. J Plant Nutr Soil Sci. 2019;182:805–14. https://doi.org/10.1002/jpln.201800398.

Souza Junior JP, Frazão JJ, Morais TCB, et al. Foliar spraying of silicon associated with salicylic acid increases silicon absorption and peanut growth. SILICON. 2020;12:00517. https://doi.org/10.1007/s12633-020-00517-y.

Souza JZ, Prado RM, Silva SLO, et al. Silicon leaf fertilization promotes biofortification and increases dry matter, ascorbate content, and decreases post-harvest leaf water loss of chard and kale. Comm Soil Sci Plant Anal. 2018;50:164–72. https://doi.org/10.1080/00103624.2018.1556288.

Stein AJ. Global impacts of human mineral malnutrition. Plant Soil. 2010;335:133–54. https://doi.org/10.1007/s11104-009-0228-2.

Subbarao GV, Ito O, Berry WL, et al. Sodium – a functional plant nutrient. Crit Rev Plant Sci. 2003;22:391–416. https://doi.org/10.1080/07352680390243495.

Svecnjak Z, Rengel Z. Canola cultivars differ in nitrogen utilization efficiency at vegetative stage. Field Crops Res. 2006;97:221–6. https://doi.org/10.1016/j.fcr.2005.10.001.

Swiader JM, Chyan Y, Freiji FG. Genotypic differences in nitrate uptake and utilization efficiency in pumpkin hybrids. J Plant Nutr. 1994;17:1687–99. https://doi.org/10.1080/01904169409364840.

Teixeira GCM, Prado RM, Rocha AMS, et al. Silicon in pre-sprouted sugarcane seedlings mitigates the effects of water deficit after transplanting. J Soil Sci Plant Nutr. 2020a;20:00170–4. https://doi.org/10.1007/s42729-019-00170-4.

Teixeira GCM, Prado RM, Oliveira KS, et al. Silicon increases leaf chlorophyll content and iron nutritional efficiency and reduces iron deficiency in sorghum plants. J Soil Sci Plant Nutr. 2020b;20:1311–20. https://doi.org/10.1007/s42729-020-00214-0.

Veloso CAC, Muraoka T. Diagnosis of macronutrient deficiency symptoms in black pepper (*Piper nigrum* L.). Sci Agric. 1993;50:232–6. https://doi.org/10.1590/S0103-90161993000200010.

Veloso CAC, Muraoka T, Malavolta E, et al. Diagnosis of macronutrient deficiencies in black pepper. Pesq Agropec Bras. 1998;33:1883–8.

Viciedo DO, Prado RM, Martinez CAH, et al. Short-term warming and water stress affect *Panicum maximum* Jacq. stoichiometric homeostasis and biomass production. Sci Total Environ. 2019a;681:267–74. https://doi.org/10.1016/j.scitotenv.2019.05.108.

Viciedo DO, Prado RM, Toledo RL, et al. Silicon supplementation alleviates ammonium toxicity in sugar beet (*Beta vulgaris* L.). J Soil Sci Plant Nutr. 2019b;19:413–9. https://doi.org/10.1007/s42729-019-00043-w.

Viciedo DO, Prado RM, Toledo RL, et al. Physiological role of silicon in radish seedlings under ammonium toxicity. Food Sci Tech. 2020a;100:10587. https://doi.org/10.1002/jsfa.10587.

Viciedo DO, Prado RM, Martinez CAH, et al. Water stress and warming impact nutrient use efficiency of Mombasa grass (*Megathyrsus maximus*) in tropical conditions. J Agron Crop Sci. 2020b;206:12452. https://doi.org/10.1111/jac.12452.

Viciedo DO, Prado RM, Martinez CAH, et al. Changes in soil water availability and air-temperature impact biomass allocation and C:N:P stoichiometry in different organs of *Stylosanthes capitata* Vogel. J Environ Manag. 2021;278:111540. https://doi.org/10.1016/j.jenvman.2020.111540.

Viégas IJM, Carvalho JG, Rocha Neto OG, et al. Carência de macronutrientes em plantas de quina. Belém: Embrapa; 1998.

Viegas IJM, Frazão DAC, Thomaz MAA, et al. Nutritional limitations for Euterpe oleracea in yellow Latosol of Para state - Brazil. Rev Bras Frutic. 2004a;26:382–4. https://doi.org/10.1590/S0100-29452004000200052.

Viegas IJM, Thomaz MAA, Silva JF, et al. Effect of omission of macronutrient and boron on growth, on symptoms of nutritional deficiency and mineral composition in camucamuzeiro plants (*Myrciaria dubia*). Rev Bras Frutic. 2004b;26:315–9. https://doi.org/10.1590/S0100-29452004000200032.

Vitorello VA, Capaldi FR, Stefanuto VA. Recent advances in aluminum toxicity and resistance in higher plants. Braz J Plant Physiol. 2005;17:129–43. https://doi.org/10.1590/S1677-04202005000100011.

Warington K. The effect of boric acid and borax on the broad bean and certain other plants. Ann Bot. 1923;37:629–72. https://doi.org/10.1093/oxfordjournals.aob.a089871.

Wen TN, Li C, Chien CS. Ubiquity of selenium containing t RNA in plants. Plant Sci. 1988;57:185–93. https://doi.org/10.1016/0168-9452(88)90124-0.

Yamamoto Y, Kobayashi Y, Devi SR et al. Aluminum toxicity is associated with mitochondrial dysfunction and the production of reactive oxygen species in plant cells. Plant Physiol. 2002;128:63–72. https://doi.org/10.1104/pp.010417.

Zakir Hossain AKM, Koyama H, Hara T. Growth and cell wall properties of two wheat cultivars differing in their sensitivity to aluminum stress. J Plant Physiol. 2006;163:39–47. https://doi.org/10.1016/j.jplph.2005.02.008.

Chapter 2
Ion Uptake by Roots

Keywords Root-ion channel · Plasma membrane · ATPases · Uptake kinetics · Transport · Redistribution

In order to study the phenomena of ion uptake by roots, it is first necessary to understand the root-ion channel. Thereafter, it is necessary to know the anatomical aspects of roots and the structures that nutrients must cover, which can be barriers or open channels to the passage of elements. The uptake process is performed by mechanisms of transference of these nutrients, which can be active (with energy expenditure) or passive (without energy expenditure). Finally, it is also important to know the external (medium) and internal (plant) factors that influence nutrient uptake by plants.

2.1 Root-Ion Channel

Before roots absorb nutrients, there must be the ion-root channel, either by ion movement in the rhizosphere soil solution (diffusion or mass flow) or by root growth, which intercepts the ion (root interception) (Barber 1966). Nutrient movement in the soil is increased under adequate water conditions (water retained with pressure between field capacity, −0.03 Mpa, and permanent wilting point, −1.5 Mpa).

Root interception: When the root develops, it finds the element in the soil solution, and its presence is required for absorption.

Normally, it can be estimated as the amount of nutrients in a soil volume (rhizospheric) equal to that occupied by roots. The root interception value is relatively low as the volume of roots in the 0–20 cm soil layer occupies approximately 1–3% of the soil volume, depending on the species.

Mass flow: The movement of the element of the soil solution, in the mobile aqueous phase, toward the root (rhizosphere) as the plant transpires. This happens due to an uninterrupted connection between molecules that evaporate through the leaf and

© Springer Nature Switzerland AG 2021
R. de Mello Prado, *Mineral nutrition of tropical plants*,
https://doi.org/10.1007/978-3-030-71262-4_2

the molecule in the soil solution; the nutrient (or ion) amount that can come in contact with the root through mass-flow transport of the solution is calculated by multiplying the volume of water transpired (or absorbed) by the plant and the nutrient concentration in the solution.

Diffusion: The nutrient crosses short distances during an aqueous stationary phase, going from a region of higher concentration to another of lower concentration on the root surface. This root-ion channel is calculated by the difference between the total absorbed by the plant and the sum of root interception and mass flow.

Ions have different transport capabilities in water and soil. It was found that mobility in the soil of NO_3^-, $H_2PO_4^-$, and K^+ ranged from 10^{-10} to 10^{-11}; 10^{-12} to 10^{-15}; and 10^{-11} to 10^{-12} m^2 s^{-1}, respectively (Malavolta et al. 1997). It is estimated that NO_3^- diffuses 3 mm per day, while K^+ and $H_2PO_4^-$ would reach 0.13 and 0.9 mm, respectively.

The three ways in which root-ion channel, root interception, diffusion, and mass flow occur are shown in Fig. 2.1a.

The relative contribution of the three root-ion channel activities in nutrient supply for vegetables was studied using maize as a test plant (Table 2.1).

We observe that mass flow is more important for N, Ca, Mg, S, and some micronutrients (B, Cu, Fe, and Mo), and while diffusion is the main contact medium for P, K, and others (Barber 1966). However, in another soil cultivated with soybean, Oliver and Barber (1966) reported that the movement of Fe, Mn, and Zn in the soil occurred by diffusion.

These processes are important as they influence to determine the location of the fertilizer in relation to the seed or plant, in order to ensure increased contact of nutrients with the root absorbents and, consequently, increased fertilization efficiency (Table 2.2).

Thus, we note that the nutrient that moves by diffusion, for example, must be located close to the root to ensure increased contact with the root. Otherwise, due to decreased movement, plant needs may not be supplied, while nutrients with increased mobility in the soil, such as in mass flow, can be applied at increased distances from the plant in broadcast fertilization by topdressing, although having increased risk of leaching. Therefore, we infer that nutrients with little mobility in the soil (immobile) and with increased mobility (mobile) have restricted and wide favorable zones for root-ion channel activity, respectively (Fig. 2.1b), with the consequence for the location of fertilization.

Only in specific cultivation situations, in areas with medium to high K and P content, consolidated direct sowing systems, and adequate liming management it is possible to apply fertilizers in the total area.

a

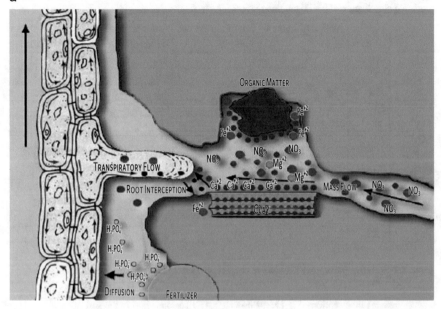

b

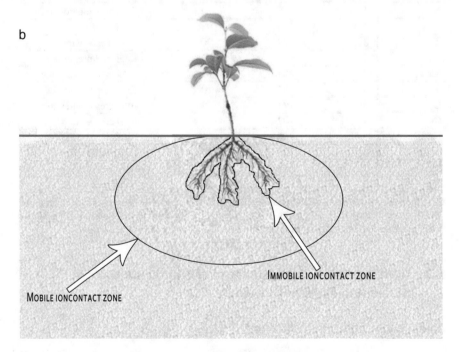

Fig. 2.1 Elements coming into contact with the root by root interception, mass flow, and diffusion (**a**) and the rhizosphere zone favorable for the channel of immobile and mobile ions (**b**)

Table 2.1 Relative contribution of root interception, mass flow, and diffusion of nutrients to maize roots in a clay loam soil (0–20 cm deep layer)

Nutrient	Uptake	Available amount	Saturation extract	Amount supplied (kg ha^{-1})		
	kg ha^{-1}	kg ha^{-1}	ppm	Interception	Mass flow	Diffusion
NO_3^-	170	–	–	2	168	0
$H_2PO_4^-$	39	45	0.5	0.9	1.8	36.3
K^+	135	190	10	3.8	35	92.2
Ca^{2+}	23	3.300	50	66	175	0
Mg^{2+}	28	800	30	16	105	0
SO_4^{2-}	20	–	–	1	19	0
H_3BO_3	0.07	1	0.20	0.02	0.70	0
Cu^{2+}	0.16	0.6	0.10	0.01	0.35	0
Fe^{2+}	0.80	6	0.15	0.1	0.53	0.17
Mn^{2+}	0.23	6	0.015	0.1	0.05	0.08
MoO_4^{-2}	0.01	–	–	0.001	0.02	0
Zn^{2+}	0.23	6	0.15	0.1	0	0,53

Table 2.2 Relative participation between contact processes and fertilizer location

Element	Contact process			Fertilizer application
	Interception	Mass flow	Diffusion	in the soil
		% of total		
N	1	99	0	Total area/topdressing
P	2	5	93	Localized/sowing
K	3	27	70	Localized/sowing
Ca	27	73	0	Total area/pre-sowing
Mg	13	87	0	Total area/ pre-sowing
S	5	95	0	Total area/topdressing
B	3	97	0	Total area/topdressing
Cu	3	97	0	Total area/sowing
Fe	13	66	21	Total area/sowing
Mn	43	22	35	Localized/sowing
Mo	5	95	0	Total area/sowing
Zn	20	20	60	Localized/sowing

Obs. Total area/topdressing refers to an application in total area ~30 days after crop emergence. Period of application may vary according to the culture, in order to meet periods of increased nutritional demand. Localized/seeding refers to an application in the sowing furrow

2.2   Root's Anatomical Characteristics and Processes of Active and Passive Uptake

Anatomical Aspects of the Root

It is important to know some anatomical aspects of the root. Thus, we will show a cross-section of the root indicating the types of existing cells (Fig. 2.2).

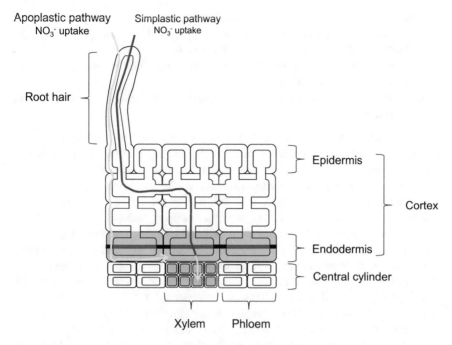

Fig. 2.2 Aspects of the anatomy of the root from a cross-section, illustrating nutrient movement through the symplast (through cell membranes) and apoplast (space between the plant cells consti-tuted by the cell wall)

The following types of cells are observed: (a) epidermis – generally a single layer of compact cells with absorbent hair; (b) cortical parenchyma (cortex) – several layers of cells spaced between them; (c) endoderm – a single layer of compact cells whose radial and transverse walls are suberin-reinforced (=Casparian strip), blocking ions passage through the walls and intercellular spaces (requiring the ion path be via symplast); (d) central cylinder – a layer of compact cells surrounding the conducting elements of phloem and xylem.

Root Nutrient Uptake Processes

It was imagined, for a long time, that elements in the soil solution were absorbed by simple diffusion, moving to a concentration gradient by going from a location of higher concentration (the external solution) to another of lower concentration (the cell sap). However, when comparing cell sap analyzes with those of the environ-ment where different species lived, it was found that, in general, the internal concen-tration of the elements was much higher than that of the external environment, with certain selectivity in element uptake. Thus, research has defined that nutrient uptake is characterized by the following factors:

- Selectivity – Consequence of the specific action of carriers acting on the membranes in the transport of solutes to the cell cytosol. Thus, ion uptake is specific and selective. Certain mineral elements are preferably absorbed.
- Accumulation – Element concentration is generally much higher in the cell sap than in the external solution.
- Genotypes – there are differences between plant species regarding uptake characteristics.

Right after the contact of the nutrient with the root (discussed earlier), the uptake process begins. Nutrient uptake by plants is defined as the entry of an element, in ionic or molecular form, into the intercellular space or in any region of the living cell of roots, such as membranes (plasmalemma). The first layer of root cells to be overcome by nutrients is the epidermis, followed by the cortex, endoderm, and the xylem. Nutrients move through the apoplastic transport pathway (ATP) or through the symplastic pathway inside the cells without crossing the plasma membrane (Fig. 2.2). This is because the plasma membranes of two adjacent cells are continuous, thanks to plasmodesmatal pores (prolongation of the cytoplasm between cells). Transport via the symplastic pathway significantly increases partitioning possibilities along the transport pathway. This can be observed in the case of P, as this ion has to travel the long symplastic pathway under deficiency. The metabolic demand along the pathway removes P from the pathway and incorporates it into the root cell metabolism, unbalancing dry matter partitioning, with increased root dry matter accumulation in relation to the shoot (Fernandes and Souza 2006).

As nutrient enters the first layer of root cells (epidermis) or other tissues, uptake occurs in two distinct phases, namely: passive and active.

Passive Phase

The passive phase corresponds to the entry of the nutrient into intercellular spaces through the cell wall or even through the plasmalemma outer surface (which is the membrane surrounding the cytoplasm) (Fig. 2.3a). In this phase, the nutrient is transported in a concentration gradient, that is, from a region of higher concentration to one of lower concentration without energy expenditure, coming directly through the transporter (the system that generates potential gradient and is the force helping to transport ions from outside to inside the cell). In this case, outside the root, nutrient concentration is higher than the concentration of the same nutrient in intercellular spaces, in the cell wall, and in the plasmalemma outer surface. These cell regions delimit the so-called apparent free space (AFS), consisting of intercellular spaces and macropores, a water free space (WFS), and micropores (Donnan free space), where cation exchange and anion repulsion occur, as the surface of these channels contain negative charges (R-COO⁻ from the cell wall fibers) that attract cations and repel anions. On the cell wall, tri- and divalent cations are attracted more strongly than monovalent cations, following the lyotropic series (Al^+ ^{3} = $H^+ > Ca^{2+} > Mg^{2+} > K^+ = NH_4^+ > Na^+$).

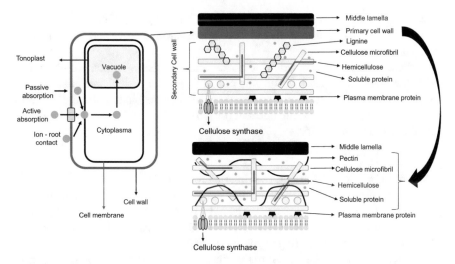

Fig. 2.3 Scheme of root-ion channel and passive uptake by a cell wall and active uptake of a nutrient by the membranes

Thus, plants containing more carboxylic groups (R-COO⁻) in the cell walls of root cells (i.e., increased cation exchange capacity – CEC) may have an increased cation reserve, which can be exchanged for H^+ or other ions and, consequently, favor its uptake into the root cells. Thus, root CEC may have influenced the sensitivity of genotypes to nutritional deficiency (cationic micronutrients).

Passive uptake occurs through several processes, such as mass flow, diffusion, ion exchange, and Donnan equilibrium. Most of these processes are similar to those that occur in the soil. This uptake mechanism is quick and reversible, that is, the nutrient in the AFS or in the cell wall can easily leave. Thus, some authors do not consider nutrients in AFS as absorbed.

Active Phase

When the nutrient reaches the plasma membrane (double layer of phospholipids encrusted with proteins) there is a barrier, as lipids in the membranes prevent the passage of ions in an aqueous solution. In this context, the uptake active phase begins, corresponding to the passage of nutrients through the plasma membrane (plasmalemma), specifically through integral proteins (Fig. 2.3b), reaching the cytoplasm and the vacuole membrane (tonoplast) until its interior. In this phase, the ion (nutrient) pathway occurs against a concentration gradient, that is, the nutrient moves from a region of decreased concentration to one of increased concentration, against the electric gradient, requiring metabolic energy expenditure. This phase corresponds to the crossing of the cytoplasmic membrane (plasmalemma) and the vacuole (tonoplast). This process is performed by an enzymatic phenomenon first described by Epstein and Hagen (1952). The theories to explain the mechanism of active uptake are not yet very clear, but it is possible to mention some main theories, namely the carrier transport theory, the Lundegardh theory, and the chemiosmotic theory.

Currently, the most accepted is the carrier transport theory, which postulates that the nutrient is transported by a specific carrier (a protein with characteristics of an enzyme) to cross the membrane, and on the inner side the binding uniting the complex (NC) would break and the nutrient (or ion) would be released. The following equation defines this theory:

$$En + C \leftrightarrow NC \leftrightarrow In + C'$$

where En = external nutrient; In = internal nutrient; C = carrier.

These carriers can be characterized as pumps (ATPases) and even channels. In plasma membranes, tonoplasts, and thylakoids, H^+-ATPases function as an H^+ pump which, triggered by ATP hydrolysis (energy+ADP^-+P), release H^+ from inside the cell to the apoplasm, generating primary active transport.

As a result, ATPase releases H^+ to the outside and makes the cytoplasm negative, generating a favorable energy situation for secondary transport via carrier (by facilitated diffusion) or another type of active transport, such as symport (when two ions pass together in opposite directions), the latter two known as cotransport. This H^+ gradient out of the cytoplasm generates an electrochemical gradient that can bind to the exit of other cations (cation antiport: H^+/Na^+) and to the entrance of cations (uniport) or anions (symport: H^+/NO_3^-).

Simultaneously, the ADP reacts with H_2O to form OH^- in the cytoplasm, so that the carrier exchanges OH^- produced by an anion. In addition, the pH gradient between the cytosol and the cell wall directly favors the reabsorption of protons into the cytosol, which may be associated with the simultaneous transport of anions. Moreover, besides ions, the H+ electrochemical gradient also provides the necessary energy for the transport of some organic compounds (Maathuis et al. 2003).

The ion pump system characterized by H^+-ATPase in the membranes, besides triggering the primary active system, generating energy for the symport and antiport nutrient transport system, can be directly involved with transport (Fig. 2.4). The integral proteins of membranes constitute the transporters. There are multiple genes that encode plasmalemma and tonoplast ATPases.

Channels are also transporters, being energy dependent and extensively studied in K uptake, as they were the first to be characterized at the molecular level. Channels are integral proteins that transport the nutrient in response to the electrochemical gradient, like K. The negative part of the protein attracts K to the channel, without water presence, and then, with the arrival of other cations in the channel, generates a series of repulsions, pumping the end of the channel and reaching the external environment. We note that channels have low specificity for solutes or nutrients.

Heavy metal transporters are members of four ATPase families. Phosphate and sulfate transporters have been identified. Micronutrient uptake systems are less well defined, in part due to problems associated with the measurement of their small fluxes, usually from small external concentrations. The genes that encode Ca channels are yet to be elucidated. Knowing the genes that encode their carriers, channels,

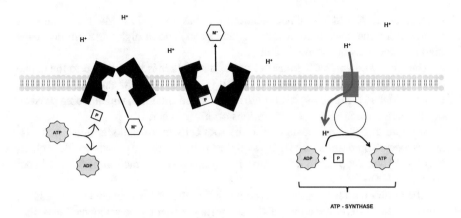

Fig. 2.4 Scheme with detail of the plasma membrane, illustrating the active nutrient uptake process through the ATP-dependent carrier and ATPase

Table 2.3 Basic aspects that characterize passive and active phases during the nutrient uptake process

Passive phase	Active phase
1. Physical or chemical process, occurs in living or nonliving systems	1. Metabolic process that occurs in a living cell
2. Not related to respiration and phosphorylation	2. It is related to respiration and phosphorylation
3. There is no direct energy expenditure for the carrier	3. There is direct energy expenditure for the carrier
4. Spontaneous	4. Nonspontaneous
5.Not influenced by inhibitors and temperature 6.Nonselective 7.Reversible	5. Influenced by inhibitors and temperature 6. Selective 7. Nonreversible

or pumps that allow nutrient uptake through studies of molecular biology could help genetic improvement to select genotypes with increased ability to absorb nutrients, having increased nutritional efficiency.

Some aspects can be highlighted when comparing passive and active uptake phases (Table 2.3).

As announced, nutrient transport through membranes can also occur through integral protein channels specialized in selective ion transport, whose action is modulated by opening and closing of the protein pore. In the membrane, several factors regulate ion transport through these channels, including light, specific hormones, and intracellular concentration of Ca^{+2} (Satter and Moran 1988). The same authors also point out that this transport is passive, as it follows the electrochemical potential gradient, being faster (a thousand times or more) than carrier transport.

Thus, the uptake kinetics of certain ions follow two phases, one related to low concentrations and the other related to high concentrations (Epstein 1972). Uptake kinetics at decreased nutrient concentrations have very low Michaelis constant (Km) (which indicates high affinity of the carrier) and at high concentration have high Km and does not seem to saturate.

Thus, it is known that K transport through membranes is attributed to the active phase by the high-affinity system, that is, it occurs at very low external concentrations; and the low-affinity system is attributed to the channels, which, being passive, occur when there are high concentrations of this element.

We note that the ion movement in the root cells to the endoderm, at short distances, occurs predominantly by the symplast pathway, as the presence of free channels (AFS) via apoplast in this region represents a small fraction of the total root volume (<10%).

After nutrients (and water) overcome these barriers and reach xylem vessels, they are transported over a long distance by root pressure (most accepted hypothesis) passively. This mechanism is called transport and will be addressed as follows.

Finally, we note that the final process of nutrient uptake by plants is vital for life on Earth, as from this moment the nutrient is part of the biosphere through consumption of that plant by humans or animals (Epstein and Bloom 2006).

2.3   Internal and External Factors Affecting Nutrient Uptake by Roots

Ionic uptake is influenced by environmental factors, that is, external factors, and by internal factors related to the plant, which can modify uptake rate, increasing or decreasing it. These factors can alter uptake efficiency (UE) by plants, which is the plant's ability to uptake nutrients per root unit. It is calculated using the following formula: UE = (total nutrient content in the plant)/(root dry matter) (Swiader et al. 1994). Mechanisms developed in plants for high uptake efficiency differ between species. Some produce an extensive root system and others have a high uptake rate per unit of root length, that is, high nutrient influx (Föhse et al. 1988). There are external and internal factors related to the plant that interfere with nutrient uptake.

External Factors

Availability

The first condition for a nutrient to be absorbed is related to being in the form available to the plant, that is, the chemical form that the plant recognizes as a nutrient (Fig. 2.5), such as boron (H_3BO_3).

For the nutrient to be absorbed, it must be in the form available in the soil solution. Thus, factors related to the passage of nutrients from the solid to the liquid phase also affect their subsequent uptake. Increased availability, measured by the

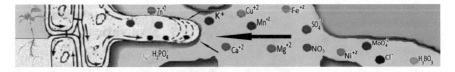

Fig. 2.5 Chemical forms of nutrients in the soil solution that can be absorbed

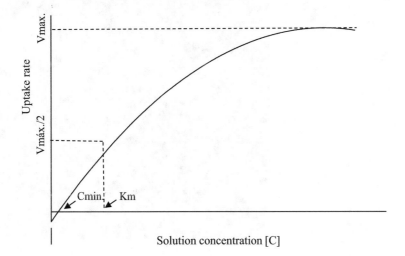

Fig. 2.6 Relationship between the ionic concentration of the solution and uptake rate according to the Michaelis–Menten equation (Km – concentration of the element that ensures ½ Vmáx. = Measurement of the affinity of a nutrient for the carrier; Cmin.– minimum initial concentration without absorption)

increased nutrient concentration in the soil solution, usually results in an increased amount absorbed per unit of time (Fig. 2.6). We note that the reaction rate or carrier action increases almost linearly at low nutrient concentrations, while the reaction rate is lower at high concentrations, as most carrier sites must be occupied.

Among factors affecting nutrient availability in the soil solution, there are pH, aeration, humidity, organic matter, temperature, and presence of other ions, among others.

We note that the most important and most studied variable aforementioned is the pH value.

- Direct effect of pH – there may be competition in the uptake process between H^+ and other cations and between OH^- and other anions for the same carrier sites in the membrane. Thus, soils with acid [H^+] or alkaline [OH^-] reactions decrease cation or anion uptake, respectively.

The acid reaction predominates in tropical soils, with a significant risk of nutritional deficiency of macronutrients such as K^+, Ca^{+2}, and Mg^{+2}. There are several factors leading to soil acidity, such as atmospheric precipitation, leaching of bases,

Fig. 2.7 Nutrients uptake
by plants illustrating the
process of cation and anion
exchanges

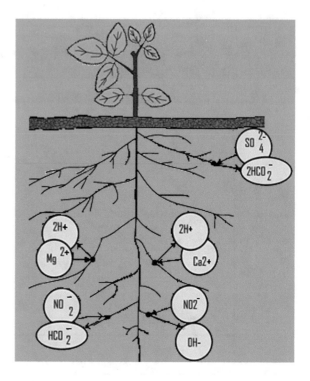

and nitrogen fertilization. In addition, the plant can acidify the medium in the rhizosphere region, as every time the plant absorbs a cation, H^+ or OH^- (if it is an anion) is released in the medium. The plant uses this device to maintain the cation–anion balance in the medium (Fig. 2.7).

In acidic soils, therefore, even decreased cation uptake tends to aggravate acidity in the rhizosphere due to this plant mechanism (H^+ release in the medium).

N, as the most absorbed nutrient, contributes the most to balance absorbed nutrients and to the release of H^+ or OH^- by the roots, besides K, Ca, Mg, P, and Fe (Marschner 1995).

- Indirect effect of pH – characterized by the fact that increased availability increases the concentration of the element in the soil solution, increasing uptake. At pH values (in water) in the range of 5.5–6.5, the availability of some nutrients is maximum (macronutrients) and not limiting for others (micronutrients) (Fig. 2.8).
- Aeration – oxygen is important during the active uptake process, which depends on the metabolic energy (ATP) originated in respiration. The compacted soil prevents O_2 flow compared to the noncompacted soil (Fig. 2.9). It was found that increased CO_2 in the root zone of soybean decreases water and nitrogen uptake and stomatal opening (Araki 2006), as well as photosynthetic activity in Pinus (Norisada et al. 2006).
- Aeration can increase nutrient availability in the soil through the activity of the soil aerobic microbiota, which oxidizes NH_4^+ to NO_3^- and S^{2-} to SO_4^{2-} from O_2.

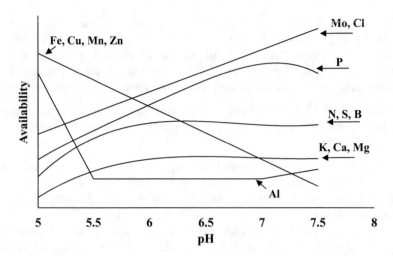

Fig. 2.8 Relationship between the pH value in the soil and nutrient availability

Fig. 2.9 O_2 flow in
noncompacted and
compacted soil

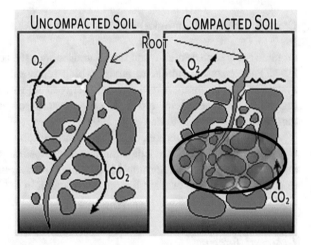

- Temperature – Uptake increases almost linearly in the 0–30 °C range with
  increasing temperature. This is explained by the fact that plant metabolic activity
  increases within certain limits, mainly due to respiratory intensity.
- Humidity – Besides influencing element availability in the soil, humidity also
  affects the uptake process, as water is the natural entry of nutrients. Rainfall or
  irrigation levels in the agricultural area can influence nutrient uptake by crops.
  However, there is not a very close relationship between water and ion uptake
  rates. K^+ and NO_3^-, for example, are absorbed much faster than water, and the
  opposite is observed for Ca^{+2}, whose uptake conditions are similar to water.
- Element – Uptake rate depends, in part, on the element considered, following the
  decreasing order:

Anions: $NO_3^- > Cl^- > SO_4^{-2} > H_2PO_4^-$;
Cations: $NH_4^+ > K^+ > Na^+ > Mg^{+2} > Ca^{+2}$

We state that hydrophobic solutes normally move more quickly through membranes at rates proportional to their solubility in lipids. Therefore, more hydrophobic solutes move more than hydrophilic solutes. Hydrophilic molecules and ions with similar lipid solubility penetrate at rates proportional to the size of the hydrated ion. Similarly, divalent cations or anions are more absorbed than trivalent cations.

• Other ions – Uptake of an element can be influenced by the presence of another element, characterizing interaction between nutrients, which will be addressed in a specific chapter.

Mycorrhizae – arbuscular mycorrhizal fungi, one of the oldest and most widespread symbioses of nature, increase the root exposure surface and element uptake a capacity, particularly P, which is usually only manifested when its concentration in the soil solution is low. It is common to inoculate coffee seedlings with mycorrhizal fungi, naturally inducing their growth, and this effect is not compensated by increasing P supply to plants (Lopes et al. 1983). This effect is only important in plants that accept mycorrhizal association.

Organic Acids and Amino Acids

Organic acids such as humic and fulvic acids induce the activity or expression of the H^+-ATPase enzyme, similar to auxin, which represents the primary H^+ transport system of the membrane, providing the energy to activate secondary uptake systems (carriers). This beneficial effect of organic acids on nutrient uptake results in increased root growth (Pinton et al. 1999).

Although amino acids have N in their composition, it is not an important source of this nutrient, but rather as a biostimulant. The biostimulant effect may be involved in the antioxidant action, hormonal production, or the transcription of genes related to nutrient absorption or assimilation, but research in field crops is lacking.

Internal Factors

Genetic Potential

The plant's genetic potential may determine whether or not it is easier to absorb a certain element. Thus, there are plant species and/or varieties that absorb and concentrate more certain nutrients, while others are inefficient in absorbing other elements. Thus, the ionic uptake process is under genetic control. Differences are manifested in several ways, such as in Km, V, and [M]min values and parameters, in the ability to solubilize elements in the rhizosphere through root excretions; and in the change of iron valence (Fe^{3+} to Fe^{2+}), which increases its uptake.

Internal Ionic State

The ion-saturated plant absorbs less than another plant with decreased ions. This is due to the fact that it reached Vmáx. and Imáx., which is the maximum uptake limit for a given nutrient.

Slightly deficient plants have increased uptake rates than nondeficient plants. Calvache et al. (1994) studied ^{32}P uptake of the solution by rice roots. According to the results, in the roots of plants with P deficit (0.0 mM P), ^{32}P uptake was 12 times higher than in the roots of nondeficient plants (0.5 and 2.0 mM P).

If the deficiency is very severe, uptake rates decrease, as irreversible metabolic damages occur.

Metabolism Intensity

The plant carbohydrate content determines its metabolic level, as these substances constitute the main respiratory substrate and the consequent energy production (ATP) to supply active uptake; that is, the higher the level of these reserves, drained from the shoot to the root, the higher the root uptake.

Transpiration Intensity

The effect of transpiration is indirect:

The transpiratory current, which carries the nutrient to the shoot in the xylem, can increase pressure, pulling the elements in the intercellular spaces and in the cell wall of the root cells.
The humidity gradient of the soil is favored with increased transpiration, increasing mass flow to the root.

Growth Intensity and Root Morphology

Uptake increases with increased intensity of plant growth. In addition, plants with well-developed roots, thinner, well-distributed, with an increased proportion of absorbent hair absorb more, especially elements that reach the root by diffusion. Thus, root morphological parameters have been widely used to evaluate nutrient uptake efficiency by plants, highlighting root area and length (Tachibana and Ohta 1983). As nutrient uptake depends, among other factors, on root morphology and the nutrient concentration in the soil solution, with the latter being a limiting factor in tropical regions, root parameters are relevant for nutrient uptake. Barber (1995) developed a mathematical model involving soil and plant variables that could explain and/or elect parameters that would most affect plant nutrient uptake. For example, important factors to predict K uptake in soybean would be the root growth rate and root radius, besides other factors (initial [K] in the soil solution; buffer power, and the diffusion coefficient of the nutrient in the soil).

2.4   Transport

Most terrestrial green plants have tissues that are far from the soil, which is a water and nutrient source. Thus, plants elaborate structures and mechanisms that transport water and nutrients inside their body. Consequently, substance transport is one of the important topics of plant nutrition (Epstein and Bloom 2006).

Transport occurs after the nutrient is absorbed by roots through epidermal cells, which is defined as the movement of the nutrient from the uptake site to any other location. There are two types of nutrient transport in plants, radial (from the epidermis to the xylem) and long distance (from the xylem to the shoot). Radial transport consists of the movement of the nutrient from the epidermal cell to the xylem vessels, occurring through the two following ways: via apoplast and/or symplast (Fig. 2.10). Nutrients can pass through intercellular spaces via apoplast pathways or move through cell walls from one cell to another until reaching the endodermis, where nutrients are blocked by Casparian strips. In the symplast pathway, nutrients travel through the interior of cells to the endoderm and may travel further inward through cytoplasmic communications between cells (plasmodium). This pathway is mandatory beyond the endoderm (Malavolta 1980).

After the nutrient reaches the xylem, there is the long-distance transport to the shoot, which occurs through a passive process for all nutrients. Long free tubes are formed after xylem cell death, allowing the transport of the nutrient solution to the shoot.

We note that the transport/translocation efficiency (TE) of nutrients by plants varies according to species and nutrient. As previously stated, TE can be calculated using the following formula: TE = ((shoot nutrient content)/(total nutrient content in the plant)) 100 (Li et al. 1991).

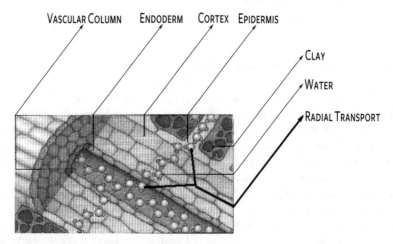

Fig. 2.10 Longitudinal section of the root illustrating the radial transport of water and nutrients by symplast pathway to the vascular column (xylem)

Studies on the transport efficiency between cultivars are important to obtain plants with increased yield and quality. In addition, a positive correlation was observed between transport efficiency and protein content in the grain of rice (Samonte et al. 2006).

Redistribution

The term redistribution refers to the transfer of nutrients from one organ to another in the same or different way from that of absorption, for example, from the leaf to the fruit in development; from an old leaf to a newer leaf; from the bark to a newer leaf.

Nutrient redistribution in the plant is a secondary process that refers to its translocation from where it was deposited by water movement in the xylem (Cerda et al. 1982) until reaching other organs via phloem vessels. Solutes (nutrients) normally tend to accumulate in organs with a higher transpiration rate (mature leaves) after absorption instead of other organs (sprouts; fruits). Plants redistribute nutrients to the different organs via phloem vessels. We note that the literature addressing solute mobility in plants comes from plant physiology, which focuses only on the movement of photosynthates, with little reference to nutrients.

The most cited theory explaining solute translocation in the phloem was created by Münch (1930), which is the pressure-flow hypothesis, with translocation from the source to the drain.

To this end, solutes go through the three following systems: diffusion in the symplast and free space; active transport across the membrane to the phloem; and passive flow through sieved tubes (pores with callose joints) (Fig. 2.11).

Nutrients with high mobility in the plant have a high concentration of this nutrient in the phloem, as immobility in the phloem is presumably caused by the inability of these elements to enter the sieved tubes. Nutrient mobility in plants (redistribution by organs in the phloem) varies from element to element, that is, the roles that the nutrient plays in the plant determine its mobility, besides interference from the culture medium and species/cultivar. Malavolta (2006) suggests the following explanations for the fact that some elements are more mobile than others: (1) permanence in ionic form and less incorporation in large molecules, favoring mobility; (2) chelate formation prevents precipitation in the vessels by OH^-, HCO_3^- or $H_2PO_4^-$; (3) the negative charge predominant in the chelate hampers cation fixation in the cell wall of the vessels; (4) increased or decreased adsorption or incorporation in the cuticle (case of Zn) or cell wall (case of B) hampers movement.

Initially, nutrient mobility was classified by Marschner (1986) as mobile, partially mobile and immobile, with further changes in this classification (Table 2.4). We note that N, P, K, Mg, and Cl are considered mobile by all authors. However, there are divergences regarding S, although visual deficiencies occur in new leaves. Recent research suggests a new classification of variable or conditional mobility, as the plant species and even the internal nutritional status can alter nutrient dynamics

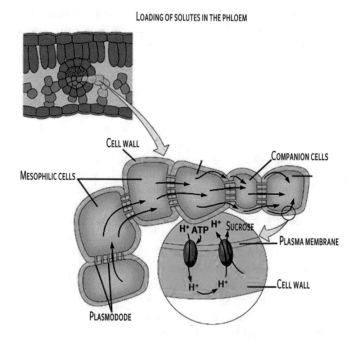

Fig. 2.11 Solute translocation (sugar and nutrients) through the phloem

Table 2.4 Classification of mobility compared only among plant nutrients

Source	Mobility classification	Mobility classification	Mobility classification
Marschner (1986)	Mobile	Mobile	Mobile
	N, P, K, Mg, Cl	N, P, K, Mg, Cl	N, P, K, Mg, Cl
Marschner (1995)	High	High	High
	N, P, K, Mg, S,Cl	N, P, K, Mg, S,Cl	N, P, K, Mg, S,Cl
Malavolta et al. (1997)	Mobile	Mobile	Mobile
	N, P, K, Mg, Cl, Mo	N, P, K, Mg, Cl, Mo	N, P, K, Mg, Cl, Mo
Welch (1999)	Mobile	Mobile	Mobile
	N, P, K, Mg, S, Cl	N, P, K, Mg, S, Cl	N, P, K, Mg, S, Cl
Epstein and Bloom (2006)	Mobile	Mobile	Mobile
	K, N, Mg, P, S, Cl, B[a]	K, N, Mg, P, S, Cl, B[a]	K, N, Mg, P, S, Cl, B[a]

[a]B is mobile in some plants (peanuts, clover, and nut-producing trees)

between plant organs. Micronutrients, like Zn and B, have restricted mobility for most crops. However, studies indicate that Zn can have mobility in the phloem, such as in wheat (Haslett et al. 2001). In other studies using the apple tree as a test plant, it was found that B translocation occurs in significant amounts in the form of a mobile organic complex (B-mannitol) in the phloem sap (Hu and Brown 1997).

The group of metallic nutrients with intermediate mobility can show very low mobility if precipitation reactions occur with oxides, hydroxides, or carbonates. However, if complexes are formed with organic acids (chelates), their mobility will be increased.

We emphasize that nutrient immobility in plants should not be thought to be absolute. All nutrients have increased or decreased mobility regarding root uptake. What happens is that, in the case of immobile nutrients, redistribution is not enough to meet the requirements or needs of new organs (leaves, branches, fruits, or roots) (Malavolta 2006).

In any case, we can infer that leaf application is not the best way to supply the nutrient in the case of immobile nutrients in the phloem, with application in the soil being the most advantageous. However, in some situations, due to low fruit transpiration, root uptake may not be sufficient to supply these fruits. In this case, localized applications have been recommended, such as Ca in tomato (fruit) to prevent apical rot, and also in apple (bitter pit).

We note that the aspect of mobility of each nutrient in the plant will be addressed in more detail in the following chapters.

The fact that part of the plant needs can be satisfied by this movement is an important consequence of redistribution in conditions not characterized by lack of supply – as long as older organs accumulate enough. Franco et al. (2005) observed in lychee seedlings that increased diameters of the hog that would originate the seedling increased nutrient accumulation in the shoot. Paredes and Primo-Millo (1988) reported that part of the need of new organs in the orange tree is satisfied by the mobilization of reserves. It is observed that the nutrients in the reserve organs meet the consumption requirements of new organs for N, P, and K, in 25–32, 12–17, and 22–29%, respectively.

In annual crops, the rate of nutrient redistribution in the reproductive phase can be a production factor, as it satisfies a significant part of the crop requirement. In wheat, there was a correlation between production and redistribution of nitrogen from the vegetative part to the grain (Xu et al. 2005). This redistribution can be affected by the environment, increasing water-deficit conditions (Xu et al. 2006). In soybeans, N biological fixation decreases considerably during flowering, and to meet the high N demand for filling the grains, N needs to be redistributed from other organs, such as the leaves that must have this reserve (5%) or from the N in the soil (Malavolta 2006). Therefore, if the N reserve in the leaves is low, grain production is likely to be affected. In rice, Souza et al. (1998) added that in the reproductive phase, the N lost in shoots corresponded to 42% and 75% of the N accumulated in the grains for cultivars IAC 47 and Piauí, respectively. The authors added that this last variety, which was not improved, showed increased redistribution efficiency as a measure of adaptation to eventual deficiency.

In sugarcane, roots can accumulate more than 30% of the applied N (Sampaio et al. 1988), showing the positive correlation between N (and S) content in the root system of sugarcane and stem production (Vitti et al. 2007). Therefore, the sugarcane stem that eventually received N application in the previous cycle may have N savings reflected in increased production, compared to that sugarcane that did not previously receive the nutrient.

Redistribution can be affected by the nutrient and nutritional status of plants. Boaretto (2006) observed that when young citrus plants were adequately supplied for boron, approximately 40% of the micronutrient in new organs came from the plant reserves, but when the plant was deficient, only 20% of the B in new organs came from the plant reserve. According to the author, this probably happened because the nutrient was in the older parts of the plant grown in deficient solution, mainly in insoluble forms, as a constituent of an organic compound (cell wall). If plants were developed in an adequate nutrient solution, the nutrient amount available in water-soluble forms (located in the apoplastic region in the form of boric acid) would increase.

Malavolta (2005) reported the quadratic decrease in the use of N and K from reserve organs during fruiting due to the leaf content of these nutrients. Thus, in plast with adequate N and K supply the redistribution is low, and fruit requirements are supplied with the reserves of the medium and not from the plant.

Therefore, saccharides, amino acids, and minerals can be transported in the phloem sap in increased quantities. In addition to these, there are proteins (~200), almost all plant phytohormones (auxin, gibberellins, cytokines, abscisic acid), RNA molecules, secondary metabolism compounds, and even chemical compounds such as insecticides, fungicides, and herbicides (Marenco and Lopes 2005). Thus, sugars predominate in the phloem sap (10–25% of mass by volume). The sugar that predominates in phloem is sucrose, although some species have other sugars, such as sugar alcohols (D-mannitol in plants of the Oleaceae family, sorbitol in *Prunus serotina* and domestic apple, and rosacea and dulcitol in plants of the Celastraceae family) (Castro et al. 2005).

According to Shelp (1988), the ratio of the concentration of a certain nutrient between new leaves and old leaves shows whether it is redistributed via phloem or not. When the ratio is much lower than 1, such as 0.5, it indicates that the nutrient is not redistributed by the phloem, being immobile, and when the ratio is higher than 1, such as 1.5, it indicates that the nutrient is redistributed via the phloem. This technique is only an indication of mobility and is not a conclusive method, such as the direct method by using the isotopic technique.

References

Araki H. Water uptake of soybean (*Glycine max* L. Merr.) during exposure to O_2 deficiency and field level CO_2 concentration in the root zone. Field Crops Res. 2006;96:98–105. https://doi.org/10.1016/j.fcr.2005.05.007.

Barber SA. The role of root interception, mass-flow and diffusion in regulating the uptake of ions by plants from soil. Viena: International Atomic Energy Agency; 1966.

Barber SA. Soil nutrient bioavailability: a mechanistic approach. New York: Wiley; 1995.

Boaretto RM, Boro (^{10}B) em laranjeira: absorção e mobilidade. Piracicaba, 120p. Tese, Centro de Energia Nucelar na Agricultura; 2006.

Calvache AM, Bernardi ACC, Oliveira FC, et al. Bioevaluation of the nutricional status of rice (*Oryza sativa* L. cv. IAC-165) and bean (*Phaseolus vulgaris* L. cv. carioca) plants using ^{15}N and [32]P. Sci Agric. 1994;51:393–8. https://doi.org/10.1590/S0103-90161994000300002.

Castro PRC, Kluge RA, Peres LEP. Manual de fisiologia vegetal: teoria e prática. Piracicaba: Agronômica Ceres; 2005.

Cerda A, Caro M, Santa Cruz F. Redistribuicion de nutrientes en limonero verna determinados por un método indirecto. Anales Edafol Agrobiol. 1982;41:697–704.

Epstein E. Mineral nutrition of plants: principles and perspectives. New York: Wiley; 1972.

Epstein E, Bloom A, Nutrição mineral de plantas: princípios e perspectivas. Maria Edna Tenório Nunes, Português editors. Londrina: Planta; 2006.

Fernandes MS, Souza SR. Absorção de nutrientes. In: Fernandes MS, editor. Nutrição mineral de plantas, vol. 1. Viçosa: Sociedade Brasileira de Ciência do Solo; 2006. p. 115–52.

Föhse D, Claassen N, Jungk A. Phosphorus efficiency of plants. I. External and internal P requirement and P uptake efficiency of different plant species. Plant Soil. 1988;110:101–9. https://doi.org/10.1007/BF00010407.

Franco CF, Prado RM, Braghirolli LF, et al. Use of pruning and different air laying diameters on plant development and nutrient accumulation in litchi. Rev Bras Frutic. 2005;27:491–4. https://doi.org/10.1590/S0100-29452005000300036.

Haslett BS, Reid RJ, Rengel Z. Zinc mobility in wheat: uptake and distribution of zinc applied to leaves or roots. Ann Bot. 2001;87:379–86. https://doi.org/10.1006/anbo.2000.1349.

Hu H, Brown PH. Absorption of boron by plants roots. In: Dell B, Brown PH, Bell RW, editors. Boron in soils and plants: reviews. Dordrecht: Kluwer Academic; 1997. p. 49–58.

Li B, Mckeand SE, Allen HL. Genetic variation in nitrogen use efficiency of loblolly pine seedlings. For Sci. 1991;37:613–26.

Lopes ES, Oliveira E, Neptune AML, et al. Effect of coffee inoculation with different species of vesicular-arbuscular mycorrhizal fungi. Rev Bras Ciên Solo. 1983;7:137–41.

Maathuis FJ, Filatov V, Herzyk P, et al. Transcriptome analysis of root transporters reveals participation of multiple gene families in the response to cation stress. Plant J. 2003;35:675–92. https://doi.org/10.1046/j.1365-313X.2003.01839.x.

Malavolta E. Elementos de nutrição de plantas. São Paulo: Agronômica Ceres; 1980. 251p.

Malavolta E. Potássio: absorção, transporte e redistribuição na planta. In: Yamada T, Roberts TL, editors. Potássio na agricultura brasileira. Piracicaba: Potafós; 2005. p. 179–238.

Malavolta E. Manual de nutrição mineral de plantas. São Paulo: Agronômica Ceres; 2006.

Malavolta E, Vitti GC, Oliveira SA. Avaliação do estado nutricional das plantas: princípios e aplicações. Piracicaba: Associação Brasileira de Potassa e do Fósforo; 1997. 319p.

Marenco RA, Lopes NF. Fisiologia vegetal: fotossíntese, respiração, relações hídricas e nutrição mineral. Viçosa: UFV; 2005.

Marschner H. Mineral nutrition of higher plants. London: Academic; 1986.

Marschner H. Mineral nutrition of higher plants. London: Academic; 1995.

Münch E. Die Stoffbewegungen in der Pflanze. Jena: Verlag von Gustav Fischer; 1930.

Norisada M, Motoshige T, Kojima K, et al. Effects of phosphate supply and elevated CO_2 on root acid phosphatase activity in *Pinus densiflora* seedlings. J Plant Nutr Soil Sci. 2006;169:274–9. https://doi.org/10.1002/jpln.200520558.

Oliver S, Barber SA. Mechanisms for the movement of Mn, Fe, B, Cu, Zn, Al and Sr from the soil to the soil to the surface of soybean roots. Soil Sci Soc Am Proc. 1966;30:468–70. https://doi.org/10.2136/sssaj1966.03615995003000040021x.

Paredes FL, Primo-Millo E, Normas para la fertilización de los agrios – Fullets Divulgación n. 5–88. València: Generalitat Valenciana; 1988.

Pinton R, Cesco S, Iacolettig G, et al. Modulation of nitrate uptake by water-extractable humic substances: involvement of root plasma membrane H^+-ATPase. Plant Soil. 1999;215:155–63. https://doi.org/10.1023/A:1004752531903.

Samonte SOPB, Wilson LT, Medley JC, et al. Nitrogen utilization efficiency: relationships with grain yield, grain protein, and yield-related traits in rice. Agron J. 2006;98:168–76. https://doi.org/10.2134/agronj2005.0180.

Sampaio EVSB, Salcedo IH, Victoria RL, et al. Redistribuition of the nitrogen reserves of 15N enriched stem cuttings and dinitrogen fixed by 90 days old sugarcane plants. Plant Soil. 1988;108:275–9. https://doi.org/10.1007/BF02375659.

Satter RL, Moran N. Ion channel in plant cell membranes. Physiol Plantarum. 1988;72:816–20. https://doi.org/10.1111/j.1399-3054.1988.tb06384.x.

Shelp BJ. Boron mobility and nutrition in brocoli (*Brassica oleracea* var. italica). Ann Bot. 1988;61:83–91. https://doi.org/10.1093/oxfordjournals.aob.a087530.

Souza SR, Stark EMLM, Fernandes MS, et al. Nitrogen remobilization during the reproductive period in two Brazilin rice varieties. J Plant Nutr. 1998;21:2049–63. https://doi.org/10.1080/01904169809365543.

Swiader JM, Chyan Y, Freiji FG. Genotypic differences in nitrate uptake and utilization efficiency in pumpkin hybrids. J Plant Nutr. 1994;17:1687–99. https://doi.org/10.1080/01904169409364840.

Tachibana J, Ohta Y. Root surface area as a parameter in relation to water and nutrient uptake by cucumber plant. Soil Sci Plant Nutr. 1983;29:387–92. https://doi.org/10.1080/00380768.1983.10434642.

Vitti AC, Trivelin PCO, Gava GJC, et al. Sugar cane yield related to the residual nitrogen from fertilization and the root system. Pesq Agropec Bras. 2007;42:249–56. https://doi.org/10.1590/S0100-204X2007000200014.

Xu Z, Yu ZW, Wang D, et al. Nitrogen accumulation and translocation for winter wheat under different irrigation regimes. J Agron Crop Sci. 2005;191:439–49. https://doi.org/10.1111/j.1439-037X.2005.00178.x.

Xu Z, Yu Z, Wang D. Nitrogen translocation in wheat plants under soil water deficit. Plant Soil. 2006;280:291–303. https://doi.org/10.1007/s11104-005-3276-2.

Chapter 3
Foliar Ion Uptake

Keywords Foliar fertilization · Air humidity · Anatomical aspects · Cuticular penetration · Mobile nutrients · Isotope

Foliar fertilization is widely used in agriculture and complements soil application to supply nutrients to the plant. In order to ensure efficient foliar fertilization, it is important to understand foliar ion uptake. For this, we must know leaf anatomical aspects, active and passive absorption processes, the external and internal factors affecting nutrient entry in the plant, and the general aspects of foliar spraying.

3.1   Introduction

It is now accepted that plant life began in water, as plants had all the necessary factors available in this habitat, where most vegetables still live today. As plants adapted to live out of water through evolution, plant parts specialized in performing certain functions. The roots specialized in fixing and absorbing nutrients; the leaves, in photosynthesis and respiration; and the stem, in solute transport, linking roots and leaves. However, shoots still have the ability to absorb nutrients.

In a brief history of foliar fertilization, we have, as follows:

1844 – Reports of application of Fe in grapevines
1874 – Application of water-diluted slurry in garden plants in Germany
1940–1945 – Great boost in ionic absorption due to radioisotope leftovers
1945 – Beginning of foliar fertilization research in Brazil by IAC (Campinas) and Esalq (Piracicaba)
1960–1970 – Large number of companies selling foliar products, often promising to replace soil fertilization

© Springer Nature Switzerland AG 2021
R. de Mello Prado, *Mineral nutrition of tropical plants*,
https://doi.org/10.1007/978-3-030-71262-4_3

Anatomical Aspects of the Leaf and Active and Passive Uptake Processes

The contact between nutrient and leaf is made mainly through foliar fertilization. In foliar fertilization, nutrients are applied in aqueous solution and need to enter the cell (cytoplasm, vacuole, or organelles) to perform their functions, as a nutrient is considered absorbed when inside the cell. For this, there are two barriers to overcome, namely the cuticle/epidermis and the membranes, plasmalemma, and tonoplast.

The epidermis and cuticle cover the upper and lower leaf surfaces. The cuticle, which is the outermost part, is of a complex chemical nature, formed by waxes, cutin, pectin, and cellulose. It is water-permeable. The epidermis provides wetting and hydrophilic properties. Foliar uptake, like root uptake, comprises a passive phase (cuticular penetration) and an active phase (cellular absorption).

(a) Passive – Consists of a non-metabolic process in which the nutrient applied to the foliar surface crosses the upper or lower cuticle (Fig. 3.1), occupying the AFS (apparent free space) formed by the cell wall, intercellular spaces, and the external surface of the plasmalemma. For cuticular uptake, solute molecules must have a diameter of less than 4–5 nanometers.

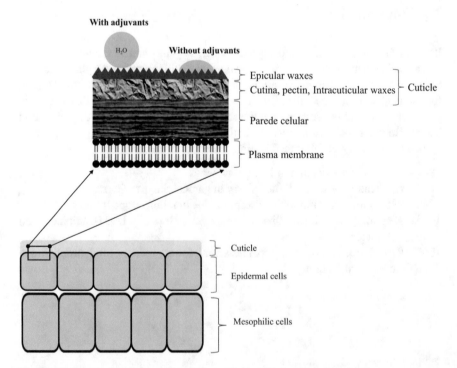

Fig. 3.1 Foliar structure from a cross-section of the leaf blade with cuticle detail

It would be the nutrient entry in the leaf apoplast. The decreasing hydrophilic order is, namely pectin> cutin> cuticular waxes. We note that, due to its structure in the form of overlapping scales and not continuous as previously thought, there is some space where solutions can pass.

We also note that some substances are capable of undoing some chemical bonds between the units of the cuticular waxes when applied to the leaf surface. The rupture of these bonds results in some openings in the cuticle, facilitating the penetration of solutions. This phenomenon is known as facilitated diffusion. Urea is an example of substance that promotes these changes, which is why it is often used in foliar spraying. Thus, urea stands out as an additive to increase the uptake rates of cations and anions (Freire et al. 1981).

(b) Active – The nutrient is effectively absorbed after overcoming the cuticle, passing through the membranes (plasmalemma and/or tonoplast) of epidermis and mesophyll cells. Afterwards, the nutrient reaches the symplast, where it is metabolized or transported between cells through cytoplasmic projections (plasmodesmata) to the phloem. Thus, nutrients are transported through long distances as in the root, although there are no Casparian strips on leaves. However, unlike roots, the phloem can be transported through apoplast on leaves. We note that this is a slow metabolic process, which occurs against a concentration gradient and requires energy supply (ATP). It is the occupation of the leaf symplast. Active nutrient uptake is moderated by a specific carrier.

In the phloem, nutrients are redistributed in forms that differ from the ones they were absorbed, such as P (hexosphosphate), N (amides), S (elemental or organic S), and micronutrients Cu, Fe, Mn, and Zn (organic, as chelates) (Malavolta 2006).

External and internal factors affecting nutrient uptake by leaves

As seen in root uptake, foliar uptake is also influenced by several external (environment) and internal (plant) factors.

External Factors

Among external factors influencing foliar uptake, the following are considered: the contact angle of the solution and leaf, the temperature and humidity, solution concentration and composition, and light.

The contact angle between solution and leaf surface is related to increased or decreased wetting of the leaf by the solution. Thus, the more the solution is spread, increased is its contact with the foliar surface and increased its possibility of uptake.

The temperature and humidity of the environment determine the drying speed of the solution applied to the foliar surface. Thus, high temperatures or low relative humidity of the air (<60%) facilitate the evaporation of the solution, contributing to decrease its permanence on the foliar surface and reduce the possibility of uptake.

The solution concentration to be applied must take into account the possibility of evaporation, and an overly concentrated solution may damage leaves. Thus, when preparing a solution, it is necessary to consider the actual conditions of its evaporation prior to its application, based on air temperature and humidity data.

The solution composition is another aspect to be considered, as each chemical element in the solution has a different uptake rate (Table 3.1) and there are relatively fast-absorbing nutrients (50% of the nutrient applied to the leaf), such as N (0.5–36 h), and slow-absorbing nutrients, such as Fe and Mo (up to 20 days). Uptake rates vary in each nutrient as data are obtained under different experimental conditions (Malavolta 1980). Recent research, such as on S, are in accordance with the data shown in Table 3.1, that is, approximately 33% of the sulfur applied to the first trifoliate leaves of bean plants was absorbed in the period of seven days (Oliveira et al. 1995). Regarding B (in citrus), researchers observed that the highest absorption efficiency occurred after 16 hours of spraying (Boaretto 2006).

There are differences in foliar uptake related to the chemical nature of the ion (cation or anion) and even to the accompanying ion. Regarding the chemical nature of the ion, we observe that the pores in the cuticle contain negative charges (polygalacturonic acids), implying increased uptake of cations in relation to anions as repulsion occurs. Thus, NH_4^+ uptake rate is higher than that of NO_3^-. Regarding the accompanying ion, studies indicate that Mg applied to apple tree leaves has increased uptake when the accompanying ion is in the form of chloride compared to nitrate or sulfate due to variation in solubility and hygroscopicity among these salts (Allen 1960).

Therefore, each nutrient has specific characteristics during uptake, with different speeds of entry into the plant and after its uptake and different mobility (nutrient transport from leaves to other organs through the phloem), varying from element to element as previously discussed (Table 3.1).

Regarding the so-called partially mobile nutrients, recent studies with citrus using the isotopic technique indicate low efficiency of these nutrients in plant nutrition. Boaretto et al. (2003), based on the results obtained, concluded that foliar fertilization with micronutrients is efficient to supply Zn, Mn, and B to sprayed leaves, but is insufficient to change the content of these micronutrients in new leaves born after foliar spraying in orange trees. The results indicate that less than 10% of the contents of Zn and Mn deposited on the foliar surface of orange trees are absorbed,

Table 3.1 Uptake rate of nutrients applied to the leaves

Nutrient	Time for 50% of total absorption
N – Urea ($CO-NH_2)_2$	0.5–36 h
P – $H_2PO_4^-$	1–15 days
K – K^+	1–4 days
Ca – Ca^{2+}	10–96 h
Mg – Mg^{2+}	10–24 h
S – SO_4^{2-}	5–10 days
Cl – Cl^-	1–4 days
Fe – Fe-EDTA	10–20 days
Mn – Mn^{2+}	1–2 days
Mo – MoO_4^{2-}	10–20 days
Zn – Zn^{2+}	1–2 days

being insufficient to increase the micronutrient contents of leaves receiving foliar fertilization. Less than 1% of the contents of Zn and Mn deposited on leaves are transported to the parts of orange trees grown after foliar fertilization, being insufficient to significantly alter leaf contents of these micronutrients in these parts. We note that the translocated amount is small (1%) and does not impress the radiographic film in contact with the new parts of leaves (Boaretto et al.2003, Fig. 3.2).

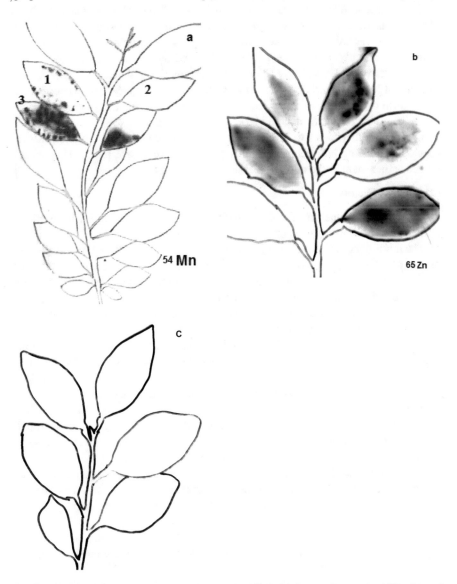

Fig. 3.2 Radioautography. Leaves 1, 2, and 3 received ^{54}Mn (**a**); Leaves that received ^{65}Zn (**b**), and New branch that developed after ^{65}Zn was applied (**c**). Leaves were outlined to locate the branch in the radiographic film

Solution pH: The solution can modify the pH of the foliar surface, changing cuticle permeability, increasing uptake rate from the beginning of the process, and increasing nutrient availability in the solution. In this sense, Rosolem et al. (1990) found that the N of solutions with decreased pH (3.0–4.0) was absorbed more quickly than that of solutions with increased pH (6.0–7.0), reaching 50% of the N applied after 5.5 and 11.5 hours, respectively.

Swanson and Whitney (1953) observed increased phosphorus uptake by bean plant leaves from solutions with decreased pH value when studying phosphate sources with variable pH. This was probably due to easier uptake of the H_2PO_4 ion found in acidic pH. Similar results were obtained by Oliveira et al. (1995), who observed increased S uptake by bean plant from sources with decreased pH values. The fastest K uptake was obtained at pH 3 and when the application was performed in the form of phosphates or citrates. For urea, the highest uptake rate occurs from pH 5–8, and the lowest from 6 and 9 (Castro et al. 2005).

Light is another factor to be considered, as it participates in the photosynthetic process, producing indispensable energy for the active uptake phase. This energy source is inexistent in the dark, decreasing uptake rates.

(a) Internal Factors

Among internal factors, that is, related to the plant, we considered the humidity of the cuticle, foliar surface, age, and ionic internal state.

The humidity of the cuticle is important for the pathway of the chemical element in the passive uptake phase. Considering that element diffusion is part of this dynamics, a minimum humidity level is indispensable for its occurrence. Dehydrated cuticles in wilted leaves are practically impermeable.

The foliar surface is an important internal factor, as the upper and lower foliar surfaces have some distinct anatomical aspects.

In the literature, it has been indicated that foliar uptake of nutrients occurs preferentially on the foliar surface, where the cuticle is thinner (such as the lower or shaded leaf surface) and the largest number of stomata are found. In this surface with an increased number of stomata there are many guard cells, which have a high amount of pores. In addition, the cuticular wax composition of guard cells is less resistant to the solute passage (Karabourniotis et al. 2001). The entry of ions through the stomatal cavity is improbable as the entry of liquids is insignificant due to its architecture (Ziegler 1987), presence of cuticle lining (although thin), and CO_2, O_2, or water vapor (positive pressure), which can also prevent the passage of solution.

Foliar age is important as there is increased cuticle development with maturation and aging of leaves, increasing solution penetration resistance and difficulty in the uptake process.

Plant age affects nutrient uptake rate and growth rate, which normally follow a sigmoidal curve where we expect increased plant response to the nutrient application in the linear part of the curve.

The ionic internal state (nutritional status), within limits, regulates the amount of elements to be absorbed, as seen for root uptake. The higher the concentration of chemical elements in the leaf cells increased the difficulty to absorb new elements.

Fig. 3.3 Nutrient application via soil and leaf

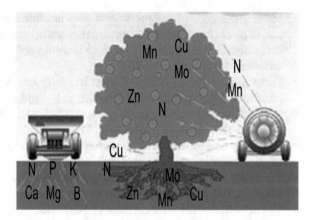

Finally, we note that plant nutrition with foliar application of nutrients should always be used as a complement to soil fertilization (Fig. 3.3).

Besides nutrients, amino acids have been indicated for foliar spraying, but there are doubts on their uptake and use in plants. Amino acids can play different roles in plants, such as stress-reducing agents, nitrogen source, and hormone precursors (Maeda and Dudareva 2012). It is important to better understand the mechanisms that amino acids use as biostimulants (hormonal and antioxidant action) and not as a source of nutrients.

General Aspects of Foliar Spraying

Foliar application with formulas like N-P-K has been negative, except for N-Urea, for the following reasons:

- Large amount of nutrients required by plants at the beginning of their development.
- Small foliar area at the beginning of the crop.
- Leaf burning problems.
- Among the many P and K forms studied, few adapt to the foliar application.
- Cost of operation.

Besides problems related strictly to compatibility, the presence of one nutrient in the solution can negatively affect the uptake of another, especially in multinutrient solutions.

Foliar fertilization has some advantages, such as:

- The high rate of utilization of the nutrients applied to the leaves by plants.
- Correction of some short-term micronutrient deficiencies.
- Possibility of applying micronutrients along with pesticides.

We note that the foliar application of nutrients requires a series of essential precautions for maximum efficiency, such as:

- Foliar application of nutrients cannot be used as a replacement of soil fertilization but as a complement, as previously stated.

- Foliar application of macronutrients does not sufficiently increase the foliar tissue, as plants have high macronutrient demands, consequently not having significant effects on production. Therefore it would not be advantageous to use this technique for these nutrients.
- Foliar application of micronutrients has their low requirement by plants as a positive point, requiring small amounts to be applied; the negative point would be the decreased mobility in the plant, that is, it will remain in increased quantity in leaves that received application. Thus, with the appearance of new leaves, there may be a repetition of deficiency symptoms. The frequency of application of micronutrients may improve their efficiency.
- The water used must be clean, as the presence of impurities such as clay can cause reactions with nutrients, reducing their action.
- The pH value of the solution must be controlled.
- The use of appropriate application technology, such as well-regulated equipment (specific spray nozzles, pressure, bar height), ensures increased homogeneity and reduced drift.
- Favorable environmental conditions, such as temperature below 30 °C, air relative humidity above 50%, wind below 3 m/s, and high probability of rain.

The use of adhesive spreader or surfactant is important to increase the solution/leaf contact surface and consequently increase uptake.

The use of humectant is important to delay drying of the solution by decreasing the deliquescence point of the solution on the foliar surface, maintaining the nutrient in ionic form for a longer time. These substances (e.g., sorbitol) are mandatory for foliar fertilization, as it is often performed in a limiting environmental condition and especially as they ensure adequate cuticular uptake of the nutrient.

References

Boaretto AE, Muraoka T, Boaretto RM. Absorption and translocation of micronutrients ([65]Zn, [54]Mn, [10]B), applied via leaf, by citrus. Laranja. 2003;24:177–98.

Castro PRC, Kluge RA, Peres LEP. Manual de fisiologia vegetal: teoria e prática. Piracicaba: Agronômica Ceres; 2005.

Freire MF, Monnerat PH, Novais RF, et al. Nutrição foliar: princípios e recomendações. Informe Agropecuário. 1981;7:54–62.

Karabourniotis G, Tzobanoglou D, Nikolopoulos D, et al. Epicuticular phenolics over guard cells: exploitation for in situ stomatal counting by fluorescence microscopy and combined image analysis. Ann Bot. 2001;87:631–9. https://doi.org/10.1006/anbo.2001.1386.

Maeda H, Dudareva N. The shikimate pathway and aromatic amino acids biosynthesis in plants. Annu Rev Plant Biol. 2012;63:73–105. https://doi.org/10.1146/annurev-arplant-042811-105439.

Malavolta E. Elementos de nutrição de plantas. São Paulo: Agronômica Ceres; 1980. 251p.

Malavolta E. Manual de nutrição mineral de plantas. São Paulo: Agronômica Ceres; 2006.

Rosolem CA, Boaretto AE, Trivelin PCO, et al. Urea absorption by cotton leaves as affected by solution pH. Pesq Agropec Bras. 1990;25:491–7.

Swanson GA, Whitney JB. Studies on the translocation of foliar applied P[32] and other radioisotopes in bean plants. Am J Bot. 1953;40:816–23. https://doi.org/10.2307/2438279.

Ziegler H. The evolution of stomata. In: Zeiger E, Farquhar GD, Cowan I, editors. Stomatal function. Stanford: Stanford University Press; 1987.

Chapter 4
Nitrogen

Keywords Symbiotic system · Nitrogen mineralization · Nitrogen uptake · Nitrogen redistribution · Nitrate reduction · Nitrogen incorporation

4.1 Introduction

When we analyze nitrogen (N) distribution in nature, we observe its predominance in the atmosphere (78.3%), being also found in the biosphere (0.27%). However, there is no such element in the lithosphere and hydrosphere. The atmosphere is the main nitrogen reservoir, reaching 82.000 t in the air surrounding 1 ha. This reservoir is practically inexhaustible, as processes (denitrification) constantly replenish the atmosphere. Despite this abundance, the N2 form in the air is not directly usable by plants, as they only recognize nitrogen in its assimilable forms ammonium (NH_4^+) or nitrate (NO_3^-). For plant nutrition, it is necessary to transform N2 gas into its assimilable forms. For this, there are three processes, namely: biological fixation, industrial fixation, and atmospheric fixation (Fig. 4.1).

The process with the greatest potential of adding nitrogen to the soil and also with the best benefit/cost ratio is the biological fixation. The free-living and symbiotic systems are the main nitrogen-fixing systems.

Free-living systems can occur in flooded rice cultivation through blue-green algae (*Azolla*), being able to fix approximately 500 kg of N ha^{-1}, and also in grasses (rice, pasture, maize, sorghum, and sugarcane). These free-living nitrogen fixers (*Azotobacter* and *Beijerinckia*) can fix approximately 30 kg ha^{-1} of N.

The symbiotic system, of greatest agricultural interest, consists of the specific association between bacteria of the genus *Rhizobium* and legumes that develop characteristic nodules. Thus, biological N_2 fixation (BNF) is a significant process that releases between 139 and 170 million tons of N per year for the biosphere, higher than the 65 million tons applied with the use of fertilizers.

In Brazil, studies with BNF started in 1963 with Dr. Joana Döbereiner (Nobel nominee in 1997) in a time when few scientists believed that this research could compete with mineral fertilizers.

Since then, most research in this area in tropical regions has been somewhat influenced by the findings of Dr. Döbereiner of EMBRAPA (Seropédica). The

© Springer Nature Switzerland AG 2021
R. de Mello Prado, *Mineral nutrition of tropical plants*,
https://doi.org/10.1007/978-3-030-71262-4_4

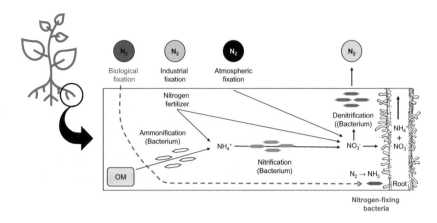

Fig. 4.1 Atmospheric nitrogen fixation processes (N_2)

Brazilian soybean improvement program, launched in 1964, was also influenced by the work of Dr. Döbereiner, among others. It has become the most successful soybean improvement program, based entirely on the BNF process. Brazil was able to compete successfully in the international market without using nitrogen fertilizers, becoming the second world producer of soybeans. This fact has represented annual savings for Brazil of more than one billion dollars in nitrogen fertilizers.

In many cases, this symbiotic system allows to suspend nitrogen fertilization in soybeans, since the system can supply approximately 60% of the N requirement. For pigeon pea, up to 90% of N comes from BNF (Valarini and Godoy 1994). We note that for biological N fixation to occur, the triple bond ($N \equiv N$) must be broken, and then three H atoms are bonded to each N, forming two NH_3 (ammonia). For this, the host plant gives carbohydrate (coming from photosynthesis) to the microorganism, which breaks the triple bond of N_2, supplying the plant with ammonia (NH_3) in exchange. The energy (ATP, electrons; H) used by the microorganism comes from compounds on photosynthesis through respiration via oxidative metabolism. Thus, biological nitrogen fixation depends on the oxidative metabolism that will supply **ATP**, **electrons**, and **H**; on an electron transport system; and, finally, on the performance of nitrogenase complex.

We note that **H** is produced during oxidative metabolism, being transferred directly to the nitrogenase complex, while electrons require a transport system (ferredoxin), where they reach the nitrogenase complex. That is, the Fe protein I (4 Fe atoms and 4 S atoms) transfers electrons to the FeMo protein II (up to 40 Fe atoms and 2 Mo atoms), with Mo and Fe as its cofactor, where N_2 binds in the presence of electrons that will break the $N \equiv N$ bond. This also requires energy (**ATP**), which originates from the oxidative metabolism powered by O_2, transported by leg hemoglobin. The presence of this protein in the nodule results in a red color inside it that is characteristic of active nodules. We note that Co is part of vitamin B12, which is required for leghemoglobin synthesis.

In the end, the ammonia (NH_3) produced in the process is transferred out of the bacterioid by diffusion and is incorporated into alpha-keto acids by the action of the enzyme glutamine synthase in the host cell (cytoplasm), forming compounds such as glutamine. Glutamine is converted into glutamate by the action of another enzyme (details on the action of these enzymes are discussed in the next item) and subsequently into urea and asparagines, which are transported via xylem to the plant shoot, where they enter the normal metabolism of N.

We highlight that the symbiotic system is a process mediated by an enzymatic complex called **nitrogenase**, with the direct participation of some nutrients such as Ca, Fe, Mo, Mg, Co, and P (ATP). Deficiency of these nutrients can decrease biological N fixation, which may cause deficiency symptoms in the legume.

Research indicates that the activity of the nitrogenase complex can benefit from application of some nutrients such as Ca, P, and Mo. N in high doses drastically reduces BNF. Waterer and Vessey (1993) demonstrated that although nitrogen is particularly inhibitory to nodule growth and nitrogenase activity, it is less harmful to the infection process. However, small N amounts in the soil (from organic matter mineralization) may favor symbiosis (Marschner 1995), possibly due to increasing photosynthesis, providing more energy for the symbiotic system.

Oxygen inhibits or disables nitrogenase activity already synthesized and in operation for being a reducing enzyme. Thus, there are mechanisms to protect the enzyme complex when NBF is occurring. Industrial fixation refers to the production of nitrogen fertilizers through industrial processes, which require a high amount of energy (1035 kJ/mol) to break the strong triple bond of the $N \equiv N$ molecule in the air through metallic iron catalysts, and depend on high temperatures ($\gg$500 °C) and pressures (200–600 atm) to combine N_2 and H_2 to produce ammonia (NH_3), the origin of several nitrogen fertilizers (urea and ammonium sulfate, among others).

Atmospheric nitrogen fixation refers to electrical discharges usual in the rainy season that unite N and O_2, forming oxides that can be decomposed or united with water, reaching the soil through the action of rain. This addition of nitrogen, considered low, can supply from 2 to 70 kg of N ha^{-1} depending on precipitation, frequency of electrical discharges, the proximity to industries that release nitrogen gases into the atmosphere, etc.

Thus, nitrogen concentration in the soil solution and its use by the plant increases with N_2 from the atmosphere through industrial, biological, or atmospheric fixation.

In the soil, nitrogen is found mostly in organic form (95%), which is nonassimilable by the plant, with the remainder nitrogen being found in assimilable mineral form, especially nitrate (NO_3^-) or ammonium (NH_4^+).

One aspect that benefits the predominance of N in organic form is that its addition in mineral form tends to pass to the organic form due to increased microbiotic activity in the soil, where a large part of the N applied in the soil is immediately absorbed by the microorganisms (incorporated into their bodies), changing to the mineral form assimilable by plants only after microorganism death.

The passage of N from organic to the mineral form, such as ammonium (NH_4^+) or nitrate (NO_3^-), is called mineralization, comprising several processes (Fig. 4.2).

N-orgânico (proteína: C-C-N-C-...) =>N-amídico (R-NH$_2$) =>N-amoniacal (NH$_4^+$) => N-nitrato (NO$_3^-$)

(aminização)   (amonificação)   (nitrificação)

Fig. 4.2 Processes involved in nitrogen mineralization

We note that the first step in the mineralization of N protein in the soil involves its hydrolysis, catalyzed by protease enzymes. The action of these enzymes creates a mixture of amino acids that undergoes a series of reactions until the first nitrogenous compound in the mineral form (NH$_4^+$) is produced. Ammonium may be subject to volatilization (NH$_4^+$ $\Leftrightarrow$ NH$_3$aq + H$^+$) as its solution concentration increases, shifting the reaction equilibrium to the right (>NH$_3$), and/or by decreasing the H$^+$ (pH > 7) concentration, since the pKa of the reaction is 9.5, that is, this is the solution pH in which the concentration of NH$_4^+$ and NH$_3$ is 50%. Therefore, a lower pH value of 7 or 9 will have, respectively, 0.4% and 36% of the total N in the NH$_3$ form. We highlight that protease activity in the soil is correlated with foliar N (Silva and Melo 2004).

Thus, knowing the factors favoring mineralization (temperature = 30 °C; humidity = 50–60% of retention capacity; aerated soil; pH > 6.0; C/N of vegetable residue <20/1) is important to maintain an increased concentration of N in mineral form in the soil solution, favoring its increased uptake by the plant.

The phenomenon of mineralization will only be effective (i.e., with predominance of mineral N) in soils with high supply of plant material, especially grasses, which have a high C:N ratio (40:80), after 15–30 days of fertilizer application. Therefore, it is important to consider this in crop management to prevent the plant from having N deficiency, even after fertilizer application.

In the study of nitrogen and other nutrients in the plant system, it is important to know all the compartments through which the nutrient is transported (soil solution, root, and shoot—leaves/fruits), that is, from the soil to its incorporation into an organic compound or as an enzyme activator, performing functions vital to enable maximum dry matter accumulation in the agricultural product (grain, fruit, etc.; Fig. 4.3).

For this, there are several processes (release from the solid to the liquid phase, root ion channel, uptake, transport, redistribution) that rule the passage of the nutrient in the different compartments to be adequately metabolized, performing their specific functions for plant life and promoting dry matter accumulation in the plant. In addition, in order to maximize the conversion of applied nutrient into agricultural production, it is necessary to supply the plant nutritional requirement, that is, to supply, in quantitative terms, the nutrients (macronutrients and micronutrients) at all stages of growth/development to achieve the expected agricultural production.

We note that each crop has a specific nutritional requirement that ensures increased conversion into agricultural product, provided that there is no nutritional disorder, whether due to deficiency or excess.

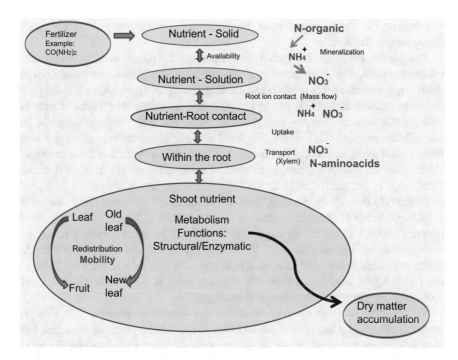

Fig. 4.3 Nitrogen dynamics in the soil–plant system, indicating the nutrient passage processes in the different plant compartments

4.2   Nitrogen Uptake, Transport, and Redistribution

Nitrogen Uptake

Before the plant absorbs nitrogen, it is necessary to link nitrogen to the root. The movement of N (as aforementioned in Chap. 2) in the soil is ruled by the mass flow phenomenon (ions moving along with water), which is responsible for more than 99% of the contact between N and root. This depends on water flow (soil–plant), which increases with the volume of water absorbed by the plant (transpiration rate), and on the concentration of the element in the soil solution. It is necessary to maintain adequate soil humidity to ensure increased contact between N and root and, consequently, increased uptake.

The uptake process begins right after contact between N and the root. The nitrogen forms of uptake by plants are, namely: N_2 (biological fixation); amino acids; urea; NH_4^+; and NO_3^- (predominating). The absorption of amino acids can occur because there are specific transporters in many species. The studies that address this subject were carried out under sterile conditions in an agar medium or nutrient solution. In the soil, the concentration of free amino acids is low due to the high competition for microorganisms. Therefore, it is estimated that the amount of nitrogen absorbed by the plant does not meet the nutritional requirement and there is a lack of research in the area.

Among the different nitrogen forms absorbed by plants noninoculated with microorganisms, the most important are nitrate (NO_3^-) and ammonia (NH_4^+). However, the nitrate form predominates during the uptake process, as it is the most abundant in the soil solution due to the high activity of the microbiota in tropical soils for nitrification ($NH_4^+ = > NO_3^-$). The NO_3^- and NH_4^+ concentrations in the soil solution are 100–50,000 and 100–2000 μmol L^{-1}, respectively (Barber 1995). We note that nitrogen absorbed as nitrate can be stored in the cell vacuole or metabolized, whereas absorbed ammonium must be completely metabolized, as it cannot accumulate in the plant due to its toxicity. When the two N forms are in similar proportions in the culture medium, we observe increased uptake of NH_4^+ compared to NO_3^- probably because this cation is absorbed through an ion channel (uniport system), without direct energy expenditure, while NO_3^- uptake usually requires a considerable amount of energy by the carrier (symport). As aforementioned, the transport system is selective, that is, nitrate has a carrier, such as NRT1 (Crawford and Glass 1998), and ammonium has another, such as AMT1 (von Wiren et al. 2000), although research has identified other carriers.

NH_4^+ uptake (uniport system) results in the release of protons (H^+) into the medium, pumped by membrane H^+-ATPases to restore the previous electrical balance, acidifying it. However, we verify the opposite when the plant absorbs NO_3^- through proton cotransport ($2H^+$), removing H^+ from the solution (Fig. 4.4). These phenomena occur as the plant seeks the maintenance of electrical neutrality inside the cytoplasm.

Thus, using ammonia as the N source acidifies the soil by proton extrusion (H^+), which associated with the inhibition of nitrification promotes an acid reaction in the rhizospheric soil. We note that decreasing the rhizosphere pH may affect the acquisition of micronutrients (Mn and Zn) and Si, decreasing the incidence of diseases (Hömheld 2005).

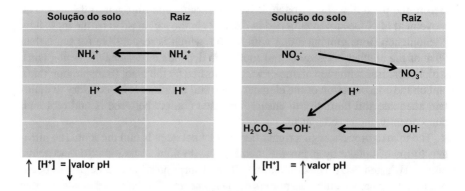

Fig. 4.4 Illustrative scheme of average variation in the pH value of the soil solution depending on whether the nitrogen source is ammonium (NH_4^+), nitrate (NO_3^-) or a mixture of both supplied to a plant

We also observe that there are some factors that can affect nitrogen uptake, such as the external and internal factors discussed previously (Chap. 2). However, there are some specific factors for N, such as:

pH Direct Effect

At an acidic pH, that is, with high H^+ concentration, there is competition with NH_4^+, inhibiting its uptake by the plant, while at a pH close to alkaline there is high OH^- concentration, inhibiting sources based on NO_3^-. Thus, regarding nitrate, the optimum pH would be below 6, as the uptake process involves H+ cotransport (Ullrich 1992). We note that it is important to balance nitrogen forms (NO_3^-, NH_4^+) to maintain a satisfactory pH value that does not affect the uptake rate of N by the plant. In addition, the use of a proportion of N in the form of ammonia saves energy for the plant, as ammonia does not need to be reduced and would go directly into carbon skeletons. However, supplying only ammonia to the plant should reduce plant dry matter, as the plant would not have the capacity to incorporate all the ammonia-N in the carbon skeletons (for lack of skeletons) on time, and with alkaline pH in the cell, ammonia would convert into NH_3 causing toxicity.

Research data in a hydroponic environment indicate the positive effect of balancing nitrogen sources. Increased ratios of nitrogen as nitrate are more advantageous for plant growth than increased proportions of N as ammonia. In forage crops, production increased when using nitrate: ammonia solution from 100:0 to 55:45, while production decreased when the proportion of nitrate/ammonia was 25:75 (Santos 2003). In passion fruit tree seedlings, it is indicated to use a maximum of 40% of N in the form of ammonia in nutritive solution with 13 mmol L^{-1} of N (Silva Junior et al. 2020). In *Apuleia leiocarpa* seedlings, the amount of ammonia must be moderate, not exceeding the 75:25 or 4:1 ratio (Nicoloso et al. 2005), while for maize the 50:50 ratio of NO_3^-:NH_4^+ performed better than 100:0 (Table 4.1). As aforementioned, the very high ammonia levels in the plant can be toxic and slow its growth (Below 2002).

Finally, the ideal ratio of nitrate to ammonium for plants depends on the species, plant age, and the average pH for crop growth (Haynes and Goh 1978).

The N form used in crops can also affect cation and anion uptake from the soil. Noller and Rhykerd (1974), while studying forage, observed increased contents of Ca, Mg, and K cations and decreased contents of P and S anions when using nitrogen in the form of nitrate (NO_3^-). Meanwhile, N in the form of ammonia (NH_4^+)

Table 4.1 Effect of N form on maize production and physiological parameters under hydroponic cultivation

Variables	NO_3^-/NH_4^+	NO_3^-/NH_4^+
	100/0	50/50
Production (t ha^{-1})	12.3	13.8
Grain number (number per plant)	652	737
N uptake by the plant (kg ha^{-1})	279	343

increased contents of P and S in plant dry matter. The different behavior for each form used is due to the balance between cation and anion.

There are also some indications to avoid nitrogen application as ammonia due to its relation with diseases. Martinez and Silva Filho (2004) reported that it is not convenient to use ammonia-N in fruit vegetables, as this ion increases the incidence of blossom-end rot.

Root Age

New roots have a high nitrogen uptake capacity, while older roots have a low capacity (Table 4.2).

Presence of Other Nutrients

In annual crops, the presence of K can increase N uptake, also serving as a counter-ion, favoring nitrate transport to the shoot. K in the shoot would transport malate to the root (Marschner 1995), which would perform decarboxylation through its oxidative metabolism, releasing HCO_3^- to the medium in exchange of nitrate uptake (Fig. 4.5).

In addition, the presence of P also increases N uptake, especially nitrate. Magalhães (1996) observed in a nutrient solution that the omission of P for 2 days reduced nitrate uptake (63%) in maize plants. In addition, in soils with low water content, the diffusion of P and consequently its absorption are impaired, influencing N uptake.

We also note that there is a negative effect of NH_4^+ on the uptake of other cations (Mg^{+2}, K^+, and Ca^{+2}) especially due to its acidifying effect on cytosol, besides competition for uptake sites.

Finally, studies on leaf removal indicate decreased N uptake by plants (Lestienne et al. 2006).

Plant age (days)	N (μmol/m root/day)
20	227
30	32
40	19
50	11
60	6
70	1
80	0.5

Table 4.2 Nutrient uptake rates on maize as a function of plant age (days)

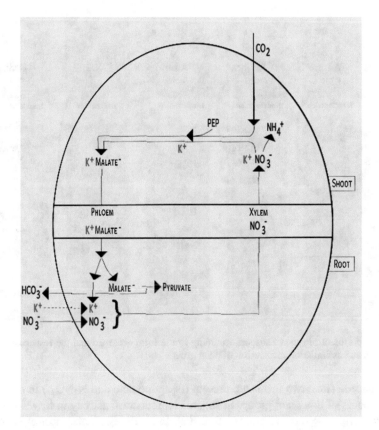

Fig. 4.5   Scheme on K circulation between root and shoot in nitrate and malate transport

Nitrogen Transport

How nitrogen is transported depends on whether it is absorbed as ammonia (NH_4^+) or nitrate (NO_3^-) and on root metabolism. Most of the NH_4^+ absorbed is generally assimilated in the roots, while NO_3^- is metabolized by roots, transported to the shoot in its original form (Mengel and Kirkby 1987), or stored in the vacuole of root cells. Therefore, nitrate (NO_3^-) absorbed in woody plants such as the rubber tree can be metabolized through nitrate reduction, as the reductase enzyme concentrates in the roots, accumulating NH_4^+. Thus, it is transported to the shoot and is incorporated into carbon skeletons, transforming into N-amino acids, as leaves concentrate enzymes GS and GOGAT. *Vaccinium macrocarpon* also reduced nitrate in the roots, and no activity of this enzyme was observed in the shoot (Dirr 1974). The presence of this enzyme in roots occurs especially in woody perennials. Assimilatory reduction does not occur in annual crops due to the absence/low activity of the enzyme in the root, and nitrate is instead transported to the shoot. In perennial herbaceous plants, part of the nitrate absorbed is metabolized in the roots and the other part is

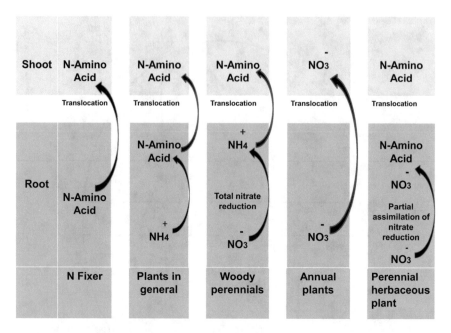

Fig. 4.6 Forms of nitrogen transport according to the form absorbed and the respective nitrate reduction and assimilation metabolism of plant groups

not reduced. Thus, two forms are present (N-amino acid and N-NO$_3^-$) in the transport process of this plant group. In sugarcane, reductase activity in leaves is much higher than in the roots (Santos et al. 2019). In N$_2$-fixing plants, most of N comes from BNF and N-amino acids (asparagine) predominate in the roots, which are transported to the shoot (Fig. 4.6).

In tropical legumes, such as cowpea, soybeans, and common beans, N transport predominates in the form of ureids (e.g., allantoin), or amides in peanuts, alfalfa, lupine, clover, peas, and lentils (e.g., asparagine and glutamine; Marenco and Lopes 2005).

In addition, the literature generally indicates that the nitrate absorbed is reduced in the roots in most species adapted to temperate climates and in conditions of low nitrate concentration in the soil. Therefore, N is transported in the ammonia form, whereas nitrate tends to be reduced in the plant shoot in species of tropical climate, without depending on the external concentration of this ion, transporting the nutrient in the same form it was absorbed (Nambiar et al. 1988). Therefore, nutrients are not necessarily transported in the ionic form absorbed in long-distance transport, being carried in organic form in low-molecular-weight compounds (N-amino acids).

In addition to amino acids, there are several low-molecular-weight organic compounds that store N (amines, ureides, etc.) and also transport organic N over long distances.

N metabolism through nitrate reduction and assimilation is discussed in the next item. We highlight that nitrogen and other nutrients are transported through the following pathway:

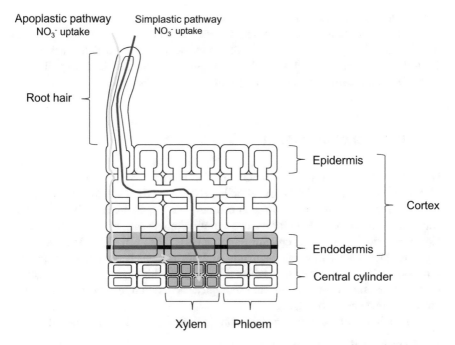

Fig. 4.7 Cross section of the root tip indicating nutrient transport via apoplast and/or symplast of cells of the epidermis, parenchyma, endoderm, and central cylinder

Epidermis = > cortical parenchyma = > endoderm = > central cylinder (xylem and phloem).

From the epidermis to the single layer of cells of the endodermis, the nutrient must overcome the cell wall and intercellular spaces (apoplast). It can also pass from one cell to another, going through the cytoplasm and its extensions between two cells (plasmodesmata), which is the symplast. Nutrient passage through the apoplast pathway and the endoderm is prevented by the presence of Casparian strips (suberin). The nutrient must be transported through symplastic pathway (active process), reaching the xylem vessels. Thus, the nutrient is transported over a long distance to the shoot through an entirely passive process (Fig. 4.7).

Nitrogen Redistribution

Nitrogen is redistributed exclusively in the form of N-amino acids, as in this phase all $N-NO_3^-$ has already been metabolized. The redistribution process occurs in the phloem, and the N-amino acid (asparagine) has high mobility, that is, if N uptake and/or transport is interrupted for any reason, the plant has the ability to mobilize N from old leaves to a new leaf or other growing organ that has high demand for this nutrient. As a result, plants with insufficient N supply (N-deficient plant) will first show deficiency symptoms in old leaves. Therefore, knowing the aspects of nutrient redistribution in plants is of significant practical importance, as it is possible to identify the characteristic symptoms of plants with nutritional deficiency in the field. We emphasize that details of the symptoms of nutritional deficiency of N are discussed in a specific item at the end (Symptoms of Deficiencies).

4.3   Participation in Plant Metabolism

Nitrogen undergoes a process of metabolization before performing its functions in plants through nitrate assimilatory reduction ($NO_3 => NH_3$), which is based on the action of reductase enzymes. Nitrogen can only participate in the metabolic pathway (GDH and GS/GOGAT) for its incorporation into carbon skeletons (from photosynthesis) in the NH_3 form, creating amino acids (...-CC-CN-CC-...) that generate proteins, coenzymes, vitamins, pigments, and nitrogenous bases that have specific functions in the life cycle of vegetables (Fig. 4.8). We highlight that N metabolism (N reduction and incorporation) and biological N fixation are the processes that most consume energy by plants.

Ammonia is formed during metabolism through the following processes:

- Nitrate reduction and biological fixation
- Photorespiration
- Metabolism of transport components, released during degradation of Asn, Arg, and ureides

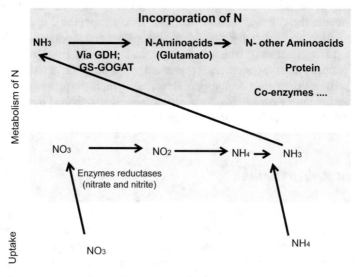

Fig. 4.8 Simplified scheme of nitrogen metabolism in plants (nitrate assimilatory reduction and nitrogen incorporation)

Nitrate Assimilatory Reduction

As electrons predominate in nitrate uptake, whose nitrogen oxidation number is +5, it is necessary to reduce its oxidation number, as the form assimilable by the plant, ammonium or ammonia (NH_4^+ or NH_3), has an oxidation number equal to −3. We note that ammonium and ammonia transform into each other in aqueous medium. The reduction requires the donation of eight electrons, which is made by the action of nitrite reductase and nitrate reductase enzymes, with increased consumption of respiratory energy. This transformation is called nitrate assimilatory reduction, being essential to incorporate nitrogen into an organic compound in the plant and to perform its various functions. The phases of nitrogen incorporation and their functions in plants are detailed in the next item.

Nitrate assimilatory reduction occurs in two phases:

(a) Reduction of nitrate to nitrite (transfer of two electrons)
(b) Reduction of nitrite to ammonia (transfer of six electrons)

Nitrate reduction requires the action of the nitrate reductase enzyme (RNO_3^-). It consists of two identical units and each unit has three prosthetic groups, namely: FAD, heme, and molybdenum complex (Campbell 1996), the latter being the active center of this complex. However, the enzyme may have other prosthetic groups (cytochrome). The activity of this enzyme is easily determined in tissue in vivo (Cazetta and Villela 2004).

This enzyme is located in the cytoplasm (meristematic cells) and obtains electrons from NADH (the main electron donor in plants) or NADPH (only in fungi and some plants), transferring these two electrons to Mo, which reduces nitrate to nitrite. NAD(P)H (coenzyme) is obtained from photosynthetic cells through glyceraldehyde phosphate (C3 plants), malate (C4 plants), or respiration, if reduction occurs in the root. The electrons pass through the R NO_3^- enzyme through the part containing the prosthetic group (heme) and the Mo complex (present in its composition) until reaching NO_3^-, which is reduced to NO_2^- (Fig. 4.9).

Experiments indicate that increased NADH increases nitrite production, that is, nitrate reduction (Viegas and Silveira 2002). In addition, the presence of elements such as Fe and especially Mo is also important for nitrate reduction. Research indicates that low activity of the nitrate reductase enzyme is observed under conditions of low nitrate and molybdenum concentrations. Increasing the nitrate concentration in the nutrient solution, associated with silicon, increased nitrate reductase activity and maize growth (Silva et al. 2020).

This enzyme can be inhibited by a high concentration of ammonium or certain amino acids/amides. Thus, nitrate reductase activity can be used to evaluate the nutritional status for nitrogen. Plants with high activity of this enzyme indicate adequate levels of nitrogen, as nitrate is a substrate of the enzyme. This diagnostic method, whether by measuring enzyme activity or metabolite content, is interesting as it is highly sensitive (high nutrient content = high enzyme activity or metabolite content). However, it has the disadvantages of high relative cost for analysis and the

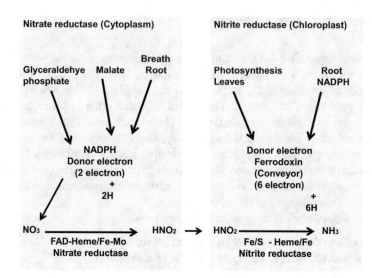

Fig. 4.9 Scheme of the reduction of nitrate and nitrite used by plants

fact that other factors may interfere with enzyme activity, besides difficult interpretation of results (lack of standards to confront data), as it is not a routine method, restricted to research.

Environmental factors promote the expression of genes that induce the production of new nitrate reductases, such as NO_2 concentration, light, and carbohydrate levels (Sivasankar and Oaks 1996).

In the second phase nitrite is reduced to ammonia, catalyzed by nitrite reductase (RNO_2), which has high activity in plants. It consists of a single polypeptide containing the two following prosthetic groups: a Fe-S cluster (Fe_2S_4) and a specialized group of heme (Siegel and Wilkerson 1989). Nitrite is a highly reactive and potentially toxic ion. This reaction occurs in the root plastids, in leaf chloroplasts, and on the outer surface of the thylakoid membrane. It requires the participation of a powerful agent, ferredoxin, which receives electrons in leaves directly from the photosynthetic electron transport chain (photosystem I) during the day; in the dark, electrons come from respiration. For the roots, the electron donor is NADPH. However, the carrier in the root is unknown, as there is no ferredoxin in the root. We note that nitrite reduction is heavily dependent on light, as ferredoxin only donates in the presence of light. In the dark, nitrite reduction occurs at very low rates.

We also note that NO_3^- may accumulate in food plants, such as forages, in the absence of Mo, which is harmful to animals. If NO_2 accumulates, it can be reduced to NO_2 in the digestive tract, which, in turn, combines with blood hemoglobin, producing a compound that cannot function as an O_2 transporter, leading the animal to suffer its deficiency.

Bloom et al. (2002) also discovered dependence between nitrate reduction in the shoot and photorespiration, that is, high CO_2 concentrations inhibit nitrate reduction.

Finally, the presence of nitrate affects carbon metabolism, as it redirects C from starch synthesis to produce organic acids or carbon skeletons necessary to incorporate N into an organic compound, a reaction that we discuss in the next item.

Nitrogen Incorporation

Once nitrogen is reduced to the NH_3/NH_4^+ form, it needs to be readily incorporated by plants that have low tolerance to this nitrogen form as it can be toxic, causing respiration inhibition, photophosphorylation decoupling in chloroplasts, and even repression of nitrogenase activity in N-fixing plants. Epstein and Bloom (2006) explain that NH_4^+ toxicity occurs due to the dissipation of transmembrane proton gradients, that is, if NH_4^+ is in high concentrations it will react with OH^- outside the membrane, producing NH_3. The NH_3 is permeable in the membrane and diffuses through it along its concentration gradient. Inside the membrane, NH_3 reacts with H^+ to form NH_4^+. The result is that both the OH^- concentration on the outside and the H^+ concentration on the inside decrease, that is, the pH gradient dissipates.

We also note that the N-NH_4^+ to be metallized can come from root uptake or from degradation of organic compounds (catabolism), coming from leaf senescence or even from seed reserves.

Regarding metabolism, nitrogen is incorporated into the first organic compound in the plant (alpha-ketoglutaric acid) and converts into **glutamic acid** (N combined with C-H-O). These reduced forms of N generate mainly amino acids, being part of proteins, enzymes, nucleic acids, and other nitrogenous compounds, including chlorophyll, a pigment that gives green color to leaves and has nitrogen in its composition.

Nitrogen incorporation in the form of ammonium into organic compounds, according to Hewitt and Cutting (1979), occurs mainly by two pathways, simultaneously.

1- **Via GDH: glutamic dehydrogenase**, which occurs in the mitochondria of leaves and roots, from the amination reaction of α-ketoglutaric acid (Fig. 4.10).
2- **Via GS-GOGAT: glutamine synthetase (GS) and glutamate synthase (GOGAT)**, discovered by Lea and Miflin (1974), which results in the amino acid glutamine from glutamic acid, followed the last reaction, which produces two molecules of the amino acid glutamate from glutamine (Fig. 4.10).

We note that the second pathway, through the two reactions, is the favorite for the introduction of NH_3/NH_4^+ in amino acids by plants. Thus, the GS enzyme is the main N incorporation pathway, which can assimilate more than 90% of the plant ammonium, as it is favored due to its increased affinity for NH_3 (Km = 50 μmol L^{-1}) compared to the GDH enzyme (Km up to 70 μmol L^{-1}), that is, even at low NH_4^+ concentrations GS is active. In addition, glutamine is the first product formed in

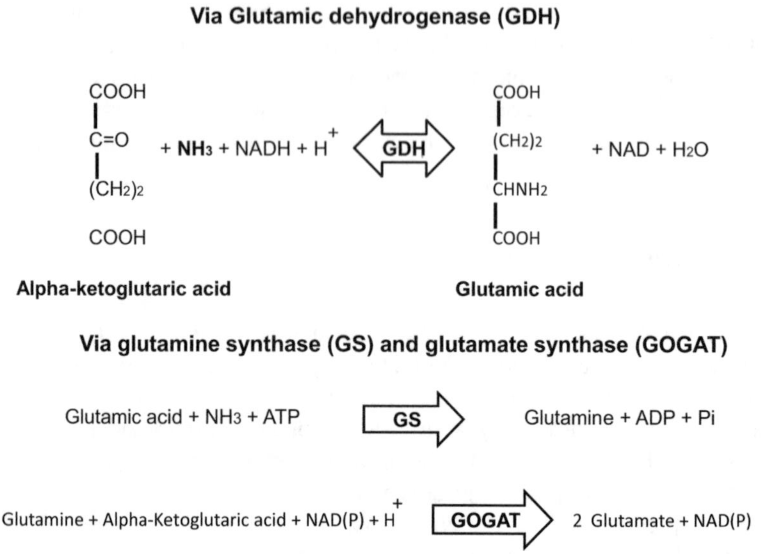

Via Glutamic dehydrogenase (GDH)

$$
\begin{array}{l} COOH \\ | \\ C{=}O \\ | \\ (CH_2)_2 \\ | \\ COOH \end{array}
\quad + NH_3 + NADH + H^+ \quad \langle\!\!\!\boxed{GDH}\!\!\!\rangle \quad
\begin{array}{l} COOH \\ | \\ (CH_2)_2 \\ | \\ CHNH_2 \\ | \\ COOH \end{array}
\quad + NAD + H_2O
$$

Alpha-ketoglutaric acid   **Glutamic acid**

Via glutamine synthase (GS) and glutamate synthase (GOGAT)

Glutamic acid + NH3 + ATP $\boxed{GS}\!\!\!>$ Glutamine + ADP + Pi

Glutamine + Alpha-Ketoglutaric acid + NAD(P) + H$^+$ $\boxed{GOGAT}\!\!>$ 2 Glutamate + NAD(P)

Fig. 4.10 Scheme illustrating the N incorporation pathway (GDH and GS-GOGAT) in plants

plants submitted to N application (NH_4^+ or NO_3^-) (Magalhães et al. 1990). There are two GSs, namely: the chloroplastic GS, which is responsible for assimilating ammonia from the N cycle of respiration, and the cytosolic GS. In addition to these two pathways, others can be observed, namely:

The reaction of $NH_3 + CO_2 + ATP$ in the presence of carbamyl kinase (Mg), which induces the formation of $NH_2 - CO - OP3H_2 + ADP$ (carbamyl phosphate).

However, the first glutamic dehydrogenase pathway is important for plants with excess NH_3, as it incorporates NH_3/NH_4 in alpha-ketoglutaric radicals, forming glutamic acid. These alpha-ketoglutaric radicals come mainly from respiration of carbohydrates in the mitochondria. Therefore, we infer that plants intensify the respiratory process as a defense mechanism, consuming more carbohydrates for the production of acid radicals that incorporate NH_4, forming amino acids. Plants in this situation produce more protein to the detriment of carbohydrates and their derivatives, resulting in a series of disorders related to excess metabolized N, such as imbalance between shoots and roots; excess vegetative growth in relation to production; decreased drought resistance; lodging; juiciness of tissues; decreased resistance to transport (fruit); and decreased sugar content.

Thus, metabolized N is incorporated into several organic compounds rich in N, besides amino acids. Several amino acids are produced in the metabolism of nitrogen, such as asparagine, which involves asparagine synthase, facilitating the transfer of amide nitrogen from glutamine to aspartate. Besides the amino acids aforementioned, we have other known amino acids such as amides (glutamine, 2 N/5C) and ureide (allantoic acid, 4 N/4C), etc.

Through the process known as transamination, with enzymes called aminotransferases or transaminases, the amino group of glutamic acid or another amino acid can be transferred to other alpha-keto acid radicals, forming other amino acids (N/C > 0.4). Once the 20 or 21 amino acids are formed, proteins are synthesized. The process occurs in the ribosomes and requires tRNA, mRNA, ATP, Mg^{2+}, Mn^{2+}, K^+, and the 20–21 amino acids to form polypeptides (proteins). According to the genetic code of DNA, different proteins and other nitrogenous compounds (nitrogenous bases, coenzymes, pigments, and vitamins) can be synthesized.

We also add that amino acids produced during the metabolism of nitrogen can also function as a hormonal system, regulating shoot and root growth. This was explained by Lam et al. (1996) who observed that under conditions of high luminosity or carbohydrate level, glutamine synthetase and glutamate synthase are stimulated while asparagine synthase is inhibited, favoring N assimilation in glutamine and glutamate, compounds rich in carbon, and stimulating root growth. In contrast, under limited light and carbohydrate conditions, glutamine synthetase and glutamate synthase are inhibited and asparagine synthase stimulated, favoring N assimilation into asparagine, a nitrogen-rich compound (stable in long-distance transport), that favors shoot growth.

We notice that nitrogen is found almost entirely in organic form (90%) in the plant after its metabolization, having mainly structural function, as component of organic compounds such as chlorophyll (Fig. 4.11). Therefore, N application linearly increases leaf chlorophyll content, such as for tomato plant from cultivar Santa Clara (Guimarães et al. 1999). Increasing chlorophyll content intensifies the green color in leaves, which is measured/estimated with a chlorophyll meter (Fig. 4.12).

Research evaluating nitrogen effects on plants usually show increased foliar area (Lin et al. 2006) and plant biomass, which is explained by the increase in the plant photosynthetic capacity, measured by CO_2 assimilation, also maintaining leaves green for a longer time. In addition, we also observe increased active photosynthesis, increasing grain production (Wolfe et al. 1988), as ears show an increased grain

Fig. 4.11 Scheme illustrating chlorophyll pigment (typical example of a natural chelate with Mg in the central part of the compound)

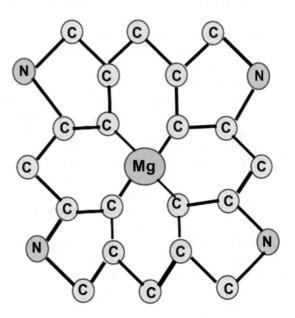

Fig. 4.12 Chlorophyll meter that estimates green color intensity corrected with chlorophyll content

number (Below 2002). The effect of N on ear size is visible (Fig. 4.13). The increased grain number is also due to the decreased abortion rate.

The effect of N to increase photosynthesis is explained by its participation in chlorophyll synthesis, as aforementioned, and also by its role in the synthesis of enzymes PEPC and RuBisCO (phosphoenolpyruvate carboxylase and ribulose 1,5 bisphosphate carboxylase-oxygenase, respectively), which participate in atmospheric CO_2 fixation. Ranjith et al. (1995), observed in sugarcane varieties. That when the foliar N content increased from 50 to 97 mmol m^{-2}, the activity of enzymes PEPC and RuBisCO and chlorophyll content practically doubled.

Yamazaki et al. (1986) observed that N is an essential factor for light to stimulate mRNA production for synthesis of PEPC and RuBisCo in maize (C4 plant). However, as N deficiency reduces growth, the use of assimilates is reduced, and increased amounts of carbon can be deviated to starch formation (Rufty et al. 1988). Photosynthesis can be seriously affected if starch accumulation in the chloroplast is excessive due to the difficulty of CO_2 in reaching RuBisCO carboxylation sites (Guidi et al. 1997).

Thus, nitrogen plays an important role in crop growth and production (Malavolta et al. 1997), participating in several physiological processes vital to the plant life cycle (Table 4.3). In adequate amounts, N favors root growth as stem growth increases foliar area and photosynthesis, increasing the flow of carbohydrates to the root (Table 4.4). Visual aspects of the barley root system indicate the beneficial effect of N, which can be observed in the part of the root where the fertilizer was in contact (Fig. 4.14). Most crops have a high response to nitrogen application. We note that irrigation is important to maximize plant response to nitrogen nutrition, such as in passion fruit tree (Fig. 4.15).

In view of environmental issues, new studies are necessary to evaluate more productive genotypes that demand less water and nitrogen. O'neill et al. (2004) observed in 12 maize hybrids that some hybrids produced approximately 27% under

Fig. 4.13 Effect of N on
the number and size of
maize ears under N
deficiency and sufficiency.
Note that the ear tips with
N deficiency have a
reduced row number

Table 4.3 Summary of the main functions of nitrogen for plants

Structural	Constituent enzymes	Processes
Amino acids and proteins	All	Ion uptake
Nitrogen bases of nucleic acids		Photosynthesis
Enzymes and coenzymes		Respiration
Vitamins		Cell multiplication and differentiation
Glyco- and lipoproteins		Inheritance
Pigments (chlorophyll)		

Table 4.4 Effect of nitrogen
on root growth of
maize hybrids

Hybrids	N applied	Root growth
	kg ha^{-1}	cm/10^2.s
P 3732	0	2.4
	227	**3.5**
B73 x Mo17	0	2.7
	227	**5.3**

water deficit and 42% under nitrogen deficit, compared to adequate water and nitrogen levels.

Calvache and Reichardt (1996) also observed for beans and wheat that deficient irrigation decreased N uptake from fertilization. We expect for any crop that N application reflects in its same proportion in foliar N content and in production (Fig. 4.16). Foliar nitrogen contents considered appropriate for the crops are presented in Chap. 19.

Fig. 4.14 Modification in
the root system of a plant
by supplying nitrate in the
middle part of the root

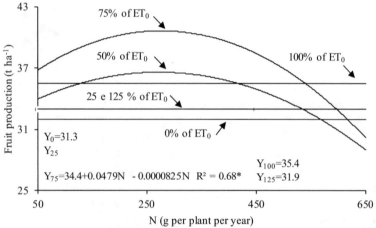

Fig. 4.15 Yellow passion fruit production as a function of nitrogen doses under irrigation depths
(% of ET0; Carvalho et al. 2000)

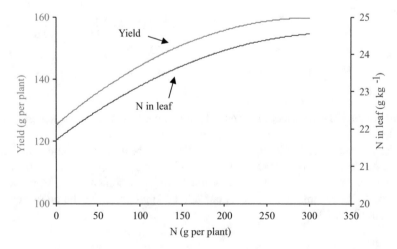

Fig. 4.16 Crop production as a function of N doses and foliar N content

4.4   Crop Nutritional Requirements

The nutritional requirement is the amount of nutrients accumulated in the entire plant during its production cycle. In order to measure the nutritional requirement (NR) of a crop, we have to consider the nutrients absorbed by the entire plant and not only by the harvested part, that is, NR (kg ha^{-1}) = nutrient content in the plant (% = g kg^{-1}x10) x dry matter of the entire plant (kg ha^{-1}). Normally, increased crop production implies increased amounts of nutrients accumulated by the plant, which are the nutritional requirement. According to Grove (1979), there is a linear relation between grain production and N accumulation in the maize shoot in different locations (Brazil, New York, Puerto Rico, Nebraska). However, increased accumulation of nutrients such as P does not indicate that plants are more or less efficient, since hybrids, through genetic adaptations, can develop and produce well with decreased nutrients, that is, plants requiring increased amounts of this nutrient may no longer be productive, and vice versa.

Nutrient Extraction and Export

Approximately 60% of nitrogen fertilizers produced are intended exclusively to meet the nutritional requirements of cereals (maize, rice, and wheat) (Ladha et al. 2005). According to the literature, we observe variation in the total N required or accumulated by crops (Table 4.5).

Therefore, as nutrient contents in the tissues and dry matter production of the crops are different, we expect that the nutritional requirement is specific to each culture.

Among crops studied, we observe that coffee and soybean are the crops that extracted the most nitrogen, reaching 253 and 181 kg for a production of 2.0 and 2.4 t, respectively. Soybeans and beans are the crops that most export N for grains at harvest, reaching 63.3 and kg t^{-1} of grains produced (Table 4.5).

However, considering the absolute need, that is, the total amount of nutrients to produce one ton of agricultural product (grains), coffee and beans demand more N, requiring 127 and 110 kg of N for each ton of product harvested, respectively.

Export (E) refers to nutrients removed from the crop along with the harvest, that is, E (kg t^{-1}) = accumulation of nutrient mobilized in the grains (kg) x dry matter of the exported part (t). Normally, part of the nutrients are exported in the product (grains) after harvesting annual crops, while the remainder stays in the agricultural area in the form of crop residues (root, stem, and leaves). In perennial plants, little of the nutrients is exported in relation to what is immobilized in the entire plant. However, in crops such as maize, the entire plant is exported for silage, except the roots, and most of the nutrients accumulated in the plant are exported.

Table 4.5 Nitrogen requirements for major crops

Crop	Plant part	Dry matter produced	N accumulated		N required to produce 1 t grains[c]
			Plant part	Total [b]	
		t ha^{-1}	kg ha^{-1}		kg t^{-1}
Annual					
Cotton	Reproductive part (seed)	1.3	29 (22.3)[1]	84	65
	Vegetative part (stem/branch/leaf)	1.7(m.s.)	49		
	Root	0.5 (m.s.)	6		
Soybean	Grains	2.4	152 (63.3)	181	75
	Stem, branches, and leaves	5.6 (m.s.)	29		
Beans	Vagem	1	47 (47)	110	110
	Stem	0.4	8		
	Leaves	1.2	53		
	Rice	0.1	2		
Maize	Grains	5.0	67 (13.4)	117	23
	Stem, leaves	4.5	50		
Rice	Grains	3	45 (15)	103	34
	Stems	2 (m.s.)	15		
	Leaves	2 (m.s.)	15		
	Husk	1	8		
	Root	1 (m.s.)	20		
Wheat	Grains	3	50 (16.7)	70	23
	Straw	3.7	20		
Perennial and/or semiperennial					
Sugarcane	Stem	100	90 (0.9)	150	1,5
	Leaves	25	60		
Coffee	Grains	2	33 (16.5)	253	127
	Trunk, branches, and leaves	–	220		

[a]Nutrient export through grains produced (kg t^{-1}): N accumulated in the grains/grain dry matter; [b]Suggests the (total) nutritional requirement by crop area for the respective yield level; [c]Suggests the relative nutritional requirement of N of the crop for production of one ton of the commercial product (grains/stems); obtained by the following formula: N accumulated in the plant (vegetative + reproductive part)/dry matter of the commercial product

Nutrient Uptake Rate

Another important point, such as the definition of the nutrient amount, is to know precisely in which periods of growth/development the crop has such nutritional requirement, as the extraction of nutrients from the soil by plants during growth varies according to the growing period. In an irrigated system (fertigation), nitrogen fertilizer application at installments, according to the crop demand, is widely used, reducing losses without increasing the cost of production. Thus, the success of nitrogen application at installments to increase fertilizer efficiency must be influenced by other factors besides the plant, such as uptake rate; the environment, that is, the cultivation system (irrigated or nonirrigated); the soil, such as its texture (sandy or medium/clayey); and management (consolidated or conventional direct sowing). Another aspect we must consider is the need to avoid gaseous losses of N applied to the soil, especially in the form of urea, through incorporation into the soil or association with other fertilizers, such as KCl, in order to decrease the pH value of the mixture.

For crops in general, in their productive stages, the curve that describes nutrient extraction as a function of time is a sigmoid, just as it occurs with the dry matter accumulation curve. Considering sugarcane seedlings, for example, nutrient accumulation is reduced when the plant is young, as well as dry matter accumulation. Afterwards there is an abrupt increase in dry matter accumulation and nutrient uptake, describing the linear part of the curve in the period from 65 to 110 days after transplanting. In the final period of seedling formation, there is stabilization, when nutrient uptake reduces (Fig. 4.17).

Considering the total amount of nutrients required to produce one ton of agricultural product (grains), coffee and beans demand more N, requiring 127 and 110 kg of N for each ton of harvested product, respectively (Table 4.5).

In maize, there is low N accumulation (9%) up to 44 days after sowing. Subsequently, the uptake rate of this nutrient is practically linear over time, reaching its maximum at 133 days of sowing. We note that maize crops indicate a beneficial response in terms of production to N application, even in the early growth stages (four to five leaves), as this nutrient has a physiological relationship to define the

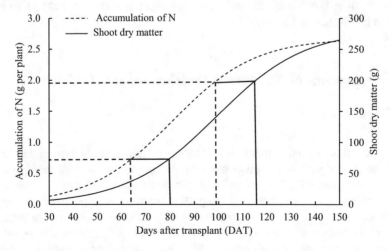

Fig. 4.17 Nitrogen uptake rate during the formation of sugarcane seedlings

potential number of grains per ear. In addition, there is a time period between N application in the soil and its availability to the plant, which is related to nutrient immobilization by the soil microbiota, as aforementioned. Thus, taking into account changes in soil N and its importance for plant physiology, the application of part of N in the initial growth of plants is interesting.

In the case of vegetables such as tomatoes, there is a slow uptake until close to 40 days, followed by a quick uptake in the period from 40 to 80 days. From that period there is slow uptake until reaching 120 days of transplanting (Fayad et al. 2002).

This information has a very important practical aspect related to the moment for nutrient application in nitrogen topdressing, which must be performed in the initial periods preceding high uptake. Thus, knowledge of the best time for nitrogen application and the use of genotypes with high ability for nutrient uptake may increase the agronomic efficiency of nitrogen fertilization. Increased efficiency of N use is important as the recovery rate of N by the crops, in the first year, peaks at 50%, decreasing drastically afterwards (<7% in six successive harvests; Ladha et al. 2005).

Finally, dividing nitrogen fertilization increases maize grain yield compared with application in just one moment (Melgar et al. 1991), especially under conditions of high loss potential, that is, in sandy soils with N doses. Reichardt et al. (1982) concluded that losses by leaching are not a problem with moderate applications in the order of 90 kg ha^{-1} of N.

Recent studies on the possibility of anticipating N application (before or at sowing) in a no-tillage system showed that it may not be advantageous under tropical conditions (central region of Brazil), characterized by the high rate of mulch decomposition, associated with high rainfall in summer and high demand for nitrogen by plants. Wolschick et al. (2003) observed for maize crop that early N application in high doses is not economically feasible, as the yield obtained with 90 kg ha^{-1} of N applied at three installments did not differ from that obtained with 180 kg ha^{-1} of N applied in advance and at sowing.

4.5 Symptoms of Nutritional Deficiencies and Toxicity

Deficiency

In nutritional disorders, as in the case of nitrogen deficiency, the typical deficiency symptom is leaf yellowing due to lack of chlorophyll, which stops being synthesized or is degraded (hydrolysis and protein redistribution) and there is proteolysis of the rubisco enzyme and other chloroplast proteins to release nitrogen to supply deficiency in new leaves. The symptom appears in older leaves due to nitrogen transport from these leaves to younger leaves or to fruit in formation. Symptoms are similar in crops with small differences (Fig. 4.18). However, over time, the deficiency worsens with chlorosis reaching all leaves, followed by necrosis.

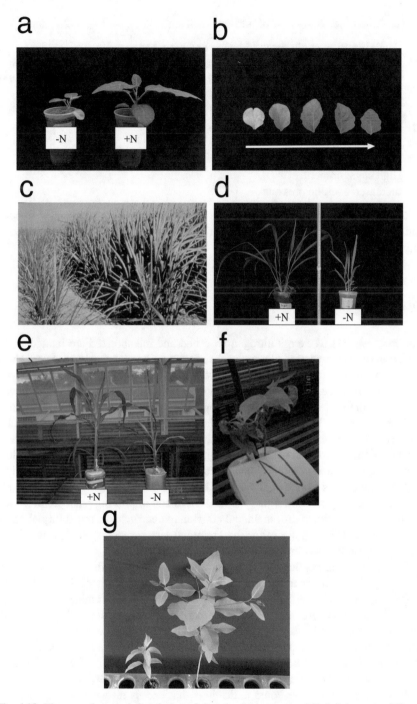

Fig. 4.18 Photos and general description of the visual symptoms of N deficiency in different crops: Scarlet eggplant (Fig. 4.18a,b), sugarcane (Fig. 4.18c), forage (Fig. 4.18d), corn (Fig. 4.18e), bean (Fig. 4.18f), eucalyptus (Fig. 4.18g)

There are other symptoms of nitrogen deficiency in plants in general, namely:

(a) Small insertion angle between leaves and branches
(b) Abbreviated maturity and senescence (due to cytokyanin reduction, which is responsible for the green color of tissues for a longer time), premature fall of leaves
(c) Decrease in flowers and dormancy of lateral buds
(d) Reduced production
(e) Small chloroplasts
(f) Low chlorophyll and protein content
(g) High sugar content
(h) Increased osmotic pressure

Given the role of resistance to auxins in adapting plants to ammoniac toxicity and N deficiency (Santos et al. 2020), further studies are needed on the interaction of N nutritional disorders and auxins.

Toxicity

The main aspects of excess nitrogen absorbed and metabolized are related to the deviation of carbohydrates to proteins, as aforementioned, promoting excess shoot growth to the detriment of the reproductive part (production), perhaps due to phytohormone imbalance. In addition, the shoot/root ratio increases, impairing the development of the root system and decreasing the capacity of plants to resist drought periods (summer), besides increasing plant lodging. In the maize crop, we also observe excess nutrient in the reproductive phase, in which the ear hair remains green. Another factor that would explain the effects of excess N to decrease production would be related to mutual shading caused by increased foliar area induced by high nitrogen doses (Stone et al. 1999). This could be minimized by increasing the Si in the system, providing rigidity to the tissues and improving plant architecture.

Increased genetic control in annual crops may contribute to the ability of small cultivars to distribute more N for the grain and less for stems and leaves, consequently responding to nitrogen fertilization without lodging.

Excess nitrogen application causes problems in the plant due to salinity stress caused by the fertilizer used, besides damaging the environment due to the high mobility of nitrate in the soil profile, which can reach the water table.

In addition, there are other negative effects of excess N on the quality of agricultural products, such as fruit, in which increased succulence decreases storage and resistance to transport; in coffee, it harms the drink quality; in sugarcane, it reduces the sucrose content. In cotton, excess N damages the opening of cotton balls. In sunflower, excess N decreases the percentage of oil (Robson 1978).

In vegetables there are some indications that excess nitrate, when reduced to nitrite, can be toxic for humans and animals using forage. In the bloodstream, N-NO2 oxidizes the ferrous ion in hemoglobin to ferric ion, forming methemoglobin. This compound prevents O_2 transport between tissues (Keeney 1982). Methemoglobin is considered moderate, that us, less than 30% of oxidized hemoglobin causes nausea and headache and more than 50% is severe and can be fatal (Boink and Speijers 2001). Maximized nitrate reduction reactions or reductase activity decreases nitrate accumulation in plants. In addition, plants (vegetables or forage) must be collected at the time of maximum light intensity, as the photosynthesis energy for the enzymatic activity of reduction is increased. Krohn et al. (2003) found that the nitrate content in lettuce decreased at 12 h (noon) compared to night (0 h) and early morning (6 h).

In summary, we have the following general symptoms of excess nitrogen:

Symptoms (visible)

(a) Dark green color
(b) Abundant foliage
(c) Lodging
(d) Delayed maturation

Other symptoms

(a) Little developed root system
(b) Decreased sugar transport to roots and tubers
(c) Increased tissue juiciness

We know that Mg deficiency can be induced by excess K or NH_4^+, especially in soils with Mg content in the medium to low limit.

We note that excess NH_4^+ in the plant that would cause toxicity can be aggravated by K deficiency, as the latter is important for the assimilation of reduced nitrogen (NH_4^+), forming amino acids (Dibb and Welch 1976). In addition, the decrease of carbon skeletons (carbohydrates) due to decreased photosynthesis (stress from other nutrients or cloudy days) or increased respiration (increased temperature) impairs nitrogen assimilation, accumulating NH_4^+. Ammonium toxicity can often include chlorosis, necrosis, and even plant death.

High N levels in the form of ammonium results in symptoms of toxicity, such as in stylo plants (chlorosis by necrotic lesions of leaves from the apex toward the base and falling of lower leaves; Amaral et al. 2000), especially in the plant shoot, where it is toxic in relatively low concentrations (Hageman and Below 1990), decoupling photophosphorylation (Trebst et al. 1960), inhibiting chlorophyll synthesis, and degrading chloroplasts (Puritch and Barker 1967) and proteins (Barker et al. 1966). Furthermore, the plant can develop ammonium detoxification mechanisms, leading to putrescine biosynthesis (Smith et al. 1990), which is a toxic product that explains necrosis in plant tissues.

Bennett and Adams (1970) observed, in cotton roots, that an increased pH value increased the toxicity of ammoniacal N. However, there are indications that Ca application may partially cancel the severity of ammonium toxicity. We know that Si attenuates ammoniacal toxicity in plants (Campos and Prado 2015, Campos et al. 2020.

References

Amaral JAT, Cordeiro AT, Rena AB. Effects of aluminum, nitrate and ammonium on the metabolic nitrogen composition and of carbohydrates in *Stylosanthes guianensis* and *S. macrocephala*. Pesq Agrop Brasileira. 2000;35:313–20. https://doi.org/10.1590/S0100-204X2000000200010.

Barber SA. Soil nutrient bioavailability: a mechanistic approach. New York: John Wiley & Sons; 1995.

Barker AV, Volk RJ, Jackson WA. Root environment acidity as a regulatory factor in ammonium assimilation by the bean plant. Plant Physiol. 1966;41:1193–9. https://doi.org/10.1104/pp.41.7.1193.

Below FE. Corn physiology, nutrition and nitrogen fertilization. Informações agronômica. 2002;99:7–12.

Bennett AC, Adams F. Concentration of NH_3 (aq) required for incipient NH_3 toxicity to seedlings. Soil Sci Soc Am J. 1970;34:259–63.

Bloom AJ, Smart DR, Nguyen DT, et al. Nitrogen assimilation and growth of wheat under elevated carbon dioxide. Proc Natl Acad Sci U S A. 2002;99:1730–5.

Boink A, Speijers G. Health effects of nitrates and nitrites, a review. Acta Hortic. 2001;20:29–33.

Calvache AM, Reichardt K. Water deficit at different growth stages for common bean (*Phaseolus vulgaris* L. cv. imbabello) and nitrogen use efficiency. Sci Agric. 1996;53:343–53. https://doi.org/10.1590/S0103-90161996000200025.

Campbell WH. Nitrate reductase biochemistry comes of age. Plant Physiol. 1996;111:355–361. https://doi.org/10.1104/pp.111.2.355.

Campos CNS, Prado RM. Use of silicon in mitigating ammonium toxicity in maize plants. Am J Plant Sci. 2015;6:1780–4. https://doi.org/10.4236/ajps.2015.611178.

Campos CNS, Silva Júnior GB, Prado RM, et al. Silicon mitigates ammonium toxicity in plants. Agron J. 2020;112:635–47. https://doi.org/10.4067/S0718-58392019000300425.

Carvalho AJC, Martins DP, Monnerat PH. Nitrogen fertilization and irrigation depths in yellow passionfruit. I. yield and fruit quality. Pesq Agrop Brasileira. 2000;35:1101–8. https://doi.org/10.1590/S0100-204X2000000600005.

Cazetta JO, Villela LCV. Nitrate reductase activity in leaves and stems of tanner grass (*Brachiaria radicans* Napper). Sci Agric. 2004;61:640–8. https://doi.org/10.1590/S0103-90162004000600012.

Crawford NM, Glass ADM. Molecular and physiology aspects of nitrate uptake in plants. Trends Plant Sci. 1998;3:389–95. https://doi.org/10.1016/S1360-1385(98)01311-9.

Dibb DW, Welch LF. Corn growth as affected by ammonium versus nitrate absorbed from soi!. Agron J. 1976;68:89–94. https://doi.org/10.2134/agronj1976.00021962006800010024x.

Dirr MA. Nitrogen form and growth and nitrate reductase activity of the cranberry. HortScience. 1974;9:347–8.

Epstein E, Bloom A. Nutrição mineral de plantas: princípios e perspectivas. Português editor: Maria Edna Tenório Nunes. Planta, Londrina; 2006.

Fayad JA, Fontes PCR, Cardoso AA, et al. Absorção de nutrientes pelo tomateiro cultivado sob condições de campo e de ambiente protegido. Hortic Bras. 2002;20:90–4.

Grove LT. Nitrogen fertility in Oxisols and Ultisols of the humid tropics. Ithaca: Cornell University; 1979.

Guidi L, Lorefice G, Pardossi A, et al. Growth and photosynthesis of *Lycopersicum esculentum* (L.) plants as affected by nitrogen deficiency. Biol Plant. 1997;40:235–44. https://doi.org/10.1023/A:1001068603778.

Guimarães TG, Fontes PCR, Pereira PRG, et al. Teores de clorofila determinados por medidor portátil e sua relação com formas de nitrogênio em folhas de tomateiro cultivados em dois tipos de solo. Bragantia. 1999;58:209–16.

Hageman RH, Below FE. Role of nitrogen metabolism in crop productivity. In: Abrol YP, editor. Nitrogen in higher plants. Research Studies: Taunton; 1990. p. 313–34.

Haynes R, Goh KM. Ammonium and nitrate nutrition of plants. Biol Rev. 1978;53:465–510. https://doi.org/10.1111/j.1469-185X.1978.tb00862.x.

Hewitt EJ, Curting CV. Nitrogen assimilation of plants: Academic Press, London; 1979. 708p.

Hömheld V. Efeitos do potássio nos processos da rizosfera e na resistência das plantas a doenças. In: Yamada T, Roberts TL, editors. Potássio na agricultura brasileira. Piracicaba: Potafós; 2005. p. 301–19.

Keeney DR. Nitrogen management for maximum efficiency and minimum pollution. In: Stevenson FJ, editor. Nitrogen in agricultural soils. Madison: American Society of Agronomy; 1982. p. 605–49.

Krohn NG, Missio RF, Ortolan ML et al. Nitrate level on lettuce leaves in function of the harvest time and leaf type sampling. Hort Bras. 2003;21:216–219. https://doi.org/10.1590/S0102-05362003000200019.

Ladha JK, Pathak H, Krupnik TJ, et al. Efficiency of fertilizer nitrogen in cereal production: retrospects and prospects. Adv Agron. 2005;87:85–156. https://doi.org/10.1016/S0065-2113(05)87003-8.

Lam HM, Coschigano KT, Oliveira IC, et al. The molecular-genetics of nitrogen assimilation into amino acids in higher plants. Annu Rev Plant Physiol Plant Mol Biol. 1996;47:569–93. https://doi.org/10.1146/annurev.arplant.47.1.569.

Lea P, Miflin B. Alternative route for nitrogen assimilation in higher plants. Nature. 1974;251:614–616. https://doi.org/10.1038/251614a0.

Lestienne F, Thornton B, Gastal F. Impact of defoliation intensity and frequency on N uptake and mobilization in *Lolium perenne*. J Exp Bot. 2006;57:997–1006. https://doi.org/10.1093/jxb/erj085.

Lin X, Zhou W, Zhu D, et al. Nitrogen accumulation, remobilization and partitioning in rice (*Oryza sativa* L.) under an improved irrigation practice. Field Crop Res. 2006;96:448–54. https://doi.org/10.1016/j.fcr.2005.09.003.

Magalhães JV. Absorção e translocação de nitrogênio por plantas de milho (*Zea mays* L.) submetidas a períodos crescentes de omissão de fósforo em solução nutritiva. Dissertação, Universidade Federal de Viçosa; 1996.

Magalhães JR, Ju GC, Rich PJ, et al. Kinetics of $^{15}NH_4$ assimilation in *Zea mays* L. Pre-liminary studies with a glutamate dehydrogenase (GDH1) null mutant. Plant Physiol. 1990;94:646–56. https://doi.org/10.1104/pp.94.2.647.

Malavolta E, Vitti GC, Oliveira SA. Avaliação do estado nutricional das plantas: princípios e aplicações. Piracicaba: Associação Brasileira de Potassa e do Fósforo; 1997. 319p.

Marenco RA, Lopes NF. Fisiologia vegetal: fotossíntese, respiração, relações hídricas e nutrição mineral. Viçosa: UFV; 2005.

Marschner H. Mineral nutrition of higher plants. London: Academic Press; 1995.

Martinez HEP, Silva Filho JB. Introdução ao cultivo hidropônico. Viçosa: Universidade Federal de Viçosa; 2004.

Melgar RJ, Smyth TJ, Cravo MS, et al. Application of nitrogen fertilizer for corn in an Oxisol in the Amazon. Rev Bras Ciênc Solo. 1991;15:289–96.

Mengel K, Kirkby EA. Principles of plant nutrition. Worblaufen-Bern: International Potash Institute; 1987.

Nambiar PTC, Rego TJ, Rao BS. Nitrate concentration and nitrate reductase activity in the leaves of three legumes and three cereals. Ann Appl Biol. 1988;112:547–53. https://doi.org/10.1111/j.1744-7348.1988.tb02091.x.

Nicoloso FT, Sartori L, Jucoski GO, et al. Mineral nitrogen sources (N-NO_3^- and N-NH_4^+) on growth of grápia (*Apuleia leiocarpa* (Vog.) Macbride) seedlings. Ciênc Rural. 2005;15:221–31. https://doi.org/10.5902/198050981839.

Noller CH, Rhykerd CL. Relationship of nitrogen fertilization and chemical composition of forage to animal health and performance. In: Mays DA, editor. Forage fertilization. Madison: American Society of Agronomy; 1974. p. 363–87.

O'neill PM, Shanahan JF, Schepers JS, et al. Agronomic responses of corn hybrids from different eras to deficit and adequate levels of water and nitrogen. Agron J. 2004;96:1660–7. https://doi.org/10.2134/agronj2004.1660.

Puritch GS, Barker AV. Structure and function of tomato leaf chloroplasts during ammonium toxicity. Plant Physiol. 1967;42:1229–38. https://doi.org/10.1104/pp.42.9.1229.

Ranjith SA, Meinzer FC, Perry MH, et al. Partitioning of carboxylase activity in nitrogen-stressed sugarcane and its relationship to bundle sheath leakiness to CO_2, photosynthesis and carbon isotope discrimination. Aust J Plant Physiol. 1995;22:903–11. https://doi.org/10.1071/PP9950903.

Reichardt K, Libardi PL, Urquiaga SC. Fate of fertilizer nitrogen in soil-plant systems with emphasis on the tropics. In: International Atomic Energy Agency (ed). Agrochemicals, Viena. 1982; pp. 277–290.

Robson RG. Production and culture. In: Carter JF, editor. Sunflower science and technology. Madison: American Society of Agronomy; 1978. p. 89–143.

Rufty TW, Huder SC, Volk RJ, et al. Alterations in leaf carbohydrate metabolism in response to nitrogen stress. Plant Physiol. 1988;88:725–30. https://doi.org/10.1104/pp.88.3.725.

Santos JHS. Proporções de nitrato e amônio na nutrição e produção dos capins aruana e marandu. Escola Superior de Agricultura Luiz de Queiroz: Dissertação; 2003.

Santos RL, Freira FJ, Oliveira ECA, et al. Nitrate reductase activity and nitrogen and biomass accumulation in sugarcane under molybdenum and nitrogen fertilization. Rev Bras Ciênc Solo. 2019;43:e0180171. https://doi.org/10.1590/18069657rbcs20180171.

Santos LCN, Barreto RF, Prado RM, et al. The auxin-resistant dgt tomato mutant grows less than the wild type but is less sensitive to ammonium toxicity and nitrogen deficiency. J Plant Physiol. 2020;252:153243. https://doi.org/10.1016/j.jplph.2020.153243.

Siegel LM, Wilkerson JQ. Structure and function of spinach ferredoxin- nitrite reductase. In: Wray JL, Kinghomn JR, editors. Molecular and genetic aspects of nitrate assimilation. Oxford: Oxford Science Publications; 1989. p. 263–83.

Silva Junior GB, Prado RM, Silva SLO, et al. Nitrogen concentrations and proportions of ammonium and nitrate in the nutrition and growth of yellow passion fruit seedlings. J Plant Nutr. 2020;43:2533–47. https://doi.org/10.1080/01904167.2020.1783299.

Silva ET, Melo WJ. Protease activity and nitrogen availability in a Latosol under orange crop. Rev Bras Ciênc Solo. 2004;28:833–41. https://doi.org/10.1590/S0100-06832004000500006.

Silva ES, Prado RM, Soares AAVL et al. Response of corn seedlings (*Zea mays* L) to different concentrations of nitrogen in absence and presence of silicon. SILICON. 2020; 12-00480 doi:https://doi.org/10.1007/s12633-020-00480-8.

Sivasankar S, Oaks A. Nitrate assimilation in higher plants: The effect of metabolites and light. Plant Physiol Biochem. 1996;34:609–620.

Smith FW, Jackson WA, Berg PJV. Internal phosphorus flows during development of phosphorus stress in *Stylosanthes hamata*. Aust J Plant Physiol. 1990;17:451–64. https://doi.org/10.1071/PP9900451.

Stone LF, Silveira PM, Moreira JAA, et al. Rice nitrogen fertilization under supplemental sprinkler irrigation. Pesq Agrop Brasileira. 1999;34:927–32. https://doi.org/10.1590/S0100-204X1999000600002.

Trebst AV, Losada M, Di A. Photosynthesis by isolated chloroplasts: XII. Inhibitors of CO_2 assimilation in a reconstituted chloroplast system. J Biol Chem. 1960;235:840–4.

Ullrich WR. Transport of nitrate and ammonium through plant membranes. In: Mengel K, Pilbeam DJ, editors. Nitrogen metabolism of plants. Oxford: Clarendon Press; 1992. p. 121–37.

Valarini MJ, Godoy R. Contribution of symbiotic nitrogen fixation on pigeon-pea yield (*Cajanus cajan* L. Millsp). Sci Agric. 1994;51:500–4. https://doi.org/10.1590/S0103-90161994000300021.

Viegas RA, Silveira JAG. Activation of nitrate reductase of cashew leaf by exogenous nitrite. Braz J Plant Physiol. 2002;14:39–44. https://doi.org/10.1590/S1677-04202002000100005.

Waterer JG, Vessey JK. Effect of low static nitrate concentrations on mineral nitrogen uptake, nodulation, and nitrogen fixation in field pea. J Plant Nutr. 1993;16:1775–1789. https://doi.org/10.1080/01904169309364649.

Von Wirén N, Gazzarrini S, Gojon A, Frommer WB. The molecular physiology of ammonium uptake and retrieval. Curr Opin Plant Biol. 2000;3:254–261.

Wolfe DW, Henderson DW, Hsiao TC, et al. Interactive water and nitrogen effects on senescence of maize. II. photosynthetic decline and longevity of individual leaves. Agron J. 1988;80:865–70. https://doi.org/10.2134/agronj1988.00021962008000060005x.

Wolschick D, Carlesso R, Petry MT et al. Nitrogen application on maize cultivated under no-tillage system in a year with normal precipitation and with "El-Niño". Rev Bras Ciênc Solo. 2003;27:461–468. https://doi.org/10.1590/S0100-06832003000300008.

Yamazaki M, Watanabe A, Sugiyama T. Nitrogen-regulated accumulation of mRNA and protein for photosynthetic carbon assimilating enzymes in maize. Plant Cell Physiol. 1986;27:443–52. https://doi.org/10.1093/oxfordjournals.pcp.a077120.

Chapter 5
Sulfur

Keywords Sulphur oxidation · Sulphur uptake · Sulphur transport · Sulphur redistribution · Assimilatory sulphate · Sulphur deficient

5.1 Introduction

Sulfur concentrations in soils range from 0.1% in mineral soils to 1% in organic soils. However, most of this element is in organic form (60–90% of the total, i.e., S-amino acids, S phenols, S carbohydrates, S-lipids, and S humus). As most of the S in the soil is organic, microbiological processes in soils are important to study this nutrient. However, organic matter is the soil S reservoir. Soils with a sandy texture and decreased organic matter content (<20 g kg^{-1}) may have decreased capacity to supply plants with this element, as each 10 g kg^{-1} would release only approximately 6 kg ha^{-1} year of S. Sulfur, like nitrogen, undergoes many transformations in the soil (immobilization/mineralization and oxidation/reduction), mainly through microorganisms and also through soil management (drainage, porosity), affecting element chemistry (oxidation state).

During mineralization, the first mineral formed is H2S, usually according to the following equation:

$$\text{Cystine} + \text{H}_2\text{O} \Rightarrow \text{acetic acid} + \text{formic acid} + \text{CO}_2 + \text{NH}_3 + 2\text{H}_2\text{S}$$

Under aerobic conditions, sulfur from H_2S is oxidized to SO_4^{-2}, whereas in anaerobic medium, H_2S produces elemental sulfur (S) (Fig. 5.1).

We observe that most of the soil sulfur is bound to an organic compound which, through microorganisms, is converted into mineral forms, becoming available to plants (SO_4^{2-}) in aerated soils. Knowledge of factors affecting the activity of the soil microbiota that favor the mineralization process (C/S < 200), besides an adequate physical management of the soil that favors increased aeration, increases the available S and, consequently, plant uptake and nutrition.

Soils receiving sulfate either through mineralization, fertilization (simple superphosphate, ammonium sulfate), or gypsum (calcium sulfate) can suffer considerable losses by leaching due to its increased mobility in the soil.

© Springer Nature Switzerland AG 2021
R. de Mello Prado, *Mineral nutrition of tropical plants*,
https://doi.org/10.1007/978-3-030-71262-4_5

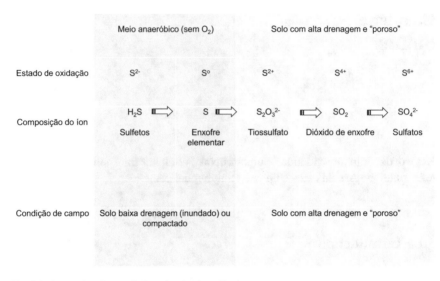

Fig. 5.1 Flow of sulfur oxidation state in the soil

When studying sulfur and other nutrients in the plant system, it is important to know all the compartments the nutrient passes in the soil solution and in the root and shoot (leaves/fruits), that is, from the soil until its incorporation in the dry matter of the agricultural product (grain, fruit, etc.) (Fig. 5.2).

There are several processes (solid–liquid phase change, root ion channel, uptake, transport, redistribution) that rule the passage of the nutrient in different compartments for it to be adequately metabolized, performing specific functions in plant life that accumulate dry matter. In order to maximize the conversion of applied nutrient into the agricultural product, we need to supply the plant nutritional requirement, that is, to supply, in quantitative terms, all growth/development stages to achieve the expected agricultural production. However, every crop has its own nutritional requirements.

5.2   Uptake, Transport, and Redistribution of Sulfur

Sulfur Uptake

Before absorption, contact between S and root occurs by mass flow, a process that depends on the rate of water movement in the soil–plant system. Thus, the root ion channel increases with increased concentrations of the element in the solution and volume of water absorbed, favoring the absorption process.

Silva et al. (2002) established the minimum dose required for each soil to reach the point where S absorbed by the plant is transported in the soil exclusively by

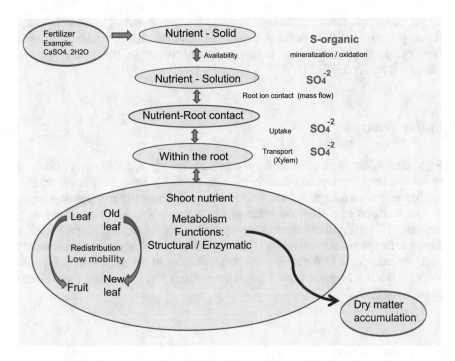

Fig. 5.2 Sulfur dynamics in the soil–plant system, indicating the passage processes of the nutrient in the different plant compartments

mass flow. The authors also determined the percentage contribution of mass flow and diffusion to the transport of S to the roots. As the mechanisms (mass flow and diffusion) act complementarily until reaching this point, only data referring to mass flow were analyzed.

The contribution of diffusion is obtained by subtracting from 100 the percentage of S transported by mass flow. A study found that S application in the soil increases the contribution of mass flow from 40% up to 100% (Silva et al. 2002).

It was observed that mass flow transported only 39 and 43% of S absorbed by the plants with the 0 mg dm^{-3} S dose in the soils of Paracatu and Lassance, respectively.

The main form of sulfur absorbed by plants is the sulfate SO_4^{2-}, absorbed by roots. However, as highlighted by Mengel and Kirkby (1978) who researched maize and sunflower, there is evidence that plants use sulfur in gaseous form (atmospheric SO_2) to meet the minimum part of their needs. In this case, plants capture S through the stomata and redistribute it to the entire plant. Sulfur can also be absorbed in organic form (amino acids cystine and cysteine), although in decreased amounts. Supply of sulfur amino acids, such as methionine, could supply plant nutrition (Miller 1947), although being of little importance in practice.

Regarding SO_4^{2-}, it is absorbed actively by roots against a concentration gradient (Legget and Epstein 1956). Smith et al. (1995) identified some sulfate transport systems, SHST1, SHST2, and SHST3. In the uptake process, sulfur is little affected by pH variations. However, there are indications demonstrating increased absorption at pH close to 6.5. In addition, luxury consumption, that is, nutrient accumulation by plants beyond their needs, does not occur in the uptake process (Simon-Sylvestre 1960).

Sulfur Transport

Right after sulfur uptake, the transport process is triggered, which occurs similarly to the absorbed form, predominantly as sulfate (SO_4^{2-}). However, a small fraction can be transported in organic form, as a fraction is metabolizable to supply this organ.

The transport of S from the epidermis to the endodermis can occur passively through AFS (cell wall and intercellular spaces) or actively through the cell interior thanks to its cytoplasmic extensions (plasmodesmata). S is only transported actively from the endoderm to the xylem due to the barrier in the endoderm (Casparian strips). After reaching the xylem, long-distance transport to the plant shoot occurs passively (Fig. 5.3).

Sulfur Redistribution

Sulfur movement in plants occurs, basically, during transport to the shoot via xylem (acropetal direction = upward from the plant base), as aforementioned. Basipetal movement of sulfur (downward toward the base), that is, redistribution, is very small and is considered S nearly immobile in the plant. Oliveira et al. (1995) observed that only 27% of the S absorbed by leaves was redistributed to the rest of the shoot and roots of bean plants. Naeve and Shibles (2005) observed that the S in soybean leaves can supply 20% of seed requirements. New leaves depend more on absorbed S than on stored S, while the opposite occurs with developing pods and the

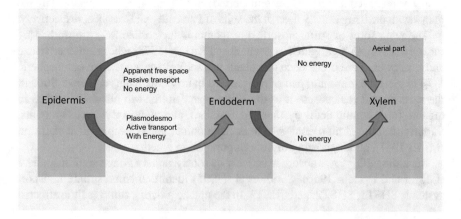

Fig. 5.3 Scheme of the sulfur transport process in plants

seeds. Thus, pods play an important role to store S for its mobilization to the seeds. The authors conclude that the development of soybean seeds depends on the S mobilized from other plant tissues.

We note that the SO_4^{2-} form is practically not detected in the phloem sap. Glutathione is the main form of reduced S transported in the phloem (Garsed and Read 1977). If there is sulfur deficiency in the soil, deficiency symptoms would appear in new leaves as the plant has difficulty to remobilize S in sufficient quantity from an old leaf, which is described in detail at the end of this chapter.

5.3 Participation in Plant Metabolism

Assimilatory Sulfate Reduction

Similar to nitrogen, sulfur is predominantly absorbed in the oxidized form SO_4^{2-} (+6) and needs to be previously reduced to the sulfide S^{2-} to be subsequently incorporated into organic compounds such as amino acids, proteins, and coenzymes, requiring energy (14 ATP) and a reducing power (transfer of eight electrons). Therefore, this reaction consumes an increased amount of respiratory energy. However, there are cases in which sulfate does not need reduction to be incorporated into some organic compounds (sulfolipids or polysaccharides).

The sulfate reduction process occurs mainly during the light period. Enzymes that participate in the reactions are located in the chloroplast membranes, with ferredoxin as electron donor. In non-chlorophyllated tissues, such as roots, the sulfate reduction process occurs in plastids with NADPH as electron donor, although being much slower than the reaction in leaves (chloroplasts).

The sulfate reduction process can be divided into the three following stages plus one last stage, with the incorporation of S into an organic compound:

(a) The first step to reduce sulfate is its activation or formation of active sulfate. Initially, SO_4^{2-} reacts with ATP (mediated by the enzyme ATP-Sulfurylase), resulting in the substitution of a pyrophosphoric radical (P-P) of the ATP by the sulfate. Thus, adenosine phospho-sulfate (APS) or active sulfate is formed. The PPi formed is hydrolyzed in inorganic phosphate via inorganic pyrophosphatase (PPi + H_2O = > 2Pi).

$$SO_4^{2-} + ATP \quad (ATP - Sulfurylase) \Rightarrow \quad APS \ + \ PPi$$
$$\text{Adenosine phospho} - \text{sulfate} \quad \text{Pyrophosphate}$$

(b) The APS produced is metabolized in the plastids, being quickly reduced or sulfated, with the predominance of reduction. The sulfuric groups of APS are transferred to a complex, carrier of sulfydryl groups, through action of APS transferases, according to the following scheme:

$$APS + Car - SH \quad (APS\,transferase) \Rightarrow \quad AMP + Car - S - SO_3H$$

Thus, SO_3^{2-} is incorporated into the SH-carrier complex (replacing H).

(c) The carrier complex ($Car-S-SO_3^{2-}H$) is reduced to S^{2-}-H, receiving electrons supplied by ferredoxin and mediated by the action of sulfite reductase.

$$Car - S - SO_3H + 6H + +6\bar{e}\ (Reductase) \Rightarrow\ Car - S - S^{-2}H + 3H_2O$$

(d) Subsequently, the radical of the SH group is transferred to acetyl serine, requiring $2\bar{e}$ originating from ferredoxin. Sulfur is incorporated into the organic compound in this stage, generating cysteine (amino acid with S) and regenerating the SH-carrier.

Car - S - SH + acetyl serine (sulfhydrase[H1] of serine) ⇒ cysteine + acetic acid + Car - SH

In summary, assimilatory sulfate reduction occurs as follows:

$$H_2SO_4 + ATP + 8H + acetyl\ serine \Rightarrow\ cysteine + acetate + 3H_2O + AMP + PPi$$

At the end of sulfate reduction the first stable compound with S, cysteine, is obtained, which can be derived to methionine, subsequently being a precursor to other organic compounds with S (other S-amino acids, proteins, sulfolipids, and coenzymes). The small amount of unreduced S (sulfate ester) is a component of sulfolipids found in all membranes, which is important for ionic transport in root cells, related to salt tolerance.

Proteins are organic compounds to which most of S (and also N) are incorporated. In proteins, approximately 34 N-atoms are normally found for each S-atom. We highlight that genetic engineering studies improving assimilation rates of S will directly affect the quality of agricultural products such as soybeans not only for increasing protein content, but also for increasing the content of essential amino acids (methionine and cysteine), improving food quality (Krishnan 2005).

Most of sulfur reduced in plants (~ 90%) is part of essential amino acids cysteine and methionine (-SH) (Salisbury and Ross 1992). These amino acids join the composition of all proteins, having structural function due to the formation of disulfide bonds (S-S), which stabilize the protein structure. They also have enzymatic function due to the participation of the sulfydryl group (-SH) as an active group of a given enzyme complex (such as urease, sulfotransferase, coenzyme A) involved in metabolic reactions and other compounds (vitamins B_1, H). The largest fraction of sulfur in plants is found in this form, already reduced.

Sulfur is also part of other organic compounds that influence several important physiological processes in plant life (Marschner 1986), such as:

Ferredoxin – low molecular weight proteins containing Fe and S, serving in redox reactions in photosynthesis; NO_3^- reduction (nitrite reductase); biological N fixation, SO_4^{-2} reduction, and glutamate synthesis.

Coenzymes (CoA) are related to the metabolism of carbohydrates, fats, and proteins during respiration.

Nitrogenase – enzyme involved in biological N fixation.

There are other compounds containing S and sulfate esters with polysaccharides as structural components of membranes. S is also part of glutathione (reducing agent), a tripeptide with antioxidant action in several detoxification processes (Droux 2004) that use dithiol <=> disulfide to participate in redox reactions. Phytoquelates are produced by polymerization of glutathione from carboxy-peptidase. Glutathione is a precursor of phytoquelatins, which detoxify heavy metals. Phytoquelatins, low molecular weight peptides (5 to 23 amino acids) with many cysteine residues, form complexes with metallic cations such as Cu, Zn, and Cd (Castro et al. 2005). Marschner (1995) has also indicated that plant nutrition with sulfur may increase tolerance to metal toxicity (Co, Cd) in contaminated areas, as plants develop mechanisms of production of small proteins (<10 kDa) rich in S (cysteine) called metallothionines, whose function is to reduce free forms of these elements in the plant tissue. In addition, the production of phytoquelates is stimulated by the increased concentration of heavy metals in the medium.

We also note that most of the S reduced in the plant (~ 90%) is in the form of glutathione, with the rest in the form of sulfydryl groups (thiol) (~ 2%), among others. Glutathione maintains ferredoxin in the reduced form and is involved in the storage and long-distance transport of reduced S.

Briefly, sulfur has different basic functions in plants (Malavolta et al. 1997) (Table 5.1).

Certain plants, such as onion, garlic, and mustard, have volatile compounds containing sulfur (isothiocyanate; sulfoxides), which contribute to the characteristic odor of these vegetables.

As sulfur participates in organic structures in plants and in various enzymatic reactions (Ergle and Eaton 1951), its deficiency can also impair protein synthesis, such as nitrogen accumulation (N-nitrate or N-organic), reducing protein content in plants and plant growth (Table 5.2).

Thus, S is closely related to N metabolism, converting N-amino acids (non-protein) into N protein. As most proteins are located in chloroplasts, where chlorophyll molecules contain prosthetic groups, there is reduced chlorophyll content in case of deficiencies, which may cause pale green color and even chlorosis.

Table 5.1 Summary of sulfur functions in plants

Structural	Enzyme constituent	Processes
Amino acids (cysteine; cystine; methionine)	Sulfydryl group -S-H and dithiol -S-S-, active in enzymes and coenzymes	Photosynthesis
Proteins (all)	Ferredoxin	Non-photosynthetic fixation of CO_2
Vitamins and coenzymes (thiamine (decarboxylation of pyruvate in the krebs cycle), biotin)		Respiration
Esters with polysaccharides (non-reduced form) is part of sulfolipids (membranes, especially thylakoids)		Synthesis of fats and proteins
		Symbiotic nitrogen fixation

Table 5.2 Effect of sulfur in the medium on some components (organic and mineral) of the cotton leaf

SO_4^{2-}	SSO_4^{2-}	Organic S	Total sugar	$N-NO_3^-$	Organic soluble N	N protein	Fresh matter
Ppm	g	%	%	%	%	%	g
0.1	0.003	0.11	0.0	1.39	2.23	0.96	13
1.0	0.003	0.11	0.0	1.37	2.21	1.28	50
10	0.009	0.17	1.5	0.06	1.19	2.56	237
50	0.100	0.26	3.1	0.00	0.51	3.25	350
200	0.360	0.25	3.4	0.10	0.45	3.20	345

Table 5.3 Dry matter production, total nitrogen accumulation, nodule matter, N and S contents, and N:S ratio in the shoot of siratro

S dose	Dry matter	Total N	Nodules	N	S	N:S ratio
kg ha^{-1}	g per pot	mg per pot	mg per pot	g kg^{-1}		
0	10.1	276	315	27	0.7	40
30	11.3	372	560	33	1.6	21
60	12.4	403	564	33	1.6	20

Table 5.4 Average response of some crops to sulfur

Crop	Increase in production
	%
Cotton	37
Soybean	24
Coffee	41
Wheat	26
Bean	28
Maize	21

In legumes that have increased protein content, sulfur can play a fundamental role in maximizing biological N fixation through nitrogenase activity with beneficial effects on production (Monteiro et al. 1983). In a study with the forage legume siratro, increased dry matter, N accumulation, and nodular matter were observed when this legume had 1.7 g kg^{-1} S and N:S ratio equal to 20 (Table 5.3). Plants grown in this soil without receiving sulfur application had 0.7 g kg^{-1} S and N:S ratio of 40.

Probably increased efficiency in the use of nitrogen by plants depends on adequate sulfur levels, indicating the importance of these nutrients in plant nutrition. Koprivova et al. (2000) reinforce that N and S assimilation are coordinated, where deficiency of one of them affects the assimilation pathway of the other.

Given the requirement of plants in relation to sulfur for the formation of amino acids/proteins and for plant physiological processes such as photosynthesis and cold resistance, research has been indicating beneficial responses to S application in some crops (Table 5.4).

This fact is more important for the process of conventional fertilization, in which sulfur has not been applied adequately (use of concentrated fertilizer with S-free formulation). In addition, improved cultivars with increased productive potential and increased S requirement may reduce soil organic matter (sulfur reserve in the soil), making sulfur a limiting nutrient in agricultural production of tropical regions.

5.4 Crop Nutritional Requirements

The study of plant nutritional requirements should address the total extraction of the nutrient from the soil, respecting extraction at each stage of plant development in order to satisfy crop nutritional needs and obtain maximum economic production.

We note that the total S content in the plant required for optimal growth/development ranges from 0.2 to more than 1% of dry matter, ranging depending on the crop and other factors that are the subject of another chapter (*Foliar diagnosis*).

In order to discuss crop nutritional requirements adequately, the two following factors are equally important: total extraction/export and accumulation (uptake rate) of the nutrient throughout cultivation.

Nutrient Extraction and Export

Total sulfur extraction occurs as a function of the content in the plant and the amount of dry matter accumulated. Therefore, it depends on production, which depends on the species, variety/hybrid, soil availability, and crop management, among others.

As for vegetable species, the amount required ranges depending on crops (Table 5.5).

From the results of different crops, we observe that total sulfur extraction ranges from 8 (wheat) to 45 kg ha^{-1} (sugarcane), considering production in one hectare. However, in absolute values of total sulfur extraction, there is an increased requirement from legumes (bean and cotton plant) and a decreased requirement from grasses (sugarcane, maize, wheat, and rice), with values of 25–26 and 0.5–4.0 kg for each ton of grains produced, respectively. We also observe differences in the nutritional requirement of soybeans for different cultivars, with the cultivar Cristalina requiring more S (20%) compared to the cultivar IAC 11 (Table 5.6). Thus, nutritional requirements for S may range depending on crop and cultivar.

The value of crop nutritional requirement is an exact recommendation of sulfur fertilization if the soil cannot supply the element and losses with fertilization are null, according to the following general equation: Fertilization = (Crop nutritional requirement) - (Provided by the soil) ÷ losses. Maize is an example, with a nutritional requirement for sulfur of 19 kg ha^{-1} (Table 5.5). However, the amount of S to be applied in the crop is greater especially due to losses, among other factors. Khan

Table 5.5 Sulfur requirements of major crops

Crop	Plant part	Dry matter produced	Accumulated S		S required to produce 1 t of grains[d]
			Plant part	Total [c]	
		t ha^{-1}	kg ha^{-1}		kg t^{-1}
Annual					
Cotton plant	Reproductive (cotton/ cottonseed)	1.3	10 (7.7)[b]	32.5	25
	Vegetative (stem/ branch/leaf)	1.7	22		
	Root	0.5	0.5		
Soybean[a]	Grains	3	6 (2.0)	23	7.7
	Branches	6	17		
Bean	Pod	1	10 (10)	26	26
	Stem	0.4	4		
	Leaves	1.2	11		
	Roots	0.1	1		
Maize	Grains	9	11 (1.2)	19	2.1
	Crop residues	6.5	8		
Rice	Grains	3	5 (1.7)	12	4
	Stems	2	3		
	Leaves	2	1		
	Husk	1	1		
	Root	1	2		
Wheat	Grains	3	3 (1.0)	8	2.7
	Straw	3.7	5		
Perennial/semi-perennial					
Sugarcane	Stems	100	25 (0.25)	45	0.5
	Leaves	25	20		
Coffee[a]	Grains	2	3 (1.5)	27	13.5
	Trunk, branches, and leaves		24		

[a]Malavolta (1980); [b]Nutrient export through the grains produced: S accumulated in the grains/grain dry matter; [c]Suggests the total nutritional requirement of the crop for the respective yield level; [d]Suggests the relative nutritional requirement of S of the crop for the production of one ton of the commercial product (grains/stems): S accumulated in the plant (vegetative + reproductive part)/dry matter of the commercial product

et al. (2006) observed that the application of 60 kg ha^{-1} of S provided adequate maize nutrition (foliar S = 4.6 g kg^{-1}) and maximum production.

Sulfur export, characterized by the nutrient leaving the crop with the harvest product, is significant. The export rate of the crops under study reached an average of 58% (Table 5.5), with cotton and beans exporting the most, 7.7 and 10 kg/t produced, respectively.

Nutritional requirements may change with the release of new cultivars/hybrids, as phytotechnicians seek to increase population densities. Nutritional requirements may also change with increased doses of fertilizers.

Table 5.6   Production and accumulation of sulfur in shoots (kg ha⁻¹) and grains (kg per t of grain) of five soybean cultivars

Soybean cultivar	Production	S – shoot	S – grain
	t ha^{-1}	kg ha^{-1}	kg per t of grain
IAC 11	3.32	34	10.2
IAC 13	2.57	28	10.9
IAC 15	3.25	34	10.5
Santa Rosa	3.05	35	11.5
Cristalina	3.27	42	**12.8**
Average	3.09	35	11.3

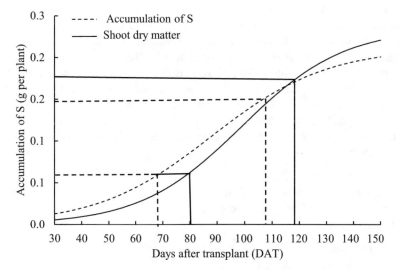

Fig. 5.4   Sulfur uptake rate by sugarcane seedlings

Uptake Rate

The study of the uptake rate of this nutrient (accumulation as a function of cultivation period) is important for determining the periods when the element is most required and for correcting deficiencies that may occur during crop development.

During the formation of sugarcane seedlings, we note that sulfur requirements range along development, being low in the initial stages. At 65 days after transplanting, the requirement increases nearly linearly until 108 days, when stabilization of S uptake begins (Fig. 5.4). Thus, sulfur (in the form of ammonium sulfate) can be applied in the formation of sugarcane seedlings at transplanting (10% of the total dose) and 65 days after transplanting, perfectly meeting the crop nutritional requirement.

5.5   Symptoms of Nutritional Deficiencies and Excesses

Chlorosis produced by sulfur deficiency is quite similar to that of nitrogen deficiency, as both have similar plant metabolism. The distinction can be detected at the
beginning, as sulfur deficiency begins in the youngest leaves due to its low mobility
in plants, while nitrogen deficiency begins in older leaves, which is attributed to the
high mobility of nitrogen. However, if symptoms worsen, chlorosis may be evenly
distributed in both new and old leaves.

In the case of legumes, S deficiency inhibits biological N fixation more than
photosynthesis, which may difficult the identification of S and N deficiency
(Fig. 5.5). Chemical analysis of leaves differentiates S and N deficiencies. We also
note that in legumes S deficiency normally occurs in new leaves when the N content
is adequate, while the opposite occurs in plants with N deficiency. S deficiency
occurs first in old leaves, indicating that its remobilization is associated with leaf
senescence induced by N deficiency (Marschner 1995).

In sulfur-deficient plants, S-SO_4^{-2} levels are decreased while N-amidic and NO_3^-
levels are increased. These analytical data contrast those of nitrogen-deficient
plants, in which soluble N levels are decreased and S-SO_4^{-2} levels are normal.

The S-$_{inorganic}$/S-$_{total}$ ratio in the tissue is considered the best indicative of the plant
nutritional status. Decreased ratios (<0.25) mean S deficiency (Marschner 1995).

Symptoms (visible)

– Chlorosis; curling of leaf margins.
– Small leaves; necrosis and defoliation.
– Short internodes; flowering reduction.
– Stunted plants.

   Other disorders.
   -        Disorders in chloroplast structure

– Decrease in photosynthetic activity.
– Increase in the soluble N/protein N ratio.

Friedrich and Schrader (1978) observed that sulfur deficiency reduced nitrate
reductase enzyme activity, nitrate accumulation, and concentrations of soluble protein and chlorophyll.

5.6   Excess Sulfur

Plants are relatively tolerant to high S-SO_4^2 levels in the soil solution. In soils with
high sulfate concentration (50 mM) problems can occur, although being confounded
with salt effects, such as decreased plant development and intense dark green color
in leaves. Premature leaf senescence may also occur.

In polluted air (0.5–0.7 mg of S-SO_2/M^3 of air) there may be toxic effects due to
the production of H_2SO_4 in the stomatal cavities, which dissociates into H^+, HSO_3^-,

Fig. 5.5 Photos and general description of visual symptoms of S deficiency in different crops. Sugarcane: Young leaves uniformly chlorotic (leaf to the left deficient and normal to the right); may develop light purple color (**a**); Aubergine: Plants stop developing, decreasing plant size (**b**); new leaves with pale yellow or soft green color (**c**). Unlike N deficiency, symptoms occur on new leaves, indicating that older tissues cannot supply S for new tissues, which depend on the nutrient absorbed by roots; Scarlet eggplant: Foliar symptoms are similar to those caused by lack of nitrogen, although appearing in the youngest leaves. Internodes shorten and decrease the size (**d**), maintaining uniform yellow color in the leaf blade (**e**); Grass of genus Panicum: Plants have a decreased number of leaves with generalized chlorosis, which start in the youngest leaves and progress to the other leaves (**f**)

capable of decoupling photophosphorylation reactions, causing leaf necrosis. Khan et al. (2006) observed that the foliar sulfur content equal to 8 g kg^{-1} was considered excessive.

References

Castro PRC, Kluge RA, Peres LEP. Manual de fisiologia vegetal: teoria e prática. Piracicaba: Agronômica Ceres; 2005.

Droux M. Sulfur assimilation and the role of sulfur implant metabolism: a survey. Photosynth Res. 2004;79:331–48. https://doi.org/10.1023/B:PRES.0000017196.95499.11.

Ergle DR, Eaton FM. Sulphur nutrition of cotton. Plant Physiol. 1951;26:639–54. https://doi.org/10.1104/pp.26.4.639.

Friedrich JW, Schrader LE. Sulfur deprivation and nitrogen metabolism in maize seedlings. Plant Physiol. 1978;61:900–3. https://doi.org/10.1104/pp.61.6.900.

Garsed SG, Read DJ. Sulfur dioxide metabolism in soybean, Glycine max var. Beloxi – II: biochemical distribution of $^{35}SO_2$ products. New Phytol. 1977;79:583–92.

Khan MJ, Khan MH, Khattak RA, et al. Response of maize to different levels of sulfur. Comm Soil Sci Plant Anal. 2006;37:41–51. https://doi.org/10.1080/00103620500403804.

Koprivova A, Suter M, Camp ROD, et al. Regulation of sulfate assimilation by nitrogen in Arabidopsis. Plant Physiol. 2000;122:737–46. https://doi.org/10.1104/pp.122.3.737.

Krishnan HB. Engineering soybean for enhanced sulfur amino acid content. Crop Sci. 2005;45:454–61. https://doi.org/10.2135/cropsci2005.0454.

Legget JE, Epstein E. Kinetics of sulfate absorption by barley roots. Plant Physiol. 1956;31:222–6. https://doi.org/10.1104/pp.31.3.222.

Malavolta E. Elementos de nutrição de plantas. São Paulo: Agronômica Ceres; 1980. p. 251p.

Malavolta E, Vitti GC, Oliveira SA. Avaliação do estado nutricional das plantas: princípios e aplicações. Piracicaba: Associação Brasileira de Potassa e do Fósforo; 1997. p. 319p.

Marschner H. Mineral nutrition of higher plants. London: Academic Press; 1986.

Marschner H. Mineral nutrition of higher plants. London: Academic Press; 1995.

Mengel K, Kirkby EA. Principles of plant nutrition. Worblaufen-Bern: International Potash Institute; 1978.

Miller LP. Utilization of DL methionine as a source of sulfur by growing plants. Contrib Boyce Thompson Inst. 1947;14:443–56.

Monteiro FA, Carriel JM, Martins L, et al. Effects of levels of Sulphur as gypsum for the growth of forage legumes in the state of so Paulo, Brazil. Bia. 1983;40:229–40.

Naeve SL, Shibles RM. Distribution and mobilization of sulfur during soybean reproduction. Crop Sci. 2005;45:2540–51. https://doi.org/10.2135/cropsci2005.0155.

Oliveira JA Jr, Rêgo IC, Scivittaro WB, Lima Filho OF, et al. Sources and additive effects on 35S foliar uptake by bean plants. Sci Agric. 1995;52:452–7. https://doi.org/10.1590/S0103-90161995000300008.

Salisbury FB, Ross CW. Plant physiology. Belmont: Wadsworth Publishing; 1992.

Silva DJ, Venegas VHA, Ruiz HA. Sulphur transport toward soybean roots in three soils from Minas Gerais state, Brazil. Pesq Agropec Bras. 2002;37:1161–7. https://doi.org/10.1590/S0100-204X2002000800014.

Simon-Sylvestre G. Les composés du soufree du sol et leur evolution – rapports avec la microflore, utilisation par les plantes. Ann Agron. 1960;3:311–32.

Smith FW, Ealing PM, Hawkesford MJ, et al. Plant members of a family of sulfate transporters reveal functional subtypes. Proc Natl Acad Sci U S A. 1995;92:9373–7. https://doi.org/10.1073/pnas.92.20.9373.

Chapter 6
Phosphorus

Keywords Phosphorus fixation · Phosphorus uptake · Phosphorus transport · Phosphorus redistribution · Assimilatory phosphorus · Phosphorus deficient

6.1 Introduction

Tropical soils usually have decreased concentration of available phosphorus and increased potential for phosphorus fixation applied via fertilizer. This context places phosphorus, along with nitrogen, as one of the most limiting nutrients for crop production in Brazil. Thus, for annual crops in soils with P considered very low (<0.6 mg dm^{-3}) and low (7–15 mg dm^{-3}) (resin extractor) (Raij et al. 1996), we expect increased responses to phosphorus application.

Increasing phosphorus in the soil is important either through mineral fertilization, providing P readily available to plants, or through organic fertilization, which is only available when soil microorganisms break the organic matter into simple forms, releasing the available inorganic phosphate ions. In the soil, 20–80% of the total P is in organic form, mainly as phytate (Raghothama 1999). After applying phosphorus in the soil, a series of transformations occur, which can remain in compartments of solid phase (labile and non-labile) and/or of liquid phase (solution). In the solid phase, labile phosphorus is weakly retained in the soil, having the function of maintaining a rapid balance with the soil solution, which does not occur with non-labile phosphorus, as the element is strongly retained in the soil. Available phosphorus is found in ionic form ($H_2PO_4^-$; HPO_4^{2-} e PO_4^{3-}) in the liquid phase. At this stage, the predominant ionic form is $H_2PO_4^-$ due to the acid reaction of Brazilian soils (~ pH = 5.5). At pH 7.1, there is an equal concentration of $H_2PO_4^-$ and HPO_4^{2-} ionic species. As the pH value decreases ([H+] increases), the presence of $H_2PO_4^-$ species increases proportionally.

As plants absorb phosphorus from the soil specifically from the solution (liquid phase), this is the form of interest in plant nutrition. Soil fertility studies address factors that increase available phosphorus, such as pH, type and quantity of clay minerals, P levels in the soil, aeration, humidity, and temperature, besides the availability of other nutrients, such as the use of silicates (Si source) (Prado and Fernandes 2001). In order to increase the efficiency of P mineral fertilization, authors indicate

© Springer Nature Switzerland AG 2021
R. de Mello Prado, *Mineral nutrition of tropical plants*,
https://doi.org/10.1007/978-3-030-71262-4_6

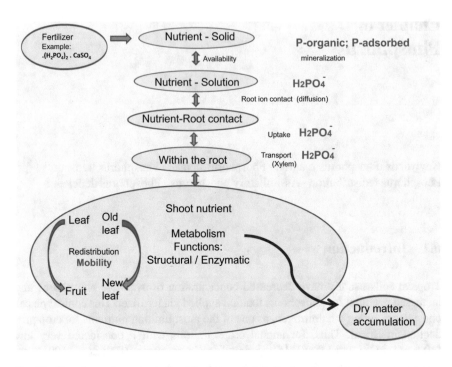

Fig. 6.1 Phosphorus dynamics in the soil–plant system, indicating the nutrient passage processes in the different plant compartments

an association with organic products with promising agronomic results (Vasconcelos et al. 2017). Factors affecting available P in the soil contribute to a better understanding of aspects related to plant nutrition.

When studying phosphorus in the plant system, it is important to know all the compartments the nutrient passes in the soil solution and in the root and shoot (leaves/fruits), that is, from the soil until its incorporation into an organic compound. In addition, it is important to know if the nutrient plays the role of enzymatic activator, contributing to vital functions that enable maximum dry matter accumulation in the agricultural product (grain, fruit, etc.) (Fig. 6.1).

6.2   Phosphorus Uptake, Transport, and Redistribution

Phosphorus Uptake

Phosphorus is absorbed from the soil solution in the $H_2PO_4^-$ or HPO_4^{2-} forms. However, as $H_2PO_4^-$ is the predominant form in the soil, it also predominates during the uptake process. Before phosphorus uptake actually occurs, this nutrient has to get in contact with the root. Specifically for phosphorus (as aforementioned in the chapter Root ion uptake), the movement of the nutrient in the soil is ruled by diffusion (ion movement against a concentration gradient), responsible for more than 94% of the contact between P and root. This P movement in the soil, characterized by a short distance, indicating the need for localized application (close to the plant root system), favors the uptake process. In addition, the diffusion process depends on water, as stated by Mackay and Barber (1985), who observed that adequate soil humidity was important to ensure sufficient P diffusion to the root and increased uptake and dry matter production (Table 6.1).

The phosphorus uptake process occurs basically in the two following phases: a passive, fast phase and a slower, active phase. Malavolta et al. (1997) observed the passive and active process in phosphorus uptake of bean plants. In the case of phosphorus ($H_2P^{32}O_4^-$), quick uptake was observed in the first 60 minutes, indicating the passive process, followed by slow uptake until completing 240 minutes, indicating the active process. The active process is slow as it needs to go against a very high concentration gradient in the root cells and in the xylem sap, which is approximately 100–1000 times greater than that of the soil solution. In addition, acidic pH (4.0) in the cell apoplast increases P uptake threefold compared to pH 6.0 (Setenac and Grignon 1985). Thus, P uptake can occur through two types of transporters, high and low affinity. One of the high-affinity P transporters is PH084 (Smith 2002). Transport through the membrane must be done through an H^+ cotransporter (Ullrich-Eberius et al. 1981).

For elements with low diffusion rate in the soil such as phosphates, plants with larger root surfaces have increased capacity to absorb the nutrient from the soil (Teo et al. 1995). For the best knowledge on P uptake, there are the mechanistic model, obtained in the laboratory and involving the plant through root surface/geometry and uptake kinetics, and also the model regarding soil nutrient requirements. In these models, the most important parameters contributing to increase P uptake by plants are, namely: 1 – root elongation rate; 2 – initial P concentration in the soil; 3 – root radius.

The literature indicates several factors that can affect the uptake process, such as external and internal factors previously discussed (*chapter root ion uptake*). In the specific case of phosphorus, external factors are important for increasing the concentration in the soil or nutrient solution, being directly related to plant uptake, as in the case of bean plant (Malavolta 1980).

External factors influencing P uptake by plants are related to soil pH, effects of other elements, temperature, and oxygen.

Table 6.1 Effect of soil moisture on P uptake and dry matter production of maize

	Adequate soil		
	Deficient (−170 kPa)	Humidity (−33 kPa)	Excess (−7.5 kPa)
P absorbed (μmol/pot)	229	381	352
Shoot (g/pot)	2.6	4.0	3.6
Roots (g/pot)	0.8	1.0	0.9

Soil (available P = 74 ppm)

Effect of pH and Other Nutrients

Soil pH value is the isolated factor that most affects soil phosphorus availability, with a pH close to 6.5 promoting the greatest availability in the soil solution (Fig. 6.2) and consequently increased plant uptake. The presence of other ions in the solution, such as magnesium, has a synergistic effect on phosphorus uptake, as Mg works as a P carrier, which is explained by ATPase activation in the membranes, contributing to uptake, and also by ATP generation in photosynthesis and respiration.

In an experiment with barley grown in the nutrient solution, it was observed that the presence of Mg along with P-labeled increased root P uptake and transport to the shoot (Malavolta 1980).

Temperature ranging from close to 0–35 °C linearly affects P uptake in most plants (Figure 6.3a) as temperature directly affects the respiration intensity of plants. Aeration through oxygenation increases P uptake in barley plants until reaching O_2 tension close to 2%, stabilizing P uptake after this value (Figure 6.3b). We observe that low oxygen tension impairs uptake but any increase above the minimum is enough to meet plant needs.

This occurs because oxygen is required for root respiration to provide energy (ATP) for absorption (active).

Effect of Temperature and Oxygen

Internal factors, related to the plant, can be summarized in those of genetic order; P and carbohydrate contents in the root; and association with microorganisms (Table 6.2).

The beneficial effects of mycorrhiza to increase P uptake by plants is a promising area with several studies. The beneficial effect on plant nutrition is maximized only in species that allow optimal root colonization rate. In addition, an imbalance in soil P availability (high content) impairs symbiosis and plant growth.

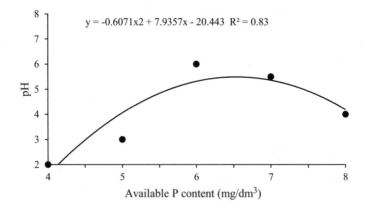

Fig. 6.2 Effect of pH value on P availability

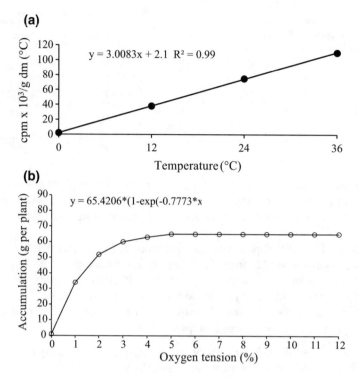

Fig. 6.3 Effect of temperature (**a**) and aeration (**b**) on P uptake by plants

Table 6.2 Some internal factors affecting phosphorus uptake by plants

Factor	Specialty
Genetic	Variation in the parameters of uptake kinetics or root morphology
	Response of soybean to P is controlled by a gene pair
	Multiple genes encode phosphate carriers
P content in the phloem	Feedback signal to regulate P uptake
P and carbohydrate content in the root	The speed, in a certain limit, increases along with the level of carbohydrates in the root.
Mycorrhizae	Increases P uptake by branched hyphae

New roots have increased phosphorus uptake capacity, while older roots have decreased capacity (Table 6.3).

Beebe et al. (2006) add that P uptake in common bean plants in a low fertility soil is affected by root architectural elements, which is an important factor to select plants that absorb P more efficiently.

The ability of plant roots to absorb P under low availability conditions is a complex process, as it depends on countless responses (Raghothama 1999), such as:

Table 6.3 Phosphorus uptake
rate by maize as a function of
plant age (days)

Plant age (days)	P (μmol/m root/day)
20	11.3
30	0.90
40	0.86
50	0.66
60	0.37
70	0.17
80	0.08

Morphological   Increased root:shoot ratio; change in root morphology and architecture; increased root hair proliferation; root hair elongation; accumulation of anthocyanin pigments; proteoid root formation; increased mycorrhizal colonization.

Physiological   Increased phosphate uptake; reduced Pi flow; increased efficiency of Pi use; mobilization of Pi from the vacuole to the cytoplasm; increased phosphorus translocation in the plant; re-translocation of Pi from old leaves; retention of more Pi in the root; secretion of organic acids, protons, and chelators; secretion of phosphatases and RNAses; change in respiration, carbon metabolism, photosynthesis, nitrogen fixation, and aromatic enzyme pathways.

Normally, when Pi availability in the soil is limited, a high-affinity uptake mechanism is activated (Schachtman et al. 1998). Very low P levels in the external medium can increase threefold the uptake rate depending on plant species. Thus, there are two mechanisms that can participate in Pi uptake, the high and low affinity. The first has low K_m (3–7 μM of Pi) with a preponderant role in soils with low P availability and the second has high K_m (50–300 μM of Pi). Only the high-affinity mechanism could be affected by the medium, which does not occur with the low-affinity mechanism as it is constitutive (Raghothama 1999).

Phosphorus Transport

The transport process is triggered after the phosphorus uptake process, occurring in a similar form to that absorbed, predominantly as $H_2PO_4^-$. However, it can also be found in low concentrations in the xylem in organic form (phosphorylcholine or carbohydrate esters).

Phosphorus Redistribution

The phosphorus redistribution process occurs exclusively in the organic form, although being also found in the phloem in inorganic form. Redistribution occurs in the phloem, and organic P is highly mobile. As it is readily mobile in the plant

Table 6.4 Foliar P content, increase in maize production, and phosphorus utilization factor as a function of modes of application of P in a soil with low P content. (Adapted from Prado et al. 2001)

P$_2$O$_5$ dose	Application Broadcast			Ridge-furrow			Double ridge-furrow		
	P [1]	Increase in production [2]	P factor [3]	P	Increase in production	P factor	P	Increase in production	P factor
kg ha^{-1}	g kg^{-1}	t ha^{-1}		g kg^{-1}	t ha^{-1}		g kg^{-1}	t ha^{-1}	
45.0	2.1	0.13	0.346	2.1	0.45	0.100	2.1	0.21	0.210
67.5	2.1	0.20	0.337	2.2	1.32	0.051	2.3	1.54	0.044
90.0	2.3	0.24	0.375	2.5	2.06	0.044	2.5	2.82	0.032
112.5	2.3	0.28	0.402	2.6	2.76	0.041	2.7	3.63	0.031
135.0	2.4	0.57	0.237	2.7	3.04	0.044	2.9	4.40	0.031
Average [4]	2.2B			2.4AB			2.4A		

[1] in the control, without P, foliar content = 2.1 g kg^{-1}. [2] Obtained in relation to average grain production (4.58 t ha^{-1}) at dose zero (absence of P application), considering the three application modes. [3] P factor: P utilization factor (kg ha^{-1} of P$_2$O$_5$/increase in production in kg ha^{-1}). [4] Same letters do not differ by Tukey test ($P < 0.05$)

upward or downward, new leaves are supplied not only by the phosphate absorbed by roots but also by phosphate originated from older leaves. P-deficient plants first demonstrate deficiency symptoms in old leaves. We estimate that approximately 60% of the P of old leaves can be remobilized via the phloem to new leaves (Malavolta et al. 1997); that is, hydrolysis of nucleic acids and phospholipids would contribute to 40–47 and 26–38%, respectively, of the total P reabsorbed from senescent leaves (Aerts 1996). Details of symptoms of P nutritional deficiency are covered at the end of this chapter.

We also note that experiments have shown that there may be some compartmentalization of P in plants (such as maize) when P is supplied in a part of the root system, which is attributed to the type of vascularization between leaves and roots (Stryker et al. 1974). Research indicates that phosphorus applied in the soil in both plant sides (double ridge-furrow) was superior to application in only one side (ridge-furrow) and application in the entire soil volume (broadcast) to increase grain production (adapted from Prado et al. 2001, Table 6.4).

6.3   Participation in Plant Metabolism

Unlike N and S that are reduced in the plant, $H_2PO_4^-$ in oxidized form remains in the form that was absorbed and transported, not exchanging valence during its metabolism in plants and remaining pentavalent (PO_4^{3-} or $P_2O_7^{4-}$). P occurs in the plant in organic and inorganic forms (accumulated in the vacuole) that are released into the cytosol according to plant needs, in order to avoid harming any important physiological process (photosynthesis) in the absence of P in the external medium.

In practice, phosphorus absorbed by cells is quickly metabolized. In approximately 10 minutes, most phosphate (approximately 80%) is incorporated into organic compounds (Jackson and Hagen 1960).

Total P in plants with adequate nutritional status occurs predominantly in inorganic form (orthophosphate), soluble in water, and found in the vacuole (approximately 75%), with the rest found in the cytoplasm and cell organelles. In P-deficient plants the inorganic form predominates, being found in the cytoplasm and chloroplasts of leaf cells, that is, the metabolic site (Foyer and Spencer 1986). In leaves, total P content can range approximately 20-fold without affecting photosynthesis. The concentration of inorganic P in the cytoplasm is regulated in a narrow range by phosphorus homeostasis, in which inorganic P of vacuoles acts as a buffer (Mimura et al. 1990). Another important part of P is in organic form, in the four following large groups:

(a) Phosphoric esters: there are more than 50, such as trioses and fructose-1-6-bisphosphate (sucrose biosynthesis) (Fig. 6.4) (intermediate compounds in the breakdown of sugars). The energetic bonds of these compounds represent the machine moving all plant metabolism. We note that most of the P contained in esters (approximately 70%) are contained in only nine compounds, in the following sequence: glucose-6-P (20%) > ATP (10%) > UDPG (uridine diphosphate and glucose) > 3-PGA (3-phosphoglyceric acid) > fructose-6-P > UDP (uridine diphosphate) > mannose-6-P; UTP (uridine triphosphate) > ADP.

(b) Phospholipids (membrane component, conferring lipid nature): chloroplasts have a quite developed membrane system (thylakoids) and represent approximately 40% of the total phospholipids of photosynthetic cells.

(c) Nucleotides:
    nitrogenated Base + Pentose +1–3 radicals of phosphoric acids, which may be free in the form of **ATP–ADP** or combined in nucleic acids, namely: **DNA** (macromolecule, responsible for carrying the genetic information of the cell) and **RNA** (involved in the translation of genetic information (via mRNA) and protein synthesis). The main function of P in the DNA is to form a structural bridge between nucleosides (= pentose + base), which are the letters of the genetic code.

    The most important pathway to its entry into organic combinations is through the esterification of an OH group of pentose bound to adenin (base set + ribose = adenosine), resulting in adenosine monophosphate, adenosine diphosphate, and adenosine triphosphate (ATP) (Figure 6.4b).

(d) Phytic acid: hexaphosphoric or inositol ester, which is the reserve substance of P in the plant.

We also highlight that the energy produced in photosynthesis and also in respiration is stored mainly in the form of ATP. Normally the cell concentration of ATP is very low, as it is renewed (resynthesized) every minute. The high-energy phosphate bond is a way to store and use large amounts of energy, from 12,000 to 16,000 calories. This energy is used by the plant in several vital processes, from solute transport through the cell membranes, promoting absorption (Fig. 6.5), and the mechanical

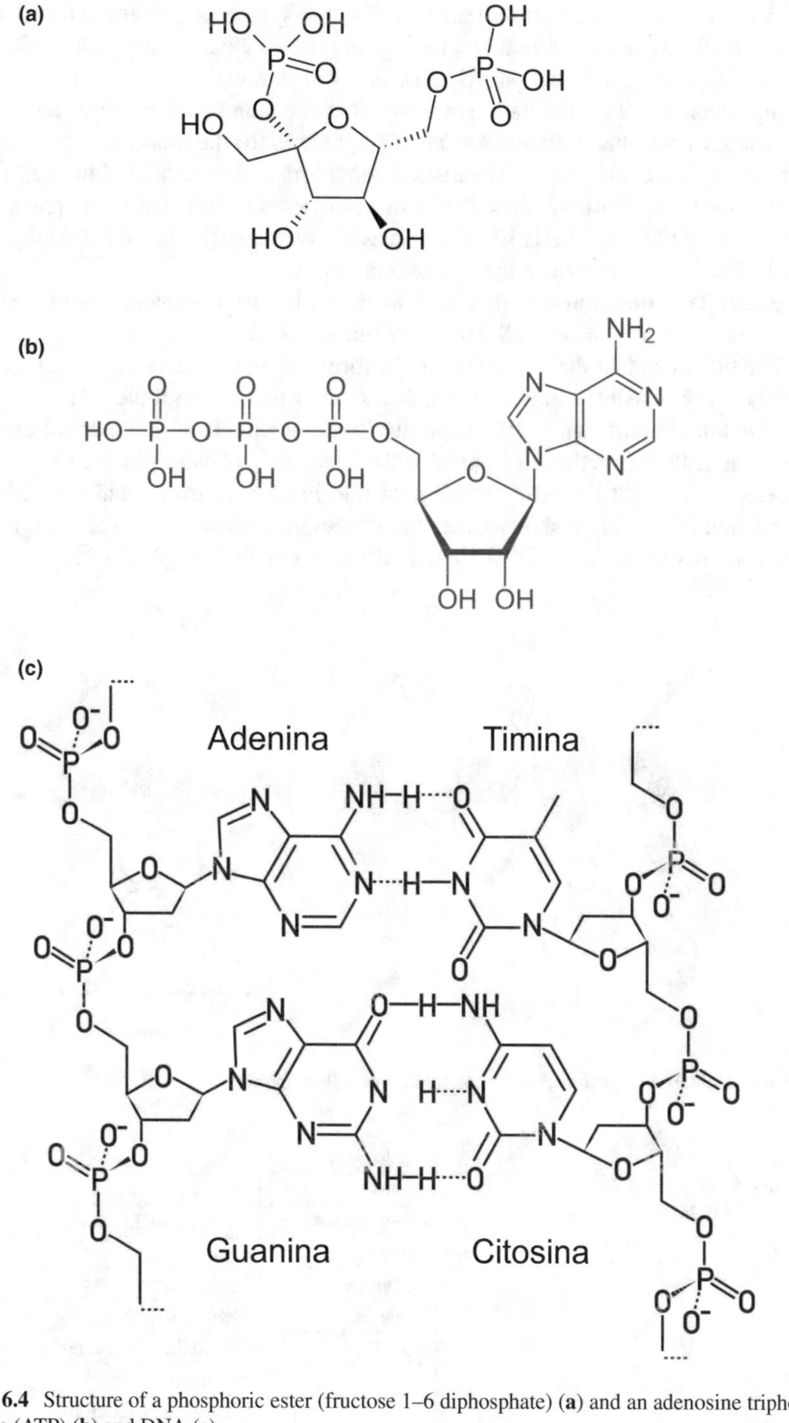

Fig. 6.4 Structure of a phosphoric ester (fructose 1–6 diphosphate) (**a**) and an adenosine triphosphate (ATP) (**b**) and DNA (**c**)

work of penetrating the root into the soil to starch synthesis (present in the grains). It is difficult to imagine if there is a biological process where phosphorus, in its different compounds, does not participate directly or indirectly.

Starch and sugar accumulation in leaves of P-deficient plants can also result from decreased export due to limitations in ATP synthesis for the transport of sugar protons through the phloem and decreased requirement in accumulation sites (Rao et al. 1990). In addition, when the Pi of the cytosol is low, the triose-phosphate produced by photosynthesis in chloroplasts is not transported to the cytoplasm to produce sucrose, being deviated to starch production.

P also participates in cell signaling, as in inositol triphosphate, modifying proteins irreversibly (Fraústo da Silva and Williams 2006).

The functions that these compounds perform are related to the structural aspect, to the energy transfer/storage process, and to other processes (Table 6.5).

Given the importance of P in plant nutrition, the effects of this nutrient are frequent on crop production in tropical soils. Thus, P application in the soil should increase its content in leaves, playing its role in plant nutrition and reflecting in production (Fig. 6.6), as shown in several studies on maize (Prado et al. 2001), passion fruit tree (Prado et al. 2005), and guava tree (Corrêa et al. 2003), etc.

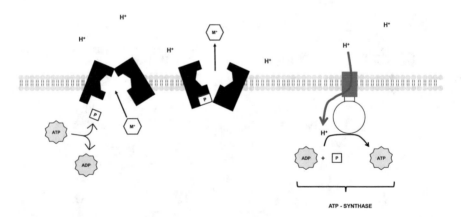

Fig. 6.5 Role of ATP in transporting solutes across cell membranes

Table 6.5 Phosphorus functions in plants

Structural	Processes
Carbohydrate esters	Energy transfer/storage
Phospholipids	Ionic uptake
Coenzymes	Photosynthesis
Nucleic acids	Synthesis (protein)
Nucleotides	Cell multiplication and division
	Inheritance
	Biological N fixation

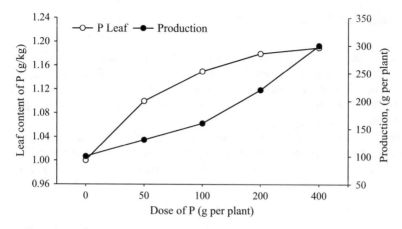

Fig. 6.6   Effect of P application on leaf content and production

Singh et al. (2006) verified in cotton that individual leaf area and water content in the fresh leaf depends on the leaf P content. The effects of P on the quality of agricultural products are scarcely studied. In peanuts, P application did not change the protein and oil content of grains (Hernandez et al. 1991; Kasai et al. 1998).

6.4   Crop Nutritional Requirements

As aforementioned, the study of crop nutritional requirements should address the total extraction of the nutrient from the soil, respecting extraction at each stage of crop development in order to satisfy crop nutritional needs and obtain maximum economic production.

We note that the total P content in the plant for optimal growth/development is 0.3–0.5% (3–5 g kg^{-1}) of dry matter. However, these values may vary depending on the crop and other factors, which are the subject of a chapter (*foliar diagnosis*).

In order to adequately discuss crop nutritional requirements, the two following factors are equally important: nutrient total extraction/export and the uptake rate of this nutrient throughout cultivation.

As aforementioned, the study of crop nutritional requirements should address the total extraction of the nutrient from the soil, respecting extraction at each stage of crop development in order to satisfy crop nutritional needs and obtain maximum economic production. The total P content in the plant for optimal growth/development is 0.3–0.5% (3–5 g kg^{-1}) of dry matter. However, these values on leaves may vary depending on crop and other factors, which are covered in Chap. 19.

Thus, in order to adequately discuss crop nutritional requirements, the two following factors are equally important: nutrient total extraction/export and the uptake rate of this nutrient throughout cultivation.

Nutrient Extraction and Export

Total extraction of phosphorus occurs as a function of P content in the plant and dry matter accumulated. Therefore, it depends on production, which depends on species, variety/hybrid, soil availability, and crop management, among others.

As for the vegetable species, there is variation in the amount required depending on the crop (Table 6.6).

From the results, we observe that total extraction of phosphorus ranged from 8.2 (a cotton plant with decreased production) to 56 kg ha^{-1} (maize). However, in absolute values of total phosphorus extraction, we observe increased requirement of

Table 6.6 Phosphorus requirements of major crops

Crop	Plant part	Dry matter produced	P accumulated Plant part	Total [3]	P required to produce 1 t of grains[4]
		t ha^{-1}	kg ha^{-1}		kg t^{-1}
Annual					
Cotton	Reproductive (cotton/cottonseed)	1.3	4 (3.1)[2]	8.2	6.3
	Vegetative (stem/branch/leaf)	1.7	4		
	Root	0.5	0.2		
Soybean[1]	Grains	3	26 (8.7)	40	13.3
	Stem/branch/leaf	6	14		
Bean	Pod	1	4 (4.0)	9.7	9.7
	Stem	0.4	0.6		
	Leaves	1.2	5		
	Root	0.1	0.1		
Maize[1]	Grains	6.4	24 (3.8)	56	8.8
	Crop residues	–	32		
Rice	Grains	3	8 (2.7)	17	5.7
	Stems	2 (m.s.)	3		
	Leaves	2 (m.s.)	1		
	Husk	1	2		
	Root	1(m.s.)	3		
Wheat	Grains	3	11 (3.7)	20	6.7
	Straw	3.7	9		
Perennial/semi-perennial					
Sugarcane	Stem	100	10 (0.1)	20	5
	Leaves	25	10		
Coffee plant[1]	Grains	2	3 (1.5)	19	9.5
	Trunk, branches, and leaves	–	16		

[1] Malavolta (1980); [2] Nutrient export through the grains produced (kg t^{-1}): N accumulated in the grains/grain dry matter; [3] Suggests the total nutritional requirement of the crop for the respective yield level; [4] Suggests the relative nutritional requirement of P of the crop for the production of one ton of the commercial product (grains/stems), obtained by the following formula: P accumulated in the plant (vegetative + reproductive part)/dry matter of the commercial product

soybean/bean legumes and decreased requirement of sugarcane/rice grasses, with values of 5.0–13.3 kg for each ton of product (stems/grains), respectively.

Regarding P export per ton of grains produced, we observe that soybean is the crop that exports the most (8.7 kg/t). However, for most crops, phosphorus is concentrated in grains. According to Staufler and Sulewski (2004), this occurs as P is related to plant metabolic processes, concentrating in the most active growth areas for being mobile and transferring most of absorbed P to the grains. Resende et al. (2006) also observed in maize that the grain was the final location of most (87%) of the absorbed P, as P in grains is related to organic reserve compounds, such as phytates. Phytates are presumably involved in regulating starch synthesis during grain filling. Thus, part of P is associated with the starch fraction and is incorporated into starch grains.

Nutrient Uptake Rate

The study of uptake rate (nutrient accumulated as a function of cultivation period) of P indicates the most demanding period. However, in practice, nutrients are not distributed over the cycle in crops grown in field conditions. Phosphorus application installments have not shown beneficial effects on nutrition and production of crops grown in field conditions, such as in pineapple crops (Teixeira et al. 2002). Localized application annually in the planting furrow proved to be efficient for maize and provided adequate residual effect (Resende et al. 2006).

In cotton crops, researchers observed slow initial uptake of P up to 30 days, reaching 13%, while reaching 40% of the total absorbed by the plant until the beginning of flowering (Mendes 1965).

We also note that phosphorus uptake by cotton plant occurs increasingly and constantly. Thus, phosphorus requires maintenance throughout the crop development phase, where P is normally applied in the fertilization at planting.

Fageria (1998) developed a greenhouse experiment with the objective of evaluating the utilization efficiency of P by 15 common bean genotypes with low, medium, and high P levels in a dark red latosol. Significant differences were found between genotypes regarding phosphorus utilization. Genotypes Rio Doce, São José, IPA9, and Roxo9 were classified as efficient and responsive.

Inter- and intraspecific differences in the capacity to use phosphorus (P) by the soil are explained, in part, by variations in root morphology and physiology, which characterize plants regarding nutrient acquisition. Föhse et al. (1988) complement that the efficiency of plants regarding P is related to their uptake efficiency, which was determined by the ratio between root system length and shoot biomass, as well as by the uptake rate per unit root length (inflow).

Machado and Furlani (2004), who studied maize cultivars in nutritive solution, found that the lowest values of Km and Cmin were good indicators of P uptake capacity of plants, being related to the greatest dry matter productions and the highest P utilization efficiency indexes.

6.5   Symptoms of Nutritional Deficiencies and Excesses

Deficiency

The most characteristic symptom of phosphorus deficiency is the appearance of dark green color in older leaves, as leaf number and leaf area expansion greatly decrease due to chlorophyll formation. Although the number of chlorophyll increases, the photosynthesis rate per unit is low. In addition, some species show a reddish or purplish color, especially along veins, due to chlorosis and necrosis of tissues. Dark color is caused by accumulation of red, blue, or purple pigments belonging to the group known as anthocyanins. These pigments are glycides formed by the reaction between sugar and a group of complex cyclic compounds, anthocyanidins. P deficiency symptoms can change depending on the species (Fig. 6.7).

P-deficient plants accumulate sugars (potential chemical energy), and have a lack of energy at the same time, as ATP compromises the biosynthesis processes of the plant. Thus, there is a marked decrease in the synthesis of RNA, starch, and lipids, decreasing the activation of amino acids required for protein-peptide binding, which causes marked protein deficiency and accumulation of soluble nitrogen compounds. This protein deficiency is responsible for decreasing plant vegetative development. In addition, P acts as a regulator of photosynthesis and carbohydrate metabolism, limiting plant development. P deficiency can cause starch accumulation in chloroplasts due to the lack of sufficient energy to translocate assimilates (sucrose transport in the phloem is a process that requires metabolic energy – ATP).

P-deficient plants induce different strategies specifically in the root system to increase P uptake (Fig. 6.8).

P-deficient plants induce genes that encode for transporters of phosphate, phosphatases, and ribonucleases, products that increase phosphate acquisition and recycling (Abel et al. 2002).

Symptoms of phosphorus deficiency, in general, can be summarized as follows:

(a) Decreased development, which occurs in most species, with the appearance of stunted plants.
(b) Dark green color in old leaves; purple color in some species (water stress and root damage can also cause purple color).
(c) Narrow-leaf insertion angle.
(d) Reduced flowering; reduced number of fruits and seeds, and delayed maturity.

The literature also indicates other effects of P deficiency, such as:

Decreased dry matter accumulation of stem or reproductive part to the detriment of other plant organs. Stem growth quickly decreases in *Stylosanthes humata* under phosphorus deficiency, although roots continue to grow as they retain more phosphorus and also due to an additional translocation of liquid of phosphorus from the stem to the roots (Smith et al. 1990). Thus, plants under deficiency have an adaptation mechanism that increases translocation of carbohydrates to the root, minimizing damage to the root system. Crusciol et al. (2005) observed that cultivars IAC 201 and IAC 202 of rice under low P availability prioritized root system development in relation to the shoot.

Fig. 6.7 Photos and general description of visual symptoms of P deficiency in different crops. Grass of genus Panicum: Plants have reduced growth and foliar area, causing production losses (**a**); Aubergine: Older yellow leaves (**b**); Maize: Dark green color in older leaves, followed by purple color at the tips and margins, also reaching the stem (**c**), leading to decrease in plant growth (plant being deficient to the right and normal to the left) (**d**); Cucumber: Lower leaves with pale green color and upper leaves with dark green color; the blade of older leaves may have internerval chlorosis with tanned appearance (**e**); Bean: With the evolution of symptoms, old leaves become necrotic and plant development decreases (**f**); Citrus: In fruit-bearing plants, fruits become rough, spongy, hollowed, and excessively acidic, with thicker albedo (**g**); Eucalyptus: In P deficiency, plants have reduced growth and leaves become reddish, especially in the margin, advancing to the center of the leaf blade (**h**)

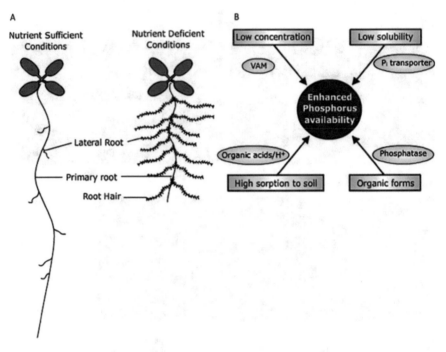

Fig. 6.8 (**a**) Plant roots exhibiting a change in response to phosphorus deficiency. Deficiency induces inhibition of primary root elongation and increases growth and density of lateral roots and root hairs. (**b**) Plants can increase phosphorus availability by secreting phosphatase, organic acids, and protons by involving Pi transporters or by association with vesicular-arbuscular mycorrhiza (VAM). (Kathpalia and Bhatla 2018)

Phytosanitary problems can occur in P-deficient plants, limiting the root system and decreasing the uptake of all nutrients. In this situation, the root system may be more susceptible to attack by root disease pathogens, with decreased microbiological activity in the rhizosphere. In addition, there may also be a delay in the physiological maturation of grains, requiring the plant to remain for an increased period in the field, increasing the chances of pest/pathogen attacks. P deficiency can increase disease incidence due to decreased physical resistance against the pathogen (decreased lignin and suberin), decreased resistance due to decrease of endogenous pesticides (alexins, glycosides, alkaloids), or due to variations in the inter- and intracellular structure in the form of accumulations of substrates that are food for pathogens (Malavolta 2006).

Excess Phosphorus

It is a rare disorder in the literature, being indicated by dark red spots in old leaves. In maize, foliar levels of P > 8 g kg^{-1} are considered excessive (Mengel and Kirkby 1987).

Fig. 6.9 Phosphorus toxicity in soybean plants grown in sandy textured soil

Several authors cite negative effects of phosphorus on the use of cationic micronutrients by plants, especially Zn and Fe. Phosphorus toxicity inhibits the growth of the soybean plant due to zinc deficiency (Fig. 6.9). The hypotheses state that P precipitation reaction and the micronutrient have their transport to the shoot reduced in the conducting vessels, causing deficiency. Another hypothesis states that dilution may affect the micronutrient with increased plant development due to P application, causing its deficiency in the plant. Therefore, excess P can induce deficiency symptoms in these micronutrients. Excess phosphorus can decrease CO_2 fixation and starch synthesis (Marschner 1986).

References

Abel S, Ticconi CA, Delatorre CA. Phosphate sensing in higher plants. Physiol Plant. 2002;115:1–8. https://doi.org/10.1034/j.1399-3054.2002.1150101.x.

Aerts R. Nutrient resorption from senescing leaves of perennials: are there general patterns? J Ecol. 1996;84:597–608. https://doi.org/10.2307/2261481.

Beebe SE, Rojas-Pierce M, Yan X, et al. Quantitative trait loci for root architecture traits correlated with phosphorus acquisition in common bean. Crop Sci. 2006;46:413–23. https://doi. org/10.2135/cropsci2005.0226.

Corrêa MCM, Prado RM, Natale W, et al. Response of guava to rates and placement of phosphate fertilizer. Rev Bras Frutic. 2003;25:164–9. https://doi.org/10.1590/S0100-29452003000100045.

Crusciol CAC, Mauad M, Alvarez RCF, et al. Phosphorus doses and root growth of upland rice. Bragantia. 2005;64:643–9. https://doi.org/10.1590/S0006-87052005000400014.

Fageria NK. Phosphorus use efficiency by bean genotypes. Rev Bras Eng Agríc Ambient. 1998;2:119–246. https://doi.org/10.1590/1807-1929/agriambi.v02n02p128-131.

Föhse D, Claassen N, Jungk A. Phosphorus efficiency of plants. I – External and internal P requirement and P uptake efficiency of different plant species. Plant Soil. 1988;110:101–9. https://doi. org/10.1007/BF02143545.

Foyer C, Spencer C. The relationship between phosphate status and photosynthesis in leaves and assimilate partitioning. Planta. 1986;167:369–75. https://doi.org/10.1007/BF00391341.

Fraústo da Silva JJR, Williams RJP. The biological chemistry of the elements: the inorganic chemistry of life. Oxford: Clarendon Press; 2006.

Hernandez FBT, Bellingieri PA, Souza ECA, et al. Adubação fosfatada e potássica em amendoim (*Arachis hypogaea* L.). Científica. 1991;19:15–27.

Jackson PC, Hagen CE. Products of orthophosphate absorption by barley roots. Plant Physiol. 1960;35:326–32. https://doi.org/10.1104/pp.35.3.326.

Kasai FS, Athayde MLF, Godoy IJ. Peanut oil and protein yield in function of phosphate fertilization and harvest time. Bragantia. 1998;57:100018. https://doi.org/10.1590/ S0006-87051998000100018.

Kathpalia R, Bhatla SC. Plant mineral nutrition. In: Bhatla SC, Lal MA, editors. Plant physiology, development and metabolism. Singapore: Springer; 2018. p. 37–81. https://doi. org/10.1007/978-981-13-2023-1_2.

Machado CTT, Furlani AMC. Kinetics of phosphorus uptake and root morphology of local and improved varieties of maize. Sci Agric. 2004;61:69–76. https://doi.org/10.1590/ S0103-90162004000100012.

Mackay AD, Barber SA. Soil moisture effects on root growth and phosphorus uptake by corn. Agron J. 1985;77:519–23. https://doi.org/10.2134/agronj1985.00021962007700040004xa.

Malavolta E. Elementos de nutrição de plantas. São Paulo: Agronômica Ceres; 1980. 251p.

Malavolta E. Manual de nutrição mineral de plantas. São Paulo: Agronômica Ceres; 2006.

Malavolta E, Vitti GC, Oliveira SA. Avaliação do estado nutricional das plantas: princípios e aplicações. Piracicaba: Associação Brasileira de Potassa e do Fósforo; 1997. 319p.

Marschner H. Mineral nutrition of higher plants. London: Academic Press; 1986.

Mendes HC. Cultura e adubação do algodoeiro. São Paulo: Instituto Brasileiro de Potassa; 1965.

Mengel K, Kirkby EA. Principles of plant nutrition. Worblaufen-Bern: International Potash Institute; 1987.

Mimura T, Dietz KJ, Kaiser W, et al. Phosphate transport across biomembranes and cytosolic phos- phate homeostasis in barley leaves. Plant. 1990;180:139–46. https://doi.org/10.1007/ BF00193988.

Prado RM, Fernandes FM. Effect of slag and limestone on the availability of phosphorus of an Oxisol planted with sugarcane. Pesq Agrop Brasileira. 2001;36:1199–204. https://doi. org/10.1590/S0100-204X2001000900014.

Prado RM, Fernandes FM, Roque CG. Effects of fertilizer placement and levels of phosphorus on foliar phosphorus content and corn yield. Rev Bras Ciênc Solo. 2001;25:83–90. https://doi. org/10.1590/S0100-06832001000100009.

Prado RM, Vale DW, Romualdo LM. Fósforo na nutrição e produção de mudas de maracujazeiro. Acta Sci. 2005;27:493–8. https://doi.org/10.4025/actasciagron.v27i3.1461.

Raghothama KG. Phosphate acquisition. Annu Rev Plant Physiol Plant Mol Biol. 1999;50:665–93. https://doi.org/10.1146/annurev.arplant.50.1.665.

Raij BV, Cantarella H, Quaggio JA, et al. Recomendações de adubação e calagem para o Estado de São Paulo. Campinas: Instituto Agronômico & Fundação IAC; 1996.

Rao IM, Fredden AL, Terry N. Leaf phosphate status, photosynthesis, and carbon partitioning in sugar beet. III. Diurnal changes in carbon partitioning and carbon export. Plant Physiol. 1990;92:29–36. https://doi.org/10.1104/pp.92.1.29.

Resende AV, Furtini Neto AEF, Alves VMC, et al. Phosphorus sources and application methods for maize in soil of the Cerrado region. Rev Bras Ciênc Solo. 2006;30:453–66. https://doi.org/10.1590/S0100-06832006000300007.

Schachtman DP, Reid RJ, Ayling SM. Phosphorus uptake by plants: from soil to cell. Plant Physiol. 1998;116:447–53. https://doi.org/10.1104/pp.116.2.447.

Setenac H, Grignon C. Effect of pH on orthophosphate uptake by corn roots. Plant Physiol. 1985;77:136–41. https://doi.org/10.1104/pp.77.1.136.

Singh V, Pallaghy CK, Singh D. Phosphorus nutrition and tolerance of cotton to water stress: II. Water relations, free and bound water and leaf expansion rate. Field Crop Res. 2006;96:199–206. https://doi.org/10.1016/j.fcr.2005.06.011.

Smith FW. The phosphate uptake mechanism. Plant Soil. 2002;245:105–14. https://doi.org/10.1023/A:1020660023284.

Smith FW, Jackson WA, Berg PJV. Internal phosphorus flows during development of phosphorus stress in Stylosanthes hamata. Aust J Plant Physiol. 1990;17:451–64. https://doi.org/10.1071/PP9900451.

Staufler MD, Sulewski G. Fósforo – essencial para a vida. In: Yamada T, Abdalla SRS, editors. Fósforo na agricultura brasileira. Piracicaba: Potafos; 2004. p. 1–12.

Stryker RB, Gilliam JW, Jackson WA. Nonuniform transport of phosphorus from single roots to the leaves of Zea mays. Physiol Plant. 1974;30:231–9. https://doi.org/10.1111/j.1399-3054.1974.tb03649.x.

Teixeira LAJ, Spironello A, Furlani PR, et al. Split application of NPK fertilizers on pineapple. Rev Bras Frutic. 2002;24:219–24. https://doi.org/10.1590/S0100-29452002000100047.

Teo YH, Beyrouty CA, Norman RJ, et al. Nutrition uptake relationship to root characteristics of rice. Plant Soil. 1995;171:297–302. https://doi.org/10.1007/BF00010285.

Ullrich-Eberius CI, Novacky A, Fischer E, et al. Relationship between energy-dependent phosphate uptake and the electrical membrane potential in Lemna gibba G1. Plant Physiol. 1981;67:797–801. https://doi.org/10.1104/pp.67.4.797.

Vasconcelos RL, Almeida HJ, Prado RM, et al. Filter cake in industrial quality and in the physiologi- cal and acid phosphatase activities in cane-plant. Ind Crop Prod. 2017;105:133–41. https://doi.org/10.1016/j.indcrop.2017.04.036.

Chapter 7
Potassium

Keywords K availability · Potassium uptake · Potassium transport · Potassium redistribution · Assimilatory potassium · Potassium deficient

Potassium is one of the most absorbed macronutrients by crops, having low availability in tropical soils, being important to N metabolism and to ensure high crop yield with quality and low disease incidence. In this chapter, we discuss initially (1) basic aspects of K in the soil; (2) K uptake, transport, and redistribution; (3) K metabolism; (4) nutritional requirements for K in crops; (5) K extraction, export, and accumulation by the main crops; and (6) K deficiency and toxicity symptoms.

7.1 Introduction

In general, tropical soils have low concentrations of available potassium, although not being as low as phosphorus. Potassium is the second most consumed nutrient in Brazilian agriculture after phosphorus.

Unlike N and S, the main form of K in the soil is mineral, which may be in the crystalline network of primary minerals – feldspars and micas (muscovite and biotite), or secondary minerals (2:1 clays, illite, and vermiculite). Soil weathering decreases minerals rich in potassium, giving rise to 1: 1 clays such as kaolinite, which have no potassium in their structure.

In addition to the structural K of minerals, there is fixed or nonexchangeable K and K-fertilizers that feed the exchangeable K compartment, besides the K solution, which can be absorbed by the plant. Nonexchangeable K comprises K adsorbed in the interlayer of 2:1 clay minerals and part of K contained in primary minerals, which are easier to weather (Mielniczuk and Selbach 1978). Thus, the nutrient in the form of exchangeable cation and in the soil solution are considered available to plants. Exchangeable levels, in general, are low in relation to total levels. However, they can be the most important reserve of available potassium in tropical soils.

We add that organic matter has exchangeable K also as part of its constitution, which is released by washing and in the mineralization process. Straw decomposition is high, benefiting the subsequent crop especially for K, releasing from 89% to

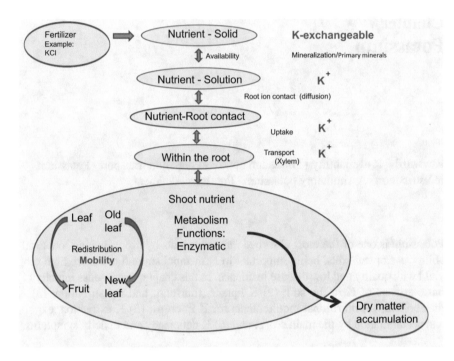

Fig. 7.1 Potassium dynamics in the soil-plant system, indicating the nutrient passage processes in different plant compartments

94% of the element contained in the straw (sugarcane) for a year (Souza Júnior et al. 2015).

In the soil, several factors affect K availability, such as clay content, temperature, and soil moistening and drying, besides pH value, which increases K availability close to 6.5.

In the study of potassium in the plant system, it is important to know all the compartments that the nutrient travels passes from the soil solution to the root and shoot (Fig. 7.1).

7.2   Uptake, Transport, and Redistribution of Potassium

Uptake

Before absorption occurs, there is contact between K and root. For this, potassium must travel from the soil to the roots. Normally, K moves in the soil through diffusion, which predominates, especially when mass flow is reduced by low water content and high clay content in the soil (Grimme 1976). However, in pots, with roots concentrated in a restricted soil volume, and in corrected soils, mass flow may be favored (Rosolem et al. 2003). Soil humidity is important to favor contact between ion and root (Mackay and Barber 1985). Therefore, adequate soil humidity increases contact and K uptake in maize, besides dry matter production (Table 7.1).

Table 7.1 Soil humidity in potassium uptake by maize and dry matter production

	Humidity level		
	Deficient (-170 kPa)	Adequate (-33 kPa)	Excess (−7.5 kPa)
K absorbed (mmol per pot)	2.9	5.0	4.6
Dry matter (g per pot)	7.8	8.7	7.3

K (Soil) = 207 mg dm^{-3}

Potassium absorbed by roots in the K$^+$ ionic form occurs through several systems (transporters and channels). K$^+$ transport across membranes occurs by carriers, such as the high affinity system (low km = 0.02 and 0.03 nM), which is very selective, such as HKT1 (Schachtman and Schroeder 1994). There are also carriers with low affinity and high Km, considered not very selective. There are other carriers which operate with high and low affinity (Fu and Luan 1998). In the first system, K uptake occurs by a proton force, i.e., H$^+$ is exchanged by K$^+$, although the K$^+$ amount entering the cell is much higher than the amount H + released into the medium. The second system has a passive process of facilitated diffusion through selective channels to the cytoplasm (Fox and Guerinot 1998).

In addition to these K uptake systems, there are channels in the plasma membrane through voltage change, which depends on proton pump activity, allowing the entry of several cations.

K uptake through low activity mechanism, that is, with high K concentration in the medium and through channels operated by the proton pump, may be affected by the presence of other ions in the medium, as these are not very selective processes. However, uptake processes that are important in plant nutrition in tropic soils (low K content) act with high affinity or low Km.

Channels are efficient in absorbing this nutrient, even if at low concentrations (10 µM) (Hirsch et al. 1998). In this process, Ca^{2+} and Mg^{2+} can interfere, which in high concentrations can inhibit potassium uptake. Normally, in the plasma membrane, K is permeable, facilitating uptake, although predominating in active form. Finally, further information regarding K uptake processes in plants can be obtained in a review by Malavolta (2005).

Transport

After absorption, K is transported to the shoot easily and quickly via xylem.

Redistribution

Internal redistribution of K is high due to its high concentration in the phloem sap, moving from old leaves to newer leaves. Usually, K is directed to meristematic tissues or growing fruits due to the fact that approximately 75% of the plant total potassium is in soluble form (K$^+$), i.e., not linked to an organic compound.

7.3   Participation in Plant Metabolism

Unlike N, potassium is not part of any organic compound in the plant, not having structural function. However, its main function in plant life is as enzyme activator. More than 60 enzymes, such as synthetases and kinases, depend on potassium for their normal activity. K is related to changes in the conformation of molecules, which increases the exposure of active sites for binding with the substrate. In general, conformation changes induced by K^+ in enzymes increase the rate of catalytic reactions (V_{max}) and, in many cases, the affinity for the substrate (K_m) (Evans and Wildes 1971).

One of the reasons explaining that plants have increased potassium requirement (usually the second most required nutrient) is the need for maintaining high potassium content in the cell cytoplasm, mainly to ensure optimal enzymatic activity, as this nutrient does not have high affinity with organic compounds (including enzymes). Another reason for the need for high K concentration in the cytosol (close to 100 mM) and in the chloroplast stroma is to maintain the neutralization of anions (soluble organic and inorganic acids and macromolecule anions) and the pH at adequate levels for cell functioning, i.e., pH of 7.0–7.5 in the cytosol and ~8.0 in the stroma (Marschner 1995).

Cations with ionic radii close to the size of potassium, such as NH_4^+, Cs^+, and Rb^+ can replace it in the activation of several enzymes, although Na^+ and Li^+, which have larger radii cannot do so, as in starch synthetase, for example (Malavolta 2006).

K-deficient plants have decreased plant metabolism with increased accumulation of soluble carbohydrates, decrease in the starch level (K activates starch synthase), and accumulation of compounds (soluble nitrogen). In addition, there are also losses in protein synthesis, as K probably activates nitrate reductase, which requires K for its synthesis. K benefits reductase activity for maintaining an optimal pH for enzymatic activity, as according to Pflüger and Wiedemann (1977), decrease in pH from 7.7 to 6.5 nearly completely inhibits the activity of this enzyme. In addition, K participates in several stages of protein synthesis, such as synthesis of ribosomes and aminoacyl-tRNA; binding of aminoacyl-tRNA to ribosomes; peptidyl-tRNA and ribosome transfers, and mRNA depolymerization after protein synthesis (Evans and Wildes 1971).

Decrease in protein synthesis accumulates basic amino acids (ornithine, citrulline, and arginine) that are decarboxylated, increasing putrescine content, a nitrogenous compound toxic to plants (Malavolta and Crocomo 1982). Citrulline is considered the best precursor for formation of putrescine (Table 7.2). We also highlight that decrease in the cytosol pH value favors the action of enzymes that participate in decarboxylation and synthesis of these compounds. In addition, putrescine biosynthesis is a mechanism for detoxification of high ammonium levels, as its accumulation precedes amine formation (Smith 1990).

Thus, putrescine accumulation can lead to necrosis of leaf margins/tips, which is common in K-deficient plants. Accumulation of polyamine putrescine prevails, although the metabolic pathway allows deployment in others (spermedine and spermine). The levels of these polyamines increase when subjected to stress, as observed for deficiency of K and other nutrients (Ca and Mg) and salt stress. In addition, the opposite also occurs, i.e., absence of stress decreases production of polyamines. At adequate levels, polyamines can be beneficial, acting on cell division in the regulation of DNA synthesis (Minocha et al. 1991).

ATPase proton pumps linked to cell membranes, which act in the ion uptake process, require Mg^{2+} for maximum activity, although increasing when associated with K^+ (Marschner 1986), except for tonoplast ATPase, which does not require K^+. This activation not only facilitates K^+ transport from the external solution through the plasma membrane, inside root cells, but also increases K importance in osmotic regulation, as ATPases in guard cell membranes pump or allow K^+ influx, turning osmotic potential more negative and causing water influx into the cell.

Potassium is important in cell expansion, which forms a large central vacuole occupying 80–90% of cell volume. There are two conditions for cell expansion to occur: ATPases, which are important for growth of meristematic cells, as they reduce the pH of cytoplasm, allowing action of enzymes that break bonds of cell wall components, influencing cell elongation; and solute accumulation to create an internal osmotic potential, also expanding the cell.

Potassium, besides being enzymatic activator, plays a physiological role in stomatal opening and closing. Barraclough and Leigh (1993) observed that the critical K content in plant tissue to exert physical function is approximately 2.7 times higher than to exert enzymatic function, although the ratio can decrease (~1.7 times) if Na is provided. The authors also added that Na did not alter the critical K content to exert enzymatic activity.

K is the main nutrient affecting plant osmotic potential, influencing from cell expansion and ion transport, as it generates high osmotic potential in the root, to stomatal opening and closing. K^+ (especially in the vacuole) influences optimal turgor in guard cells, as it increases their osmotic potential, resulting in water uptake from guard and adjacent cells, increasing turgor and stomatal opening, as shown in the diagram below (Fig. 7.2). In addition, plants with turgid cuticle have increased physical resistance against mechanical damage or microorganisms.

Thus, reduced water loss by plants adequately supplied with potassium is provided by decreased respiration rate, which depends on cell osmotic potential and stomatal opening and closing. Plants adequately supplied with K have increased water-use efficiency, as in the case of maize (Neiva 1977). Thus, in period of water

Component	+K	−K
	mmol/g dry matter	
Arginine	72	115
Citrulline	118	377
Ornithine	45	117
Agmatine	20	29
N-carbamylputrecin	26	92
Putrescine	114	1000

Table 7.2 Amino acids, amines, total N, protein, and potassium in sesame leaves (38 days old) influenced by K level

Fig. 7.2 Scheme of
stomatal movement
influenced by K

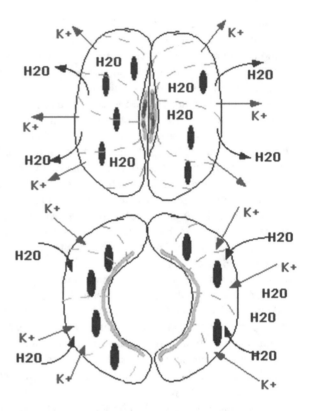

deficit, plants with adequate K supply maintain more water in the tissues compared to deficient plants (Fig. 7.3, Table 7.3).

Maintaining increased water content in the cell also favors metabolic reactions and leaf orientation, maintaining increased light interception, although this effect can be replaced by other cations, such as Na (Malavolta 1980).

Water deficit or salt stress induces stomatal closure as plants produce the ABA hormone in the root, which is transported via xylem to leaves, inducing K^+ efflux from stomatal guard cells. The effect of ABA is caused by activation of Ca^{2+} channels in guard cell membranes, increasing cytoplasmic Ca and depolarizing these membranes. As a result, voltage-dependent anion passages are activated and the plasma membrane is induced from K^+ conduction state to anion conduction state, which decreases membrane potential and induces K^+ outflux of guard cells and stomatal closing (Hendrich et al. 1990).

Controlling stomatal opening/closing is also important for photosynthesis rate, as stomatal opening does not occur regularly in K-deficient plants, reducing CO_2 entry (Steineck and Haeder 1978). Thus, application can increase CO_2 assimilation by leaves and foliar area (Table 7.4).

Chloroplasts contain approximately half of foliar K. K increases CO_2 diffusivity in mesophilic cells and also activates RuBP carboxylase, increasing photosynthetic activity. However, all nutrients affecting the biochemistry of photosynthesis increase

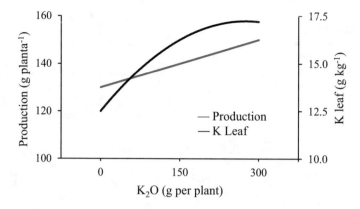

Fig. 7.3 Effect of potassium doses on foliar K content and production

Table 7.3 Water-use efficiency by maize crop in relation to potassium nutrition

K_2O	Water use	Transpiration coefficient	Water use efficiency
kg ha⁻¹	L m⁻²	L per kg of dry matter	L kg⁻¹ grain
0	375	276	573
465	330	203	379
930	320	198	360
1395	317	171	323
D.M.S.5%	5	40	54

tCO_2 concentration in stomatal cells, which may induce their closure. Finally, K is primarily responsible for H⁺ flux, induced by light, through thylakoid membranes (Tester and Blatt 1989) and for the establishment of trans-membrane pH gradient, required for ATP synthesis (photophosphorylation), analogous with ATP synthesis in the mitochondria.

We also add that plants with adequate potassium supply have shown increased water retention in the tissues and tolerance to climatic stress (droughts and frosts). As for frosts, K can increase the freezing point of tissues by inducing more solute content in cells, mitigating damage from low temperature (Grewal and Singh 1980).

Research on potato crop indicated that potassium application, besides increasing production and leaf content, decreased freeze-damaged leaves (Table 7.5).

Several studies have attributed photosynthate transport in the phloem to potassium, which is an active process that requires energy from ATPase activity in membranes. K favors the active passage of photoassimilates through the membranes of sieved tubes and also favors the passive flow of solutes inside the tubes, as it keeps a high pH, facilitating sucrose transport. Transport of solutes or photosynthates is important at all stages of crop development. In the reproductive phase, often, most of nutrients that are drained into the developing fruits do not come from the roots but from the leaves or stem.

Table 7.4 Effect of
potassium on CO_2
assimilation and leaf area
growth in maize grown in
nutrient solution

K in solution	CO_2 assimilation	Foliar area
$\mu g\ cm^{-3}$	$mg/cm^2 \times h$	cm^2
15	22.6 (100)	51.9 (100)
45	28.8 (127)	51.9 (100)
135	32.8 (145)	53.6 (103)
400	35.3 (156)	59.8 (115)

Light intensity $=7.5$ lúmen/cm^2

Table 7.5 Effect of potassium application on tuber production, foliar content, and percentage of leaves damaged by freezing in potatoes

K doses (kg ha^{-1})	Tuber production (t ha^{-1})	Foliar K (mg g^{-1} of dry matter)	Freeze-damaged leaves (%)
0	2.39	24.4	30
42	2.72	27.6	16
84	2.87	30.0	7

In sugarcane, after 90 minutes, 50% of the photosynthesized compounds (^{14}C) were exported from the leaf to other organs and 20% of the total were already in the reserve tissues (stem) of plants adequately supplied with potassium. In a situation of deficiency, even after 4 h, transport rates were much lower (Hartt 1969, cited by Marschner 1986). In eucalyptus, it was found that K application increased carbohydrate transport from leaves to the wood, consequently resulting in increased accumulation of cellulose and hemicellulose in the wood (Silveira and Malavolta 2003).

K-deficient plants can decrease cell wall synthesis (in stems), predisposing the crop to winds, causing lodging.

Due to its different roles in the plant, potassium has direct effects on the production of most crops. Carvalho et al. (1999) observed that the dose of 434 g per passion fruit plant increased fruit production, while maximum juice production was obtained when 562 g per plant/year were applied. This K amount was 30% higher than the amount that provided the highest fruit yield. Increased production of passion fruit seedlings was associated with the K content in the shoot of 39 g kg^{-1} and in the roots of 20 g kg^{-1} (Prado et al. 2004). Therefore, K application increased its content in leaves and, consequently, increased crop yield (Fig. 7.3).

The role of K in transport can also affect protein content in grains, increasing their quality. This occurs because K transports N for protein synthesis in grains (Blevins 1985).

Potassium can also increase soybean quality by increasing isoflavone. Isoflavone is considered a functional food component for human health, reducing the rate of diseases (cancer, cardiovascular diseases, and osteoporosis). Bruulsema (2001) found that K application (101 kg ha^{-1}) in soybean increased the element in leaves and the content of K and isoflavones (by 21%) in the grain, besides increasing production.

It is possible to obtain high soybean production along with high concentration of isoflavones without any significant decline in oil and protein concentrations (Yin and Vyn 2005).

7.4   Nutritional Requirements of Major Crops

The study of crop nutritional requirements should address the total extraction of the nutrient from the soil, respecting extraction at each stage of crop development in order to satisfy crop nutritional needs and obtain maximum economic production.

The total K content in the plant for optimal growth/development is 2–5% (20–50 g kg^{-1}) of dry matter. However, these values may vary depending on crop and other factors, which are covered in Chap. 19.

Thus, in order to adequately discuss crop nutritional requirements, the two following factors are equally important: nutrient total extraction/export and the uptake rate of this nutrient throughout cultivation.

Nutrient Export and Extraction

Total extraction of potassium occurs as a function of K content in the plant and dry matter accumulated. Therefore, it depends on production, which depends on species, variety/hybrid, soil availability, and crop management, among others.

As for the vegetable species, there is variation in the amount required depending on the crop. In some species K is the most accumulated macronutrient, such as in tomato (Prado et al. 2011) and cauliflower (Alves et al. 2011).

From different crops, we observe that total extraction of potassium ranged from 39 (wheat) to 257 kg ha^{-1} (maize) (considering production in one hectare) (Malavolta et al. 1997). However, in relative values of potassium extraction in kg per ton produced, there is increased requirement for coffee (116 kg) and decreased requirement for grasses such as wheat (13 kg) and sugarcane (1.6 kg) (Table 7.6).

Regarding K export in the product harvested by area, we observe that sugarcane is the one that exports the most (65 kg ha^{-1}), considering that the whole harvested part is exported, while annual crops, such as beans and soybeans, are the ones that export the most per ton produced, 22 and 19 kg t^{-1}, respectively. Therefore, it is important to restitute fertilization for these crops.

We also note that even within the same species there are considerable differences in the amount of K absorbed and exported by crops. In citrus, Bataglia and Mascarenhas (1977) found that fruits of Tahiti lime and Natal orange have contrasting exports per fruit box, ranging from 48 to 102 g of K_2O, respectively.

Uptake Rate

The study of uptake rate (nutrient accumulated as a function of cultivation period) is important for determining the periods when the elements are most required and to correct deficiencies that may occur during plant development.

Table 7.6 Potassium requirements of major crops

Crop	Plant part	Dry matter produced t ha^{-1}	Accumulated K Plant part kg ha^{-1}	Total[c]	K required to produce 1 t of grains[d] kg t^{-1}
Annual					
Cotton	Reproductive part (cotton/cottonseed)	1.3	24 (18.5)[b]	66	50.8
	Vegetative part (stem/branch/leaf)	1.7	39		
	Root	0.5	3		
Soybean[a]	Grains (pods)	3	57 (19)	115	38.3
	Stem, branches and leaves	6	58		
Bean	Pod	1	22 (22)	92	92.0
	Stem	0.4	11		
	Leaves	1.2	57		
	Root	0.1	2		
Maize[a]	Grains	6.4	30 (4.7)	257	40.2
	Crop residues	–	237		
Rice	Grain	3	13 (4.3)	111	37.0
	Stems	2	60		
	Leaves	2	12		
	Husk	1	6		
	Root	1	20		
Wheat	Grains	3	12 (4.0)	39	13.0
	Straw	3.7	27		
Perennial/semi-perennial					
Sugarcane	Stems	100	65 (0.65)	155	1,6
	Leaves	25	90		
Coffee plant[a]	Grains	2	52 (26)	232	116
	Trunk, branches, and leaves	–	180		

[a]Malavolta (1980)
[b]Nutrient export through the grains produced (kg t^{-1}): K accumulated in the grains/grain dry matter
[c]Suggests the total nutritional requirement of the crop for the respective yield level
[d]Suggests the relative nutritional requirement of K of the crop for production of 1 ton of the commercial product (grains/stems), obtained by the following formula: K accumulated in the plant (vegetative + reproductive part)/dry matter of the commercial product

In the cotton tree crop, Mendes (1965) obtained the uptake rate accumulated throughout the plant development cycle, which lasted 150 days (Table 7.7). The author observed slow initial uptake, followed by 50% during flowering (initial phase) and reaching 73% (final phase). In the initial dehiscence phase, uptake reached 86%, completing with 100% in the final phase, at 130 days. We note that right at the beginning of flowering half of the K required by the crop was already absorbed, indicating that by that time, most of the K should already have been applied in topdressing.

Table 7.7 Effect of potassium application in stallments in dry matter accumulation of leaves + stems at flowering and maize stem breakage

Days after application			Plant part	Dry matter	Stem breakage
Planting	30	45			
kg de K_2O per ha				g per plant	%
90	0	0	Leaves	33	
45	45	–	Leaves	30	
30	30	30	Leaves	27	
90	0	0	Stems	65	10
45	45	–	Stems	65	13
30	30	30	Stems	50	31

In the case of potassium, excessive stallments can often result in decreased dry matter production. In the case of maize, there is a high rate of K accumulation in the first 30–40 days, suggesting that application should precede this period, in the initial development phase (Table 7.7). K application in two or three stallments was the worst treatment compared to total application at planting for both dry matter accumulation and lodging rate.

In clayey soil with moderate doses of potassium, the authors did not verify benefits of potassium application in stallments. Silva et al. (2002), who studied potassium fertilization (50 kg K_2O ha^{-1}) in rice under different K application periods (sowing until panicle differentiation), observed that the highest yield was obtained with K application at sowing.

Utilization Efficiency of K by Plants

Fageria (2000) studied the response of 15 rice genotypes in highlands to K treatments with 0 and 200 mg kg^{-1} (high level) in the soil. Rice genotypes indicated differences in grain production and K utilization. Agronomic, physiological, and utilization efficiency were highly correlated with grain production. Based on grain production (greater than the average of genotypes), the low K level (efficient) and K agronomic efficiency (greater than the average of genotypes) (responsive), genotypes Rio Paranaíba, L141, and Guarani were classified as efficient and responsive.

7.5   Symptoms of Nutritional Deficiencies and Excesses

Deficiency

K deficiency leads to changes in different levels, from the beginning, affecting metabolism dynamics, to biochemical, molecular, subcellular, and cellular levels, until reaching tissues. Malavolta (1984) indicates weakening of cell membranes of the outermost layers, chloroplast isolation in the form of rods or spindles, grease presence, destruction of chloroplast, deterioration of mitochondria, swelling of

Fig. 7.4 Symptom of potassium deficiency in quinoa leaf

proplastids, thinning of the protoplastid matrix, and subsequent failures in the differentiation of conducting tissues, and loss of exchange activity.

Symptoms of K deficiency in crops in general are characterized by marginal chlorosis and leaf necrosis, initially in the oldest leaves (Figs. 7.4, 7.5). In some crops, K deficiency develops dark green or bluish-green leaves, similar to P deficiency. There is less translocation of carbohydrates from the shoot to the root, reducing root growth.

K deficiency in grasses can accumulate Fe in the nodes at the plant base, causing Fe deficiency symptoms in younger leaves (Malavolta 1980).

Highly productive maize hybrids have high capacity to transport photosynthates from the stem to fill grains. Thus, in crops adequately supplied with K, there is increased production of photosynthates to supply grain filling and ensure adequate levels in the stem. Thus, the stem is weakened in K-deficient plants, with decreased cell wall thickness and possible collapse of parenchymal tissue, which results in increased lodging rate. In cotton, K deficiency affects reproductive development, decreasing carbohydrate transport to cotton balls (Read et al. 2006).

K-deficient plants have decreased cytosol pH, increased activity of some hydrolases (β-glycosidase) or oxidases (polyphenol), where soluble nitrogen compounds accumulate, and sugar accumulation, changing cell chemical composition. In addition, the cell wall is thinned, making the plant more vulnerable to attack by pathogens (Mengel and Kirkby 1987).

Premature leaf senescence (cotton tree) may occur due to metabolic changes caused by K deficiency (Wright, 1999). As potassium acts on osmotic regulation and resistance of plants to water stress, plants with adequate K nutrition have decreased levels of abcysic acid (ABA), a phytohormone that accelerates leaf senescence (Beringer and Trolldenier 1979).

Plants subject to K deficiency develop a series of control mechanisms, such as high-affinity uptake. There are some molecules that signal low K+ level in plants, including active oxygen and phytohormone species (auxin, ethylene, and jasmonic acid) (Ashley et al. 2006).

Fig. 7.5 Photos and general description of visual symptoms of K deficiency in different crops. Maize: Chlorosis starts in older leaves, occurring in the tips and margins, progressing to dryness and necrosis ("burning") (**a**); Bean: Lower leaves show pale green color and upper leaves show darker green color. Leaflets of older leaves may have marginal chlorosis, which evolves to necrosis, decreasing plant development (**b**); Aubergine: Lesser plant development, with decreased leaf number and height (**c**); yellowish spots in the leaf margin, which are initially pale yellow and become tanned as they spread and coalesce. They may curl up (**d**); Grass of genus Brachiaria: K-deficient plants with shorter stature, with appearance of marginal chlorosis and necrosis in the tips of older leaves (**e**); Scarlet eggplant: Symptoms are characterized initially by stunting of plant growth and older leaves become deformed (**f**). As damage progresses, the leaves become chlorotic with dried margins due to tissue necrosis (**g**); Eucalyptus: Drying of leaf margins of older leaves and shorter stature compared to plants without deficiency (**h**)

Excess

Although plants have luxury consumption of K, characteristic symptoms are not known. However, in plants subjected to excess K, symptoms are confounded with damage caused by salinity, which is high in the main potassium fertilizers. In maize, foliar contents of $K > 55$ g kg^{-1} are considered excessive (Mengel and Kirkby 1987). Increased K content in plants can occur due to induced calcium and magnesium deficiency. On the other hand, increased K doses can induce deficiency of other nutrients, such as Mg. Excess KCl over many years, especially in areas with low rainfall, induces sufficient salinity to decrease soil microbiota activity.

References

Alves AU, Prado RM, Corrêa MAR, et al. Cauliflower cultivated in substrate: progress of absorption of macro and micronutrients. Ciênc Agrotec. 2011;35:45–55. https://doi.org/10.1590/S1413-70542011000100005.

Ashley MK, Grant M, Grabov A. Plant responses to potassium deficiencies: a role for potassium transport proteins. J Exp Bot. 2006;57:425–36. https://doi.org/10.1093/jxb/erj034.

Barraclough PB, Leigh RA. Critical plant K concentrations for growth and problems in the diagnosis of nutrient deficiencies by plant analysis. Plant Soil. 1993;155:219–22. https://doi.org/10.1007/BF00025023.

Bataglia OC, Mascarenhas HAA. Absorção de nutrientes pela soja. Campinas: Instituto Agronômico; 1977.

Beringer H, Trolldenier G. Influence of K nutrition on the response to environmental stress. In: Ipi, editor. Potassium research review and trends. Bern: Ipi; 1979. p. 189–222.

Blevins DG. Role of potassium in protein metabolism in plants. In: Mund-Son RD, editor. Potassium in agriculture. Madison: ASA/CSSA/SSSA; 1985. p. 413–24.

Bruulsema T. Potássio aumenta a produção de isoflavona na soja. Informações Agronômicas. 2001;94:5.

Carvalho AJC, Rodrigues MGV, Lima AA, et al. Nitrogen and potassium effects on yield and quality of yellow passion fruit, under irrigation. Rev Bras Frutic. 1999;21:333–7. https://doi.org/10.1590/S0100-29452003000200019.

Evans HJ, Wildes RA, Potassium and its role in enzyme activation. In: Proceedings of the 8th Colloquium of the International Potash Institute, Bern; 1971. p. 13–39.

Fageria NK. Potassium use efficiency of upland rice genotypes. Pesq Agropec Bras. 2000;35:2115–20. https://doi.org/10.1590/S0100-204X2000001000025.

Fox TC, Guerinot ML. Molecular biology of cation transport in plants. Annu Rev Plant Physiol Plant Mol Biol. 1998;49:669–96. https://doi.org/10.1146/annurev.arplant.49.1.669.

Fu HH, Luan S. AtKUP1: a dual-affinity K^+ transporte from Arabidopsis. Plant Cell. 1998;10:63–73. https://doi.org/10.1105/tpc.10.1.63.

Grewal JS, Singh SN. Effect of potassium nutrition on frost damage and yield of potato plants on alluvial soils of the Punjab (India). Plant Soil. 1980;57:105–10. https://doi.org/10.1007/BF02139646.

Grimme H. Soil factors of potassium availability. In: Ghosh AB, Hasan R, editors. Potassium in soils, crop and fertilizers. New Delhi: Indian Society of Soil Science; 1976. p. 144–63.

Hendrich R, Busch H, Raschke K. Ca^{2+} and nucleotide dependent regulation of voltage dependent anion channels in plasma membrane of guard cells. EMBO J. 1990;9:3889–92.

Hirsch RE, Lewis BD, Spalding EP, et al. A role for the AKTI potassium channel in plant nutrition. Science. 1998;280:918–21. https://doi.org/10.1126/science.280.5365.918.

Mackay AD, Barber SA. Soil moisture effects on potassium uptake by corn. Agron J. 1985;77:524–7. https://doi.org/10.2134/agronj1985.00021962007700040005x.

Malavolta E. Elementos de nutrição de plantas. São Paulo: Agronômica Ceres; 1980. 251p.

Malavolta E. O potássio e a planta. Piracicaba: Potafós; 1984.

Malavolta E. Potássio: absorção, transporte e redistribuição na planta. In: Yamada T, Roberts TL, editors. Potássio na agricultura brasileira. Piracicaba: Potafós; 2005. p. 179–238.

Malavolta E. Manual de nutrição mineral de plantas. São Paulo: Agronômica Ceres; 2006.

Malavolta E, Crocomo OJ. O potássio e a planta. In: Potafos, editor. Potássio na agricultura brasileira. Piracicaba: Potafós; 1982. p. 95–162.

Malavolta E, Vitti GC, Oliveira SA. Avaliação do estado nutricional das plantas: princípios e aplicações. Piracicaba: Associação Brasileira de Potassa e do Fósforo; 1997. 319p.

Marschner H. Mineral nutrition of higher plants. London: Academic; 1986.

Marschner H. Mineral nutrition of higher plants. London: Academic; 1995.

Mengel K, Kirkby EA. Principles of plant nutrition. Worblaufen/Bern: International Potash Institute; 1987.

Mielniczuk J, Selbach PA. Potassium supply capacity of six soils in Rio Grande do Sul. Rev Bras Ciênc Solo. 1978;2:115–20.

Minocha R, Minocha SC, Komamine A, et al. Regulation of DNA synthesis and cell division by polyamines in Catharanthus roseus suspension cultures. Plant Cell Rep. 1991;10:126–30. https://doi.org/10.1007/BF00232042.

Neiva LCS, Influência do potássio sobre a economia de água de quatro cultivares de arroz submetidos a déficit hídrico. Dissertação, Universidade Federal de Viçosa; 1977.

Pflüger RE, Wiedemann RDER. The influence of monovalent cations on the reduction of nitrate in Spinacia oleracea L. Z Pflanzenphysiol. 1977;85:125–33. https://doi.org/10.1016/S0044-328X(77)80286-9.

Prado RM, Braghirolli LF, Natale W, et al. Potassium application on the nutricional status and dry matter production of passion fruit cuttings. Rev Bras Frutic. 2004;26:295–9. https://doi.org/10.1590/S0100-29452004000200028.

Prado RM, Santos VHG, Gondim ARO, et al. Growth and nutrient absorption by Raisa tomato cultivar grown in hydroponic system. Semina. 2011;32:17–28. https://doi.org/10.5433/1679-0359.2011v32n1p19.

Read JJ, Reddy KR, Jenkins JN. Yield and fiber quality of upland cotton as influenced by nitrogen and potassium nutrition. Eur J Agron. 2006;24:282–90. https://doi.org/10.1016/j.eja.2005.10.004.

Rosolem CA, Silva RH, Esteves JAF. Potassium supply to cotton roots as affected by potassium fertilization and liming. Pesq Agropec Bras. 2003;38:635–41. https://doi.org/10.1590/S0100-204X2003000500012.

Schachtman DP, Schroeder JI. Structure and transport mechanism of a high-affinity potassium uptake transporter from higher plants. Nature. 1994;370:655–8. https://doi.org/10.1038/370655a0.

Silva TRB, Soratto RP, Ozeki M, et al. Potassium management in upland rice sprinkler irrigation. Acta Sci. 2002;24:1455–560. https://doi.org/10.4025/actasciagron.v24i0.2401.

Silveira RLVA, Malavolta E. Production and chemical characteristics of young Eucalyptus grandis wood progenies as affected by potassium rates in the nutrient solution. Sci Florestalis. 2003;63:115–35.

Smith TA. Plant polyamine: metabolism and function. In: Flores HE, Arteca RN, Shanon JC, editors. Polyamine and ethylene: biochemistry, physiology and integration. Rockville: American Society of Plant Physiology; 1990. p. 1–23.

Souza Júnior JP, Flores RA, Prado RM, et al. Release of potassium, calcium and magnesium from sugarcane straw under different irrigation layers. Aust J Crop Sci. 2015;9:767–71.

Steineck O, Haeder HE. The effect of potassium on growth and yield components of plants. In: Resumos do 11 congress international of the potash institute. Bern: International Potash Institute; 1978. p. 165.

Tester M, Blatt MR. Direct measurement of K+ channels in thylakoid membranes by incorporation of vesicles into planar lipid bilayers. Plant Physiol. 1989;91:249–54. https://doi.org/10.1104/pp.91.1.249.

Wright PR. Premature senescence of cotton (Gossypium hirsutum L.) – pre- dominantly a potassium disorder caused by an imbalance of source and sink. Plant Soil. 1999;211:231–9. https://doi.org/10.1023/A:1004652728420.

Yin X, Vyn TJ. Relationships of isoflavone, oil, and protein in seed with yield of soybean. Agron J. 2005;97:1314–21. https://doi.org/10.2134/agronj2004.0316.

Chapter 8
Calcium

Keywords Ca availability · Calcium uptake · Calcium transport · Calcium redistribution · Assimilatory calcium · Calcium deficient

Calcium (Ca) in general is the third macronutrient most absorbed by crops, being limited for specific crops and important to ensure high crop yield with quality and low disease incidence. In this chapter, we will discuss initially (i) basic aspects of Ca in the soil; (ii) Ca uptake, transport, and redistribution; (iii) Ca metabolism; (iv) nutritional requirements for Ca in crops; (v) Ca extraction, export, and accumulation by the main crops; and (vi) Ca deficiency and toxicity symptoms.

8.1 Introduction

The primary origin of calcium (Ca) are rocks, as it is found in minerals such as dolomite, calcite, feldspars, and antibolics, and also occurs in sedimentary and metamorphic rocks. In acidic soils, these minerals are weathered, and calcium is partly lost by leaching. The Ca in the soil is adsorbed in soil colloids or as organic matter component. Under increased pH conditions, Ca can precipitate as carbonates, phosphates, or sulfates, with decreased solubility. Ca considered available to plants is that adsorbed to colloids (exchangeable) and present in the soil solution (Ca^{2+}). Ca^{2+} levels in the solution of acidic soils are quite low. Corrective materials such as limestone (calcium carbonate) are used in these soils, which, besides correcting acidity, also provide calcium. In cultivated soils, in general, calcium is not a limiting factor for most crops, similar to N and P. Liming improves soil fertility, reflecting on nutrition (foliar Ca) and production (Prado et al. 2007). Several factors affect Ca availability in the soil, such as pH value, with increased Ca availability with pH close to 6.5.

When studying calcium in the plant system, it is important to know all the compartments the nutrient passes in the soil solution and in the root and shoot (leaves/fruits); that is, from the soil until its incorporation into an organic compound or as enzyme activator, contributing to vital functions that enable maximum dry matter accumulation in the agricultural product (grain, fruit, etc.) (Fig. 8.1).

© Springer Nature Switzerland AG 2021
R. de Mello Prado, *Mineral nutrition of tropical plants*,
https://doi.org/10.1007/978-3-030-71262-4_8

8.2   Uptake, Transport, and Redistribution of Calcium

Uptake

The pathway of calcium from the soil to the roots occurs not only by mass flow but also with contribution of root interception, characterized by null pathway in the soil, with the root contacting Ca during its growth. Calcium application in the soil should occupy the largest volume, with broadcast application (incorporated in the 0–20 cm layer) to increase the chances of the root having contact with the nutrient, favoring uptake.

Although the soil solution has a high Ca concentration, being, for example, tenfold higher than that of K, Ca uptake rate is lower than that of K, as Ca is absorbed only in young roots, in which endoderm cell walls are not suberized without Casparian strips (Clarkson and Sanderson 1978).

In the Ca isotope uptake process, we clearly observed quick uptake in the first 60 min, indicating the passive process, followed by slow uptake until completing 240 min, characterizing the active process (Malavolta et al. 1997). We know that most of Ca is absorbed passively, following water entry, as the internal concentration of the element is not much higher than the external, as with K. If the external Ca concentration is low, active uptake prevails. Therefore, depending on the

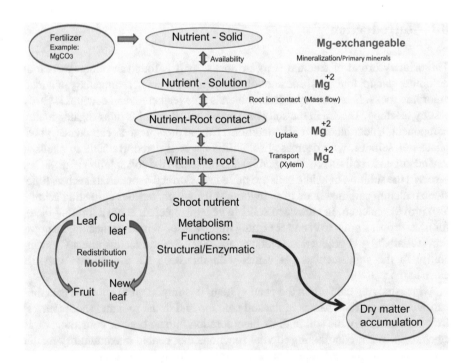

Fig. 8.1 Calcium dynamics in the soil–plant system, indicating nutrient passage processes in different plant compartments

external concentration, Ca uptake can be passive or active (Malavolta 2006). The author adds external factors that could decrease Ca uptake besides its external concentration, which are the presence of other ions at high concentration (NH_4^+, K^+, Mg^{+2}, Al^{+3}, Mn^{+2}), possibly causing deficiency, and the accompanying ion. Uptake follows the decreasing order $Cl \rightarrow NO_3 \rightarrow SO_4^{-2}$ e $Cl^- > NO_3^- > SO_4^{-2}$.

Soil acidity (high Al) can affect Ca uptake by plants. In this sense, Marschner (1986) indicates that aluminum can inhibit calcium uptake, mainly as a result of blocking or competition in exchange sites. Foy (1984) verified that Al toxicity can be manifested as an induced Ca deficiency as a result of reduced nutrient transport in the plant, collapsing growth points at pH values <5.5. Al and Ca antagonism is perhaps the most limiting factor for calcium uptake.

Transport

After uptake, Ca is transported until reaching the xylem, with subsequent passive transport to the shoot (through transpiratory current).

Calcium is translocated in the plant along with water, being affected by transpiration rate. Organs with increased transpiration rate receive greater amounts of Ca. In organs with decreased transpiration, such as new leaves or fruits, calcium transport depends on environmental conditions favoring the development of root pressure (Bradfield and Guttridge 1984). Root pressure exists when transpiration is reduced to a lower rate than the rate of water entering roots, as occurs at night or in periods of high relative humidity (François et al. 1991). Through root pressure, positive pressure develops in the xylem, causing liquid flow in its interior, which translocates calcium to the organs with difficult transpiration. Root pressure usually results in gutting (Tibbitts and Palzkill 1979).

Thus, Ca deficiency can occur in organs with low transpiration rate, leading to physiological disorders. These disorders can occur in new leaves, such as tipburn in lettuce (Collier and Tibbitts 1982), or in fruits, such as internal collapse in mango. In order to prevent these disorders, it is necessary to favor increased Ca flow in these organs. Factors inhibiting the development of root pressure such as drought, wind, and high salinity promote the appearance of these disorders (Collier and Tibbitts 1982). Bradfield and Guttridge (1984) prevented calcium deficiency in hydroponic tomatoes by maintaining high air humidity at night and low electrical conductivity of the nutrient solution. There are also other factors, such as the high rate of organ growth (high luminosity, photoperiods, and nitrogen fertilization).

Redistribution

Ca is only transported in plants via xylem, from roots to shoots, with the opposite being very reduced. This occurs as transport in the phloem occurs through cell cytoplasm, which has low Ca concentration, in the order of $0.1–10$ μM (Raven 1977).

Low Ca concentration in the cytosol is attributed to the low general permeability of membranes and action of membrane transporters that remove the nutrient, placing it in the apoplast or in the endoplasmic reticulum, chloroplast, and vacuoles (Evans et al. 1991). Ca in phloem form insoluble salt complexes, forming oxalate or phosphate, restricting Ca redistribution (Clark 1984). Most of the plant Ca is in the form of calcium pectates, constituting the middle lamella of cell walls, as these structures have a large amount of binding sites (R-COO⁻) for Ca. In addition, Ca in plants is also found in the form of calcium salts of low solubility, such as carbonate, sulfate, phosphate, silicate, citrate, malate, and oxalate. We observe that unlike potassium, Ca has low solubility (wheat: 1.8%, potato: 4.6%, and tomato: 7.7%) and phloem concentration, having very restricted mobility in the plant (Malavolta et al. 1997).

As a consequence of near immobility of Ca in plants, foliar applications are not adequate to correct eventual nutritional disorders, such as in lettuce crops (Johnson 1991).

8.3   Participation in Plant Metabolism

Differently from other nutrients, the highest Ca proportion is in the apoplast, extracellular space, where it is strongly retained in cell wall structures (30–50% of the total Ca of the plant), besides being found in the external surface of plasmalema. Inside cells, it is concentrated in the vacuole/mitochondria and to a lesser extent in the cytoplasm. Most of Ca in the vacuole is in the form of calcium oxalate, being responsible for the cation–anion balance, maintaining a low concentration of this element in the cytoplasm.

One of the main functions of calcium is in the plant structure as part of the cell wall, increasing mechanical resistance of tissues, and neutralizing organic acids in the cytosol. The cell wall is quantitatively the largest product of plants, constituting its actual structure.

Normally, the contact surface between cells increase with their growth, also increasing the need for Ca supply (calcium pectate) to form pectin, elongating the cell wall until reaching its final size, where lignin will be deposited, rigidifying the cell wall. In addition, Ca from calcium pectate is also part of the middle lamella (space between two adjacent cells), which has the function of "cementing", that is, binding neighboring cells (Natale et al. 2005). The cell wall of guava fruits with Ca application presented organized middle lamella (Fig. 8.2), while the medium lamella was unstructured or absent in the cell wall of fruits without Ca application (Fig. 8.3).

The effect of Ca on the organization of the middle lamella influences the texture, firmness, and ripeness of fruits (Hanson et al. 1993) and reduce the rate of vitamin C degradation, ethylene and CO_2 production, and postharvest disease incidence (Conway and Sams 1983). Increased Ca in the guava fruit increased fruit firmness and decreased water loss (Prado et al. 2005a, Fig. 8.4), improving postharvest quality and ensuring a longer storage period. Similar results were obtained by Prado et al. (2005b) on star fruit.

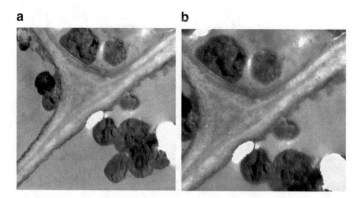

Fig. 8.2 Transmission electron micrographs of guava fruit with detail of the cell wall with Ca application (limestone). (**a**). 10,000 X; (**b**). 120000 X

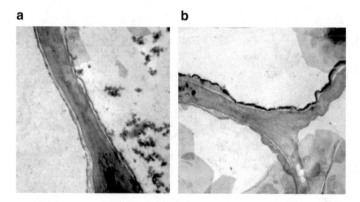

Fig. 8.3 Electron micrographs of guava transmission with detail of the cell wall without the application of Ca (limestone). (**a**). 10,000 X; (**b**). 120000 X

In the cell membrane, Ca is important for binding phosphate/carboxylic groups of phospholipids and stabilizing proteins, especially peripheral proteins (Marschner 1995), besides activating ATPases. Ca is important for membrane stability and selective uptake of ions.

As aforementioned, Ca is part of pectin through calcium pectates, being required for mitotic elongation and division of cells, reflecting in root growth. In the absence of Ca supply, root growth ceases in few hours (Fig. 8.5) and may die. When Ca is complexed with indoleacetic acid, the cell wall may be less rigid, breaking the bond and leaving the wall more elastic, and stimulating cell growth (Rains 1976). There are indications that indoleacetic acid is involved in the transport of Ca to the apical regions of the plant (root or branch). Thus, decreased auxin levels cause Ca deficiency in these tissues.

In order to maintain optimal root growth, it is necessary to maintain minimum concentrations of Ca in the soil between 2.5 and 8.0 mmol$_c$ dm^{-3} (Adams and Moore

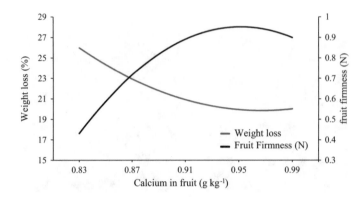

Fig. 8.4 Relationship between calcium content in the fruit pulp, mass loss, and firmness of guava after 8 days of storage at room temperature

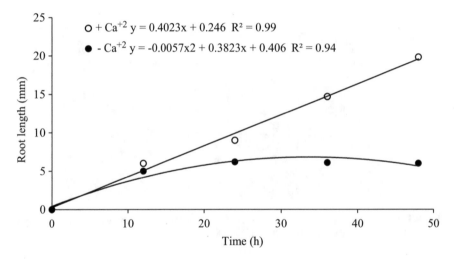

Fig. 8.5 Effect of calcium on growth of primary roots of common bean (Ca^{+2} in the solution =+/− 2 mM)

1983). Ca concentration in the soil can be increased through liming, resulting in increased root growth in cotton (Rosolem et al. 2000). Similar results were obtained in other crops, such as guava (Prado and Natale 2004a).

This indicates, however, that for good development of the plant root system, it is necessary that Ca occupies the entire soil volume, which concentrates most of the root system, especially soil subsurface layers (below 20 cm). As Ca is immobile in the plant, there is no Ca transfer from the root supplied with this element to Ca-deficient roots. Agricultural gypsum (calcium sulfate: high solubility and mobility in the soil profile) is an important Ca source for subsurface layers.

In addition, calcium sulfate has been used to minimize saline effects. Bolat et al. (2006) verified that salinity damages membrane permeability and electrolyte efflux, and addition of Ca sulfate reduced such effects on the membrane. The presence of Ca^{2+} in plants grown in saline environments can promote better control of root ion uptake, such as decreasing Na^+ (Hansen and Munns 1988) and increasing K^+, acting on cell membrane, and maintaining appropriate levels of these ions in photosynthetic tissues (Lacerda et al. 2004). Ca^{2+} also promotes the accumulation of organic solutes, such as proline (Colmer et al. 1996) and glycinebetaine (Girija et al. 2002), which allow the establishment of osmotic balance in the cytoplasm that is more compatible with cell metabolism, decreasing salt stress.

One of the key functions of calcium is to maintain the structural integrity of membranes of several organelles. When there is deficiency, the membranes begin to leak, cell compartmentalization is broken, and the binding of Ca to the pectin in the wall is affected. Pectate degradation is mediated by the action of polygalacturonase enzyme, which is drastically inhibited by high Ca concentrations (Table 8.1).

Corroborating this fact, polygalacturonase activity increases in Ca-deficient plants. The typical deficiency symptom of Ca is the disintegration of the cell wall, with collapse of petiole tissues and of the youngest parts of the stem. In addition to the collapse of stem, there may also be increased disease incidence as the efflux of low molecular weight organic compounds from cells is a food source for parasites, aggravating disease severity. Moreover, in the process of plant infection, parasites produce pectolytic enzymes that dissolve the middle lamella. Ca presence inhibits these enzymes, as in tomato, in which Ca decreased *Fusarium* wilt severity (Table 8.2).

The concentration of cytoplasmic Ca is relatively low (<0.1 μM), playing an important enzymatic role as it activates the action of the Ca-calmodulin coenzyme, as without Ca this enzyme has no catalytic activity. Ca-calmodulin is required for the activity of a series of other enzymes (phospholipases, nucleotides, membrane ATPase-Ca, glutamate decarboxylase, and NAD kinase) and for the synthesis of α-amylase, which has the role of breaking starch in seed germination (cereals). However, high Ca levels in the cytoplasm can cause unwanted reactions, such as formation of insoluble salts (Ca-ATP; Ca-phosphates, and callose) and closure of plasmododesmata, reducing the radial transport of ions. According to Rengel and Zhang (2003), production of callose may play an important role in the expression of Al toxicity. Increased Ca in the cytoplasm can inhibit even certain important enzymes, such as PEP carboxylase, phosphatases, and fructose-1.6-bisphosphatase (sucrose synthesis in cytosol). High Ca concentration does not occur due to the action of ATPases in the membranes that remove Ca from the cytoplasm. These

Table 8.1 Effect of Ca on the hydrolysis of a pectate by the polygalacturonase enzyme

Ca^{2+}	Amount of galacturonic acid released
mg L^{-1}	μ mol per 4 h
0	3.5
40	2.5
200	0.6
400	0.2

Table 8.2 Effect of Ca on *Fusarium* wilt severity in tomato plants after inoculation

Ca^{2+}	Ca concentration in plant sap	Disease
$\mu g\ mL^{-1}$		%
0	72	100
50	219	92
200	380	80
1000	1081	9

ATPases are activated by Ca-calmodulin (a low molecular weight compound). Most (90%) of calmodulins are located in the cytosol. In the roots, they are mainly associated with the plasma membrane (via microtubules), especially in the first millimeters of the apex, a region of high metabolic activity occupied by the cap, meristem, and the root elongation zone.

Ca is part of the structure of only one enzyme (amylase), while Ca only activates or is part of the synthesis of other enzymes. During germination, Ca activates enzymes (phospholipases) that degrade lipid bodies in cotyledon membranes (Paliyath and Thompson 1987).

Ca is also indispensable for the germination of pollen grain and for pollen tube growth, which may be due to its role in cell wall synthesis or plasmalema functioning, with high Ca content detected at the apex of growing pollen tubes. In addition, phytates are degraded by phytases for germination of the pollen grain, which are activated by Ca (Scott and Loewus 1986). Moreover, development of the pollen tube is chemotopically oriented by extracellular Ca. Thus, Ca content in the flowers of coffee plants was, in order of magnitude, approximately 1.8 times greater than the content in leaves and branches in relation to the cultivar Catuaí Amarelo, and 1.4 times greater in relation to the cultivar Mundo Novo (Malavolta et al. 2002).

In biological N_2 fixation by legumes, root nodules need more Ca than the plant. Once nodules are formed, fixation and plant growth occur normally, with relatively low concentrations of the element.

Calcium plays an important role in osmoregulation. Stomatal movements are typical processes that regulate cell turgor due to changes in osmotic potential in neighboring cells, guard cells or tissues. These changes are promoted mainly by flows of potassium, chloride, and malate as active osmotic components. Aabscisic acid (ABA) action in stomatal closure depends on calcium concentrations in the leaf epidermis, which are usually much higher than Ca concentrations of other cells. ABA-induced activation in calcium channels and quick increase in cytosolic Ca^{2+} concentrations seem to block proton pumps and open channels for anions. Both events lead to loss of turgor in guard cells and stomatal closure, which can provide the plant with defense against temperature stresses and anaerobiosis (Atkinson et al. 1990).

Lately, specialized literature has discussed another important function of Ca, acting as secondary messenger in the conduction of signals for the response of plants to environmental factors, changing growth metabolism and plant development. External (light, gravity, and mechanical) and internal (hormones) stimuli act on Ca^{2+}

Table 8.3 Calcium functions in plants

Structural	Enzyme activator	Processes
Pectate (medium lamella)	ATPases	Structure and functioning
Carbonate	α amylase	of membranes
Oxalate	Phospholipase D	Ion uptake
Phytate	Nuclease	Reactions with plant hormones and enzyme activation (via calmodulin)
Calmodulins		
		Secondary messenger

transport mechanisms by modifying their concentration in the cytoplasm. The stimulus is a message that is carried by Ca^{2+} as a secondary messenger. When the cell perceives the message, Ca^{2+} is discharged from its reservoirs, such as the apoplast, mitochondria, and endoplasmic reticulum in the cytosol (Malavolta et al. 1997). We note that for Ca to perform this function as a messenger for signal transduction (light and touch), it is necessary to maintain a low concentration in the cytoplasm (0.1–0.2 µM) (Trewavas and Gilroy 1991), although a transitory increase in the concentration of cytosolic Ca^{2+} is necessary (Mansfield et al. 1990).

Finally, four groups of calcium-sensitive proteins in plants have been described, namely: (1) calcium-dependent protein kinases; (2) calmodulins; (3) other calcium-associated proteins with EF-hand motifs; and (4) calcium-associated proteins without EF-hand motifs. Indirectly, only those in groups 1 and 2 are related to the induction of stress-regulating genes (Reddy 2001). We note that mechanisms through which Ca acts to reduce stress effects lack conclusive information. However, the latest research indicates that this nutrient serves as a messenger in many development processes in response to stress. Plants under stress, whether biotic or abiotic, introduce a high concentration of Ca in the cytosol, losing the function of Ca of warning the plant of the harmful agent so that the plant can develop defense mechanisms in time, minimizing damage to its growth.

In summary, Ca has structural function and work as enzyme activator (Malavolta et al. 1997), influencing various processes in plants (Table 8.3).

8.4   Crop Nutritional Requirements

The study of crop nutritional requirements, as aforementioned, reflects the extraction of the nutrient from the soil throughout the production cycle. Normally, with adequate liming, besides correcting soil acidity, bases such as Ca are supplied. Thus, with the use of the adequate V (%) [soil base saturation: (K+Ca+Mg) x100 / (K+Ca+Mg+H+Al)] for the crop, we expect that Ca nutrition in the plant is also suitable, for example, in passion fruit seedlings (V = 56–58% and Ca of the shoot = 7.4–12.8 g kg^{-1}) (Prado and Natale 2004b, 2005), guava (V = 65% and Ca of the shoot = 7.8 g kg^{-1}) (Prado et al. 2003), and also in crops such as rubber tree (V = 57% and foliar Ca content = 8 g kg^{-1}) (Roque et al. 2004). We note that there are different responses of crops to V% and Ca nutrition. We also add that even within the same species, such as maize, there may be different responses to base saturation values regarding cultivar (Prado 2001), a fact also observed in soybean (Andrade et al. 2020).

Total Ca content in the plant may range from 0.1 to 0.5% (1–5 g kg^{-1}), although reaching 10% (100 g kg^{-1}) in old leaves. These values may range depending on the crop and other factors that are the subject of Chap. 19.

Information on the mechanisms of Ca efficiency by plants is poorly understood. According to Caines and Shennan (1999), the relationship between efficient utilization of Ca and plant growth is very complex and may involve several physiological controls, such as the internal retranslocation capacity of compartmentalized Ca in membranes and cell storage organs (endoplasmic reticulum, chloroplasts, and vacuole) (Caines and Shennan 1999). In addition, Ca inactivation due to binding and/or precipitation in the form of oxalate or calcium phosphate have been suggested as a cause for low efficiency of nutrient use (Behling et al. 1989).

Behling et al. (1989) added that the high Ca uptake efficiency of a tomato strain was due its ability to maintain a high proportion of total Ca in soluble form and to maintain growth and metabolism in all plant parts, even under low Ca concentration in their tissues. In turn, the low efficiency of other tomato strain was associated with its high concentrations of insoluble Ca in the tissues of the shoot. In grafted coffee, Tomaz et al. (2003), who studied different genotypes under different combinations of rootstock and canopy, verified that the H 514-5-5-3 strain benefited from Ca utilization efficiency and dry matter production only by one rootstock Mundo Novo.

Nutrient Extraction and Transport

Total extraction of calcium occurs as a function of Ca content in the plant and dry matter accumulated. Therefore, it depends on production, which depends on species, variety/hybrid, soil availability, and crop management, among others.

As for the vegetable species, there is variation in the amount required depending on the crop.

From different crops, we observe that total extraction of potassium ranged from 39 (wheat) to 257 kg ha^{-1} (maize).

From results of different crops, we observe that total Ca extraction ranged from 7 (wheat) to 142 kg ha^{-1} (coffee). However, in relative values of calcium extraction in kg per ton produced, there is increased requirement for coffee (71) and beans (54) and decreased requirement for grasses such as sugarcane (1.0) and wheat (2.3). However, when comparing only annual crops, we note that legumes require much more than grasses. In legumes, Ca is most required for nodule formation. Once formed, the requirement decreases (Table 8.4).

Regarding nutrient export per ton of grains produced, legumes (3.3–8.5 kg t^{-1}) export much more Ca than grasses (0.06–0.33 kg t^{-1}). However, we noted that generally the amounts exported by crops are relatively low.

Nutrient Uptake Rate

The study of uptake rate (nutrient accumulated as a function of cultivation period) is important to determine the periods when elements are most required. However, as Ca is supplied with liming at preplanting, information on the uptake rate is little applicable. The opposite occurs if cultivation is performed with supply of a nutrient solution, such as cultivation in containers.

In the cotton crop, Mendes (1965) obtained accumulative uptake rate in the entire plant development cycle (150 days). Slow initial uptake (up to 30 days) (19%) was observed, with values close to 50% (initial phase) to 75% (final phase) obtained in the flowering phase. We note that by the end of flowering, approximately 75% of the Ca required by the crop has already been absorbed, indicating that this nutrient must be available to the plant at the very beginning of development, being supplied in the form of liming, applied at preplanting (3 months before sowing).

8.5 Symptoms of Nutritional Deficiencies and Excesses

Deficiency

Symptoms of Ca deficiency occur initially in meristematic regions (growth points) and on new leaves (Fig. 8.6). The most frequently reported symptoms are as follows:

1. Whitish color on leaf margins.
2. Irregular leaf shapes, with tearing of margins and gelatinous appearance in the tips of leaves.
3. Internerval necrotic spots on leaves.
4. Shoot death from the tips, which can cause plant tillering.
5. Root growth is severely affected.
6. Decreased fruiting.
7. Decreased seed production.
8. Petiole collapse.

At cellular level, Ca deficiency causes the appearance of polyploid or constricted nuclei, binucleated cells, and amitotic divisions, paralyzing plant growth, especially root growth, which darkens reaching necrosis.

Rates of leaf senescence can be altered in Ca-deficient plants. Senescence is a consequence of membrane lipid peroxidation by increased free oxygen radical

Table 8.4 Calcium requirements of major crops

Crop	Plant part	Dry matter produced	Ca accumulated Plant part	Ca accumulated Total (3)	Ca required to produce 1 t of grains(4)
		t ha^{-1}	kg ha^{-1}		kg t^{-1}
Annual					
Cotton	Reproductive (cotton/cottonseed)	1.3	11 (8.5)(2)	61	46.9
	Vegetative (stem/branch/leaf)	1.7	49		
	Root	0.5	1		
Soybean(1)	Grains (pods)	3	10 (3.3)	70	23.3
	Stem/branch/leaf	6	60		
Bean	Pod	1	4 (4.0)	54	54.0
	Stem	0.4	8		
	Leaves	1.2	40		
	Root	0.1	2		
Maize(1)	Grains	6.4	0.4 (0.06)	36	5.6
	Crop residues	–	35.6		
Rice	Grains	3	2 (0.66)	25	8.3
	Stems	2	4		
	Leaves	2	12		
	Husk	1	2		
	Root	1	5		
Wheat	Grains	3	1 (0.33)	7	2.3
	Straw	3.7	6		
Perennial/semi-perennial					
Sugarcane	Stems	100	60 (0.6)	100	1.0
	leaves	25	40		
Coffee(1)	Grains	2	7 (3.5)	142	71
	Trunk, branches, and leaves	–	136		

(1) Malavolta (1980); (2) Nutrient export through the grains produced (kg t^{-1}): Ca accumulated in the grains/grain dry matter; (3) Suggests the total nutritional requirement of the crop for the respective yield level; (4) Suggests the relative nutritional requirement of Ca of the crop for production of 1 ton of the commercial product (grains/stems), obtained by the following formula: Ca accumulated in the plant (vegetative + reproductive part)/dry matter of the commercial product

levels. The protective effect of calcium (and cytokines) occurs due to its action in inhibiting lipoxygenase enzyme activity, which degrades membranes, and also in inhibiting polygalacturonases that degrade the cell wall. Early senescence begins with degradation of membranes and cell wall, associated with increased ethylene production.

Coffea arabica plants under Ca^{2+} deficiency conditions decreased chlorophyll and soluble protein contents (Ramalho et al. 1995).

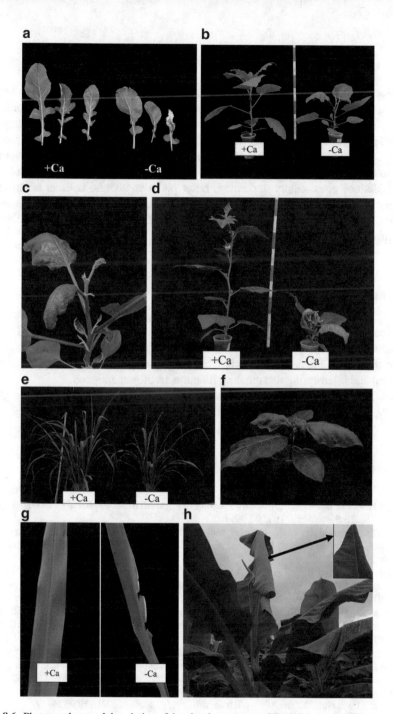

Fig. 8.6 Photos and general description of the visual symptoms of Ca deficiency in different crops. Garden rocket: New leaves have tanned margins that evolve to necrosis, the leaf blade decreases, and leaves deform (Fig. 8.6a); Aubergine: The shoot apical region has its development paralyzed, younger leaves are curved, and folding downwards (Fig. 8.6b); internerval chlorosis (broad bands) and necrosis appear on upper leaves with laceration of the leaf blade and death of the growth region, causing the appearance of leaves adjacent to the petiole (Fig. 8.6c); and Scarlet eggplant: Plant growth decreases and new leaves become deformed and curved downwards. Intense chlorosis appears between veins, evolving to necrosis until leaves dry (Fig. 8.6d); irregular deformations in the leaf blade of the apical region hamper plant development (Fig. 8.6e); Grass of genus Panicum: Leaves become cut in the marginal regions of the leaf blade, the foliar area decreases, and leaves narrow (Fig. 8.6f). Cut regions along the leaf margin turn yellow (Fig. 8.6g); Banana tree: The apical region deforms and the leaf blade is characterized by surface undulations (Fig. 8.6h)

Normally, organs with increased transpiration receive more Ca, such as leaves due to their larger specific surface compared to fruit. Certain organs such as fruits receive most of the water, via phloem, with near Ca absence. Some physiological disorders occur in plant tissues with decreased transpiration, that is, which receive little Ca through transpiratory flow, triggering Ca deficiency symptoms. Thus, there are some anomalies in certain crops, such as apical rot of tomato and melon fruits, rot of lettuce apex, celery darkening, bitter pit in apple, and internal collapse of mango, among others. Calcium deficiency in tomato plants decreased with foliar spraying of silicon (7.1 mM) in the plant reproductive phase (Alonso et al. 2020), requiring further studies to evaluate fruit apical rot.

Excess

In excess, calcium is highly tolerated by plants, reaching about 10% Ca in old leaves without toxicity symptoms. However, at subcellular level, excess Ca increases its concentration in the cell cytoplasm, which may precipitate P, decreasing ATP production and interrupting its signaling function, as aforementioned.

We note that excess Ca could eventually induce magnesium or potassium deficiency, especially if their soil concentration is medium to low.

References

Adams F, Moore BL. Chemical factors affecting root growth in subsoil horizons of coastal plain soils. Soil Sci Soc Am J. 1983;47:99–102.

Alonso TAS, Barreto RF, Prado RM, et al. Silicon spraying alleviates calcium deficiency in tomato plants, but Ca-EDTA is toxic. J Plant Nutr Soil Sci. 2020;183:0055. https://doi.org/10.1002/jpln.202000055.

Andrade CA, Patinni IRG, Pantaleão AA, et al. Physiological response and earliness of soybean genotypes to soil base saturation conditions. J Agron Crop Sci. 2020;206:806–14. https://doi.org/10.1111/jac.12439.

Atkinson CJ, Mansfield TA, Mcainsh MR, et al. Interactions of calcium with abscisic acid in the control of stomatal aperture. Biochem Physiol Pflanz. 1990;186:333–9. https://doi.org/10.1016/S0015-3796(11)80228-3.

Behling JP, Gabelman WH, Gerloff GC, et al. The distribution and utilization of calcium by two tomato (*Lycopersicon esculentum* Mill.) lines differing in calcium efficiency when grown under low Ca stress. Plant Soil. 1989;113:189–96. https://doi.org/10.1007/BF02280180.

Bolat I, Kaya C, Almaca A, et al. Calcium sulfate improves salinity tolerance in rootstocks of plum. J Plant Nutr. 2006;29:553–64. https://doi.org/10.1080/01904160500526717.

Bradfield EG, Guttridge CG. Effects of night-time humidity and nutrient solution concentration on the calcium content of tomato fruit. Sci Hortic. 1984;22:207–17. https://doi.org/10.1016/0304-4238(84)90054-2.

Caines AM, Shennan C. Growth and nutrient composition of Ca^{2+} use efficient and Ca^{2+} use inefficient genotypes of tomato. Plant Physiol Biochem. 1999;37:559–67.

Clark RB. Physiology aspects of calcium, magnesium and molybdenum deficiencies in plants. In: Adams F, editor. Soil acidity and liming. Madison: American Society of Agronomy; 1984. p. 99–170.

Clarkson DT, Sanderson J. Sites of absorption and translocation of iron in barley roots. Tracer and microautoradiographic studies. Plant Physiol. 1978;61:731–6. https://doi.org/10.1104/pp.61.5.731.

Collier GF, Tibbitts TW. Tipburn of lettuce. Hortic Rev. 1982;4:49–65.

Colmer TD, Fan TWM, Higashi RM, et al. Interactive effects of Ca^{2+} and NaCl stress on the ionic relations and proline accumulation in the primary root tip of *Sorghum bicolor*. Physiol Plant. 1996;97:421–4. https://doi.org/10.1111/j.1399-3054.1996.tb00498.x.

Conway WS, Sams CE. Calcium infiltration of Golden Delicious apples and its effect on decay. Phytopathology. 1983;73:1068–71. https://doi.org/10.1094/Phyto-73-1068.

Evans DE, Sally-a B, Williams LE. Active calcium transport by plant cell membranes. J Exp Bot. 1991;42:285–303.

Foy CD. Physiological effects of hydrogen, aluminum and manganese toxicities in acid soil. In: Adams F, editor. Soil acidity and liming. Madison: Soil Science Society American; 1984. p. 57–97.

François LE, Donavan TJ, Maas EV, et al. Calcium deficiency of artichoke buds in relation to salinity. HortScience. 1991;26:549–53.

Girija C, Smith BN, Swamy PM. Interactive effects of sodium chloride and calcium chloride on the accumulation of proline and glycinebetaine in peanut (*Arachis hypogaea* L.). Environ Exp Bot. 2002;47:1–10. https://doi.org/10.1016/S0098-8472(01)00096-X.

Hansen EH, Munns DN. Effect of $CaSO_4$ and NaCl on mineral content of Leucaena leucocephala. Plant Soil. 1988;107:101–5. https://doi.org/10.1007/BF02371549.

Hanson EJ, Beggs JL, Beaudry RM. Applying calcium chloride postharvest to improve highbush blueberry firmness. HortScience. 1993;28:1033–4. https://doi.org/10.21273/HORTSCI.28.10.1033.

Johnson JR. Calcium nutrition and cultivar influence incidence of tipburn of collard. HortScience. 1991;26:544–6. https://doi.org/10.21273/HORTSCI.26.5.544.

Lacerda CF, Cambraia J, Oliva MA, et al. Calcium effects on growth and solute contents of sorghum seedlings under NaCl stress. Rev Bras Ciênc Solo. 2004;28:289–95. https://doi.org/10.1590/S0100-06832004000200007.

Malavolta E. Elementos de nutrição de plantas. São Paulo: Agronômica Ceres; 1980. 251p

Malavolta E, Favarin JL, Malavolta M, et al. Nutrients repartition in the coffee branches, leaves and flowers. Pesq Agropec Bras 2002;37:1017–1022. https://doi.org/10.1590/S0100-204X2002000700016.

Malavolta E. Manual de nutrição mineral de plantas. São Paulo: Agronômica Ceres; 2006.

Malavolta E, Vitti GC, Oliveira SA. Avaliação do estado nutricional das plantas: princípios e aplicações. Piracicaba: Associação Brasileira de Potassa e do Fósforo; 1997. 319p

Mansfield TA, Hetherington AM, Atkinson CJ. Some current aspects of stomatal physiology. Annu Rev Plant Physiol Mol Biol. 1990;41:55–75. https://doi.org/10.1146/annurev.pp.41.060190.000415.

Marschner H. Mineral nutrition of higher plants. London: Academic Press; 1986.

Marschner H. Mineral nutrition of higher plants. London: Academic Press; 1995.

Mendes HC. Cultura e adubação do algodoeiro. São Paulo: Instituto Brasileiro de Potassa; 1965.

Natale W, Prado RM, Morô FV. Anatomical modifications in the cell wall of guava as influenced by calcium. Pesq Agrop Brasileira. 2005;40:1239–42. https://doi.org/10.1590/S0100-204X2005001200012.

Paliyath F, Thompson JE. Calcium and calmodulin regulated breakdown for phospholipid by microsomal membranes from bean cotyledons. Plant Physiol. 1987;83:63–8. https://doi.org/10.1104/pp.83.1.63.

Prado RM. Base saturation and corn hybrids under no-tillage system. Sci Agric. 2001;58:391–4. https://doi.org/10.1590/S0103-90162001000200024.

Prado RM, Natale W. Effect of the liming on the nutrition and the development of the guava root system. Pesq Agrop Brasileira. 2004a;39:1007–12. https://doi.org/10.1590/S0100-204X2004001000008.

Prado RM, Natale W. Application of industrial waste (silication) in red Argisoil of passion fruit cuttings. Acta Sci. 2004b;26:387–93. https://doi.org/10.4025/actasciagron.v26i4.1714.

Prado RM, Natale W. Effect of application of calcium silicate on growth, nutritional status and dry matter production of passion fruit seedlings. Rev Bras Eng Agríc Ambient. 2005;9:185–90. https://doi.org/10.1590/S1415-43662005000200006.

Prado RM, Corrêa MCM, Cintra ACO, et al. Response of guava plants to basic slag application as corrective of soil acidity. Rev Bras Frutic. 2003;25:160–3. https://doi.org/10.1590/S0100-29452003000100044.

Prado RM, Natale W, Silva JAA. Liming and quality of guava fruit cultivated in Brasil. Sci Hortic. 2005a;104:91–102. https://doi.org/10.1016/j.scienta.2005.03.001.

Prado RM, Natale W, Corrêa MCM, et al. Liming and postharvest quality of carambola fruits. Braz Arch Biol Technol. 2005b;48:689–96. https://doi.org/10.1590/S1516-89132005000600003.

Prado RM, Natale W, Rozane D. E. Soil liming effects on the development and the nutritional status of the carambola tree and its fruit yielding capacity. Commun Soil Sci Plant Anal. 2007;38:493–511. https://doi.org/10.1080/00103620601174536.

Rains DW. Mineral metabolism. In: Bonner J, Varner JE, editors. Plant biochemistry. New York: Academic Press; 1976. p. 561–98.

Ramalho JC, Rebelo MC, Santos ME, et al. Effects of calcium deficiency in *Coffea arabica*. Nutrients changes and correlation of calcium levels with some photosynthetic parameters. Plant Soil. 1995;172:87–96. https://doi.org/10.1007/BF00020862.

Raven JA. H^+ and Ca^{2+} in phoem and symplast: relation of relative immobility of the ions to the cytoplasmic nature of the transport paths. New Phytol. 1977;79:465–80. https://doi.org/10.1111/j.1469-8137.1977.tb02229.x.

Reddy ASN. Calcium: silver bullet in signalling. Plant Sci. 2001;160:381–404. https://doi.org/10.1016/S0168-9452(00)00386-1.

Rengel Z, Zhang WH. Role of dynamics of intracellular calcium in aluminium toxicity syndrome. New Phytol. 2003;159:295–314. https://doi.org/10.1046/j.1469-8137.2003.00821.x.

Roque CG, Prado RM, Natale W, et al. Nutritional status and productivity of rubber tree on soil with surface applied lime. Pesq Agrop Brasileira. 2004;39:485–90. https://doi.org/10.1590/S0100-204X2004000500011.

Rosolem CA, Giommo GS, Laurenti RLB. Root growth and mineral nutrition of cotton cultivars as affected by liming. Pesq Agrop Brasileira. 2000;35:827–33. https://doi.org/10.1590/S0100-204X2000000400021.

Scott JJ, Loewus FAA. calcium-activated phytase from pollen of *Lilium longiflorum*. Plant Physiol. 1986;82:333–5. https://doi.org/10.1104/pp.82.1.333.

Tibbitts TW, Palzkill DA. Requirement for root pressure flow to provide adequate calcium to low-transpiring tissue. Commun Soil Sci Plant Anal. 1979;10:251–7. https://doi.org/10.1080/00103627909366892.

Tomaz MA, Silva SR, Sakiyama NS, et al. Efficiency of uptake, translocation and use of calcium, magnesium and sulphur in young *Coffea arabica* plants under the influence of the rootstock. Rev Bras Ciênc Solo. 2003;27:885–92. https://doi.org/10.1590/S0100-06832003000500013.

Trewavas AJ, Gilroy S. Signal transduction in plant cells. Trends Genet. 1991;7:356–61. https://doi.org/10.1016/0168-9525(91)90255-O.

Chapter 9
Magnesium

Keywords Mg availability · Magnesium uptake · Magnesium transport · Magnesium redistribution · Assimilatory magnesium · Magnesium deficient

Magnesium (Mg) is one of the most absorbed macronutrients by crops, although being essential to metabolism, with structural function in chlorophyll composition and in enzymatic processes, ensuring high crop yield. In this chapter, we will discuss initially (i) basic aspects of Mg in the soil; (ii) Mg uptake, transport, and redistribution; (iii) Mg metabolism; (iv) nutritional requirements for Mg in crops; (v) Mg extraction, export, and accumulation by the main crops; and (vi) Mg deficiency and toxicity symptoms.

9.1 Introduction

Magnesium (Mg) has its primary origin in the soil, in igneous, metamorphic, and sedimentary rocks. The main minerals containing Mg are biotite, dolomite, chlorite, serpentine, and olivine. Mg is part of clay mineral structure (illite, vermiculite, and montmorillonite). However, after soil weathering exchangeable Mg remains adsorbed to colloids and being a component of the soil organic matter. Exchangeable (5–10% of the total content) and soil solution forms are considered available to plants.

In soils cultivated with increased application of potassium fertilizers and in crops with increased K requirement (coffee, cotton, citrus, and banana), there may be Mg deficiency induction, especially in soils with decreased Mg content. Another situation is the indiscriminate use of calcitic limestone with low magnesium content (<5% MgO), which may cause deficiency problems in plants as this element is not found in conventional fertilizers. There are indications that the Ca:Mg ratio in the soil must be balanced. Munoz Hernandez and Silveira (1998) verified that low Ca:Mg ratio in the soil (2 1 or 3:1) provided better maize growth compared with a high ratio (4:1 or 5:1).

As with other nutrients, increase in the soil pH value close to 6.5 increases Mg availability in the soil.

© Springer Nature Switzerland AG 2021
R. de Mello Prado, *Mineral nutrition of tropical plants*,
https://doi.org/10.1007/978-3-030-71262-4_9

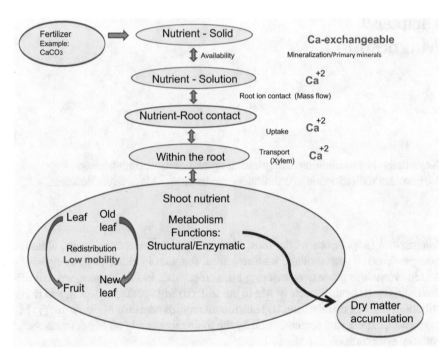

Fig. 9.1 Magnesium dynamics in the soil–plant system indicating the nutrient passage processes in the different plant compartments

When studying magnesium in the plant system, it is important to know all the compartments the nutrient passes in the soil solution and in the root and shoot (leaves/fruits), that is, from the soil until its incorporation into an organic compound or as enzyme activator, contributing to vital functions that enable maximum dry matter accumulation in the agricultural product (grain, fruit, etc.) (Fig. 9.1).

9.2   Uptake, Transport, and Redistribution of Magnesium

Uptake

The pathway of magnesium from the soil to the roots occurs by mass flow (85% of the total). This movement depends on the water dynamics in the soil–plant system, moved by plant transpiration.

Magnesium uptake process (passive and active) in form of Mg^{2+} is widely studied, as high Ca^{2+} and especially K^+ concentrations in the medium can inhibit its uptake by ionic competition, which may cause deficiency in plants ($Mg \times Ca \times K$ ratios are discussed in Chap. 20). Interaction between Mg and other cations is important, as normally the uptake rate of this ion is relatively low. According to

Marschner (1986), Mg has a relatively large hydrated radius (0.480 nm) and very high hydration energy, causing low Mg affinity for binding sites on the plasma membrane.

Transport

After absorption, Mg^{2+} is transported (active and passive) until reaching the xylem, subsequently transported passively to the area under action of the transpiratory current.

Redistribution

Contrary to what occurs with Ca^{2+} and similarly to what occurs with K^+, Mg^{2+} is mobile in the phloem. As most of the plant Mg is found in soluble form, this explains its redistribution in plants.

9.3   Participation in Plant Metabolism

Among the main functions of magnesium in plants, we highlight its participation in the central constitution of chlorophyll, corresponding to 2.7% of the molecular weight, and also as an enzyme activator.

Structural (Chlorophyll) and Enzyme Activity

Approximately 20% of foliar total Mg is found in chloroplasts, which represent approximately 5% of the total volume of a mature leaf cell. Mg in the chloroplasts is distributed with 20% being part of chlorophylls and the rest in ionic form. In the dark, it is found in intrathylakoid spaces, and when chloroplasts are illuminated, the element is secreted to the stroma, where it activates several enzymes (Castro et al. 2005).

Mg^{2+} deficiency decreases chlorophyll synthesis and, consequently, photosynthetic rate. However, quantitative evaluation demonstrates that the photosynthetic rate decreases more than chlorophyll content, indicating that Mg deficiency affects photosynthesis in other steps, besides participating in pigment composition from covalent bonds. We also note that a small part of Mg accompanies Ca in cell wall formation.

Magnesium and potassium are the most important elements regarding enzyme activation. Currently, it is known that Mg^{2+} acts as enzyme activator of ribulose diphosphate regeneration reactions, which is a CO_2-accepting compound, that is, the sugar that accepts photosynthetically fixed CO_2, which occurs at the beginning of the Calvin cycle in the chlorosplast. Thus, Mg^{2+} affects modulation of rubisco in the chloroplast stroma (Pierce 1986). The activity of this enzyme is highly dependent on Mg^{2+} and pH. The binding of Mg^{2+} with the enzyme increases both its affinity (Km) with the CO_2 substrate as its maximum speed (Vmax) of reaction (Sugiyama et al. 1968). When participating in reactions involving CO_2 fixation, the lack of Mg can inhibit photosynthesis even with chlorophyll presence. In addition, the nutrient provides the maintenance of optimal pH for enzyme activity. Portis Junior and Heldt (1976) verified that Mg^{2+} increase in the stroma from 3 to 5 mM would be sufficient to increase RuDPase affinity for CO_2 to the point that its fixation could proceed at maximum rate. Ribulose diphosphate production or regeneration, which is a carbon skeleton that receives fixed CO_2, increases with Mg enzyme action.

Considering the sum of its effects, it is possible to understand the reason for the marked effect of Mg deficiency on photosynthetic rate. As photosynthesis is one of the main ways of capturing metabolic energy in adenosine triphosphate (ATP) form, it is interesting to establish a parallel between Mg functions and the energy metabolism of plants. In leaf cells of green plants, at least 25% of the total protein is located in the chloroplasts, consisting mainly of rubisco, explaining why Mg^{2+} deficiency affects chloroplast size, structure, and function, including electron transfer in photosystem II (McSwain et al. 1976). For its ionic characteristics, Mg^{2+} forms a bridge between ATP or adenosine triphosphate (ADP) molecules and enzyme molecules, enabling energy transfer from radical phosphates to different reactions of organic synthesis. In addition, the ATPase substrate is not only ATP but also ATP complexed with Mg (Mg-ATP), which is another relationship of Mg^{2+} and energy metabolism.

Protein Synthesis

The first step in protein synthesis is the amino acid activation reaction. Phosphate radicals are incorporated into different amino acids that pass from activated form (aminoacyl), necessary for tRNA binding, to subsequent incorporation into the polypeptide chain, maintaining the necessary configuration for protein synthesis. Mg^{2+} is necessary for energy transfer to amino acids through phosphorylating enzymes. Mg as cofactor benefits several biochemical reactions in plants, uniting N and P complexes. There is evidence that Mg^{2+} acts in the aggregation of ribosome subunits (Cammarano et al. 1972), which are the particles responsible for the formation of the polypeptide chain. Another important enzyme in N metabolism is glutamate synthetase (GS), which acts in the important assimilation pathway of NH_3–GS/ GOGAT, also activated by Mg^{2+} in chloroplasts. In addition, Mg is required for the action of nuclear RNA polymerase (Castro et al. 2005). Thus, the proportion of N

Table 9.1 Functions of magnesium in plants

Structural	Enzyme activator	Processes
Chlorophyll	Acetic thiokinase	Ion uptake
	Pyruvic kinase	Photosynthesis
	Hexokinase	Respiration
	Enolase	Energy storage/transfer
	Isocitric dehydrogenase	Organic synthesis
	Pyruvate decarboxylase	Electrolytic balance
	Ribulose carboxylase	Ribosome stability
	Phosphopyruvate synthetase	
	Glutamyl synthetase	
	Glutamyl transferase	

protein decreases and nonprotein N increases in Mg-deficient plants (Haeder and Mengel 1969).

In summary, Mg functions in plants are shown in Table 9.1.

9.4   Crop Nutritional Requirements

Normally, with adequate liming, besides correcting soil acidity, Mg is provided by crops.

Total Mg content in the plant can range from 0.15 to 0.35% (1.5–3.5 g kg^{-1}). These values may range depending on the crop and other factors that are the subject of Chap. 19.

In order to adequately discuss crop nutritional requirements, two factors are equally important: Mg extraction and uptake absorption rate throughout cultivation.

Nutrient Extraction and Export

The crops that most extracted Mg per area were sugarcane and corn, 52 and 48 kg ha^{-1}, respectively, while wheat and rice extracted only 9 kg ha^{-1} (Table 9.2).

The bean plant exported the most Mg for grains (5.0 kg ha^{-1}), showing the importance of monitoring this crop for periodic replacement of this nutrient via corrective material (limestone with MgO content > 5.0%), as it is the Mg source with the lowest cost.

Among annual crops, considering the need for Mg per ton of grain produced, legumes (8.7–18.5 kg/t) were less efficient than grasses (3–7.5 kg/t). Regarding nutrient export by the grains produced by annual crops, we note that legumes (2–5 kg/t) export more than grasses (1.3–2.0 kg/t).

Table 9.2 Magnesium requirements of major crops

Crop	Plant part	Dry matter produced t ha^{-1}	Mg accumulated Plant part kg ha^{-1}	Total^c	Mg required to produce 1 t of grains^d kg t^{-1}
Annual					
Cotton	Reproductive (cotton/cottonseed)	1.3	5 (3.8)^b	12.7	9.8
	Vegetative (stem/branch/leaf)	1.7	7		
	Root	0.5	0,7		
Soybean[a]	Grains (pods)	3	6 (2.0)	26.0	8.7
	Stem/branch/leaf	6	20		
Bean	Pod	1	5 (5.0)	18.5	18.5
	Stem	0.4	1		
	Leaves	1.2	12		
	Root	0.1	0.5		
Maize[a]	Grains	6.4	10.0 (1.6)	48.0	7.5
	Crop residues	–	38.0		
Rice	Grains	3	4 (1.3)	9	3.0
	Stems	2	1		
	Leaves	2	2		
	Husk	1	1		
	Root	1	1		
Wheat	Grains	3	6 (2.0)	9	3.0
	Straw	3.7	3		
Perennial/semi-perennial					
Sugarcane	Stems	100	35 (0.35)	52	0.5
	Leaves	25	17		
Coffee[a]	Grains	2	3 (1.5)	33	16.5
	Trunk, branches, and leaves	-	30		

[a]Malavolta (1980); [b] Nutrient export through the grains produced (kg t^{-1}): Mg accumulated in the grains/grain dry matter; [c] Suggests the total nutritional requirement of the crop for the respective yield level; [d]Suggests the relative nutritional requirement of Mg of the crop for production of 1 ton of the commercial product (grains/stems) obtained by the following formula: Mg accumulated in the plant (vegetative + reproductive part)/dry matter of the commercial product

Different cultivars may have different abilities for Mg uptake, transport, and utilization. In grafted coffee, Tomaz et al. (2003), who studied different genotypes under different rootstock and canopy combinations, found that the H 514-5-5-3 strain benefited from Mg utilization efficiency and dry matter production by Mundo Novo IAC 376-4 and Apoatã LC 2258 rootstocks.

Barbosa (1978) used tomato cultivars Kadá, Ângela, Manalucie, and Maçã de Ibirité in three assays and observed that the cultivar Ângela was the first to show Mg deficiency symptoms when cultivated under a lack of Mg; concluding that its decreased efficiency can be attributed to increased root retention, decreased translocation to leaves, and increased redistribution of the nutrient from lower to upper leaves.

Nutrient Uptake Rate

In field crops where liming is normally performed three months before sowing, the nutritional requirement for Mg by the crop should be met throughout the production cycle. Under these conditions, uptake rate would not define the period for Mg application. However, for didactic purposes, it is important to know the time the crop most requires Mg, such as for maize, where Mg uptake is considered slow up to 59 days (12th leaf), reaching only 16% of the total, and from this period, uptake accelerates with uptake rate peaks between 12th leaf and bolting and in milky ripening and grain formation stages (Flanery 1987).

In perennial crops such as coffee plant, Mg accumulation by flowers of cultivars Catuaí Amarelo and Mundo Novo represents 52% in relation to the total extracted by plant parts (flowers, leaves, and branch), indicating that Mg must be applied before flowering in this crop (Malavolta et al. 2002).

9.5 Symptoms of Nutritional Deficiencies and Excesses

Deficiency

Mg deficiency can occur by low concentration in the crop medium or by competition with other cations, such as K. As the Mg in the plant is mobile in the phloem, deficiency initially appears in older leaves through internerval chlorosis, which may be accompanied by yellowish spots that can unite forming ranges along leaf margins that become red or with other pigmentation (Fig. 9.2). Chlorosis starts with spots that later unite and spread to the tips and margins of leaves.

Mg deficiency symptoms may vary depending on the species and/or cultivar.

In leaves, mesophyll cells near vascular bundles retain chlorophyll for a longer period than parenchyma cells, which can delay chlorosis (Epstein 1972).

Nonstructural carbohydrates (starch and sugars) accumulation is a common characteristic of Mg^{2+}-deficient plants. In bean plant, carbohydrate accumulation in leaves is related to decreased carbohydrate content in drain regions, as occurs in roots and pods (Fischer and Bussler 1988). Limiting the carbohydrate supply to the roots greatly impairs their growth.

Fig. 9.2 Photos and general description of visual symptoms of Mg deficiency in different crops. Maize: Older leaves turn yellow in the margins and subsequently between veins, with the appearance of stretch marks (Fig. 9.2a); as damage evolves, chlorotic regions become necrotic; and the symptom progresses to new leaves, causing chlorosis in all leaves (Fig. 9.2b). Aubergine: Plant growth decreases and internerval chlorosis appears, which progresses from the center to the margins of leaves, subsequently affecting newer leaves (Fig. 9.2c). As deficiency progresses, pale spots with irregular contour, near whitish, appear on the leaf blade. Afterwards, spots dry as a result of necrosis (Fig. 9.2d). Scarlet eggplant: Mg deficiency inhibits plant growth, becoming short; chlorosis appears between veins of older leaves (Fig. 9.2e). Associated with internerval chlorosis, the tissue becomes deformed as a result of tissue degradation, which may dry with evolution (Fig. 9.2f). Citrus: Chlorophyll begins to disappear in older leaves between the main vein and the margin, progressing usually outwards, leaving the figure of a wedge at the leaf base. However, it can progress inwards, causing a yellow wedge, which may reach the entire leaf with a golden bronze color. Leaves fall prematurely, causing descending death of new branches (Fig. 9.2g)

Tewari et al. (2006) found that the lowest dry matter production of blackberry plants under Mg deficiency was associated with decreased leaf carbohydrate content, photosynthetic pigment, and photosynthetic rate. In addition, H_2O_2 and activity of antioxidant enzymes (peroxidase and dismutase) increased to decrease oxidative stress.

Mg deficiency can induce decarboxylation of amino acids, forming putrescine (Basso and Smith 1974), which is also described for K, possibly explaining leaf necrosis, as putrescine is a nitrogenous compound toxic to plants.

Mg-deficient plants often have delayed reproductive phase (Taiz and Zeiger 2004). Carbohydrate translocation from shoot to root decreases, reducing root growth.

Forage plants with low levels of this macronutrient can promote low levels of serum magnesium in animals, causing tetany.

Mg deficiency can increase or decrease diseases depending on crop and pathogen. A specific defense mechanism against diseases enhanced by Mg includes increased tissue resistance to degradation by pectolytic enzymes of pathogens (Huber and Jones 2013).

Excess

Excess magnesium supply results in deposition of the element in the form of different salts in cell vacuoles. There are few studies on the harmful effects on plant development and production.

Kobayashi et al. (2005) studied rice plants and *Echinochloa* species grown in nutrient solution with excess Mg (30mM) in the form of $MgCl_2$ and $MgSO_4$ for 20 days after transplanting. Dry matter production reduced from 33 to 67%. Mg, K, and Cl uptake increased and Ca uptake decreased in both crops, affecting more the *Echinochloa* species than rice plants. The authors concluded that rice plants were more tolerant to excess Mg than *Echinochloa*, and tolerance was related to Ca deficiency.

The greatest amount of Mg application inhibits Zn uptake, as these are elements that have similar valence, ionic radius, and hydration degree (Kabata-Pendias and Pendias 1984).

Mass et al. (1969) observed in barley that inhibition between Mg and Mn is non-competitive. This fact is important in areas that received increased amounts of limestone with increased Mg content, which impairs Zn and Mn uptake by the two following factors: increasing the pH and decreasing its availability in the soil, besides interacting with uptake process.

In isolated chloroplasts, photosynthesis is inhibited by 5 mmol L^{-1} of Mg in the external solution. Inhibition is caused by decreased K influx. Photosynthesis inhibition can occur due to high Mg concentrations in the metabolic pool in plants under drought stress (Vitti et al. 2006).

References

Barbosa DS. Comportamento de tomateiro (*Lycopersicum esculentum*, Mill) cultivados sob diversos níveis de magnésio em solução nutritiva. Dissertação, Universidade Federal de Viçosa (1978)

Basso LC, Smith T. Effect of mineral deficiency on amine formation in higher plants. Phytochemistry. 1974;13:875–83. https://doi.org/10.1016/S0031-9422(00)91417-1.

Cammarano P, Felsani A, Gentile M, et al. Formation of active hybrid 80-S particles from subunits of pea seedlings and mammalian liver ribossomes. Biochim. Byophis. Acta. 1972;281:625–42. https://doi.org/10.1016/0005-2787(72)90160-8.

Castro PRC, Kluge RA, Peres LEP. Manual de fisiologia vegetal: teoria e prática. Piracicaba: Agronômica Ceres; 2005.

Epstein E. Mineral nutrition of plants: principles and perspectives. New York: Wiley; 1972.

Fischer ES, Bussler W. Effects of magnesium deficiency on carbohydrates in *Phaseolus vulgaris*. Zeitschrift für Pflanzenernährung und Bodenkunde. 1988;151:295–8.

Flanery RL. Exigências nutricionais do milho em estudo de produtividade máxima. Informações Agronômicas. 1987;37:6–7.

Haeder HE, Mengel K. The absorption of potassium and sodium in dependence on the nitrogen nutrition level of the plant. Landw Forsch. 1969;23:53–60.

Huber DM, Jones JB. The role of magnesium in plant disease. Plant Soil. 2013;368:73–85.

Kabata-Pendias A, Pendias H. Trace elements in soils and plants. Boca Raton: CRC Press; 1984.

Kobayashi H, Masaoka Y, Sato S. Effects of Excess magnesium on the growth and mineral content of rice and Echinochloa. Plant Prod Sci. 2005;8:38–43. https://doi.org/10.1626/pps.8.38.

Malavolta E. Elementos de nutrição de plantas. São Paulo: Agronômica Ceres; 1980. 251p

Malavolta E, Vitti GC, Oliveira SA. Avaliação do estado nutricional das plantas: princípios e aplicações. Piracicaba: Associação Brasileira de Potassa e do Fósforo; 1997. 319p

Malavolta E, Favarin JL, Malavolta E, et al. Nutrients repartition in the coffee branches, leaves and flowers. Pesq Agropec Bras. 2002;37:1017–22. https://doi.org/10.1590/S0100-204X2002000700016.

Marschner H. Mineral nutrition of higher plants. London: Academic; 1986.

Mass EV, Moore DP, Mason B. Influence of calcium and magnesium on manganese absorption. Plant Physiol. 1969;44:796–800. https://doi.org/10.1104/pp.44.6.796.

Mcswain BD, Tsujimoto HY, Arnon DI. Effects of magnesium and chloride ions on light-induced electron transport in membranes fragments from a blue-green alga. Biochim. Biophys Acta. 1976;423:313–22. https://doi.org/10.1016/0005-2728(76)90188-2.

Munoz Hernandez RJ, Silveira RI (1998) Effect of base saturation, Ca:Mg ratios in soil and levels of phosphorus on the mineral nutrition and dry matter production of corn (*Zea mays* L). Sci Agric. 1998;55:79–85. https://doi.org/10.1590/S0103-90161998000100014.

Peaslee DE, Moss DN. Photosynthesis in K and Mg-deficient maize (*Zea mays* L.) leaves. Soil Sci Soc Am J. 1966;30:220–3. https://doi.org/10.2136/sssaj1966.03615995003000020023x.

Pierce J. Determinants of substrate specificity and the role of metal in the reaction of ribolosebisphosphate carboxylase/oxygenase. Plant Physiol. 1986;81:943–5. https://doi.org/10.1104/pp.81.4.943.

Portis Junior AR, Heldt HW. Light-dependent changes of the Mg^{2+} concentration in the stroma in relation to the Mg^{2+} depending of CO_2 fixation in intact chloroplasts. Biochim Biophys Acta. 1976;449:434–46. https://doi.org/10.1016/0005-2728(76)90154-7.

Sugiyama T, Nakayama N, Akazawa T. Structure and function of chloroplast proteins. V. Homo- tropic effect of bicarbonate in RuBP carboxylase relation and the mechanism of activation by magnesium ions. Arch Biochem Biophys. 1968;126:737–45. https://doi.org/10.1016/0003-9861(68)90465-7.

Taiz L, Zeiger E. Fisiologia vegetal. Porto Alegre: Artmed; 2004.

Tewari RK, Kumar P, Sharm PN. Magnesium deficiency induced oxidative stress and antioxidant responde in mulberry plants. Sci Hortic. 2006;108:7–14. https://doi.org/10.1016/j.scienta.2005.12.006.

Tomaz MA, Silva SR, Sakiyama NS, et al. Efficiency of uptake, translocation and use of calcium, magnesium and sulphur in young *Coffea arabica* plants under the influence of the rootstock. Rev Bras Ciênc Solo. 2003;27:885–92. https://doi.org/10.1590/S0100-06832003000500013.

Vitti GC, Lima E, Cicarone F. Cálcio, magnésio e enxofre. In: Fernandes MS, editor. Nutrição mineral de plantas. Viçosa: Sociedade Brasileira de Ciência do Solo; 2006. p. 299–325.

Chapter 10
Boron

Keywords B availability · Boron uptake · Boron transport · Boron redistribution · Assimilatory boron · Boron deficient

Boron is not one of the micronutrients that most limits crop yields due to its low soil availability. Thus, tropical crops have a high productive response to its application. In this chapter, we will discuss initially (i) basic aspects of B in the soil; (ii) B uptake, transport, and redistribution; (iii) B metabolism; (iv) nutritional requirements for B in crops; (v) B extraction, export, and accumulation by the main crops; (vi) B deficiency and toxicity symptoms.

10.1 Introduction

Soil organic matter is the main B source for plants. Soils with low organic matter content and/or low mineralization of organic matter (humidity, temperature, etc.) may have critical B concentration in the critical soil for adequate plant nutrition. In addition, regions with high rainfall and/or sandy soils can leach B from the soil solution, causing deficiency.

Thus, it is important to control factors affecting B availability in the soil, in order to maintain adequate B levels in the soil. Whenever B concentration is low (<0.20 mg dm^{-3}) or even medium (0.20–0.60 mg dm^{-3}) (hot water extraction) (Raij et al. 1996), there is potential response of plants in general to B application, which may range depending on crop nutritional requirement.

When studying boron in the plant system, it is important to know all the compartments the nutrient passes in the soil solution and in the root and shoot (leaves/fruits), that is, from the soil until its incorporation into an organic compound or as enzyme activator, contributing to vital functions that enable maximum dry matter accumulation in the agricultural product (grain, fruit, etc.) (Fig. 10.1).

© Springer Nature Switzerland AG 2021
R. de Mello Prado, *Mineral nutrition of tropical plants*,
https://doi.org/10.1007/978-3-030-71262-4_10

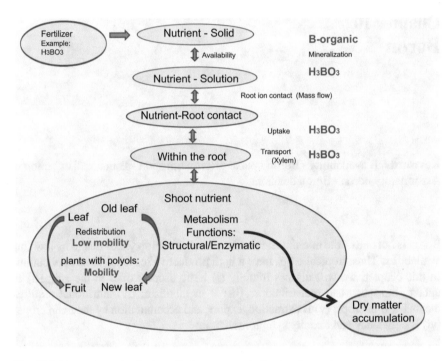

Fig. 10.1 Boron dynamics in the soil–plant system, indicating the nutrient passage processes in the different plant compartments

10.2 Uptake, Transport, and Redistribution of Boron

Uptake

Contact between B and root occurs basically through mass flow, which is affected by the plant transpiratory rate. Boron is absorbed via roots in the H_3BO_3, $H_2BO_3^-$, and $B(OH)_4^-$ forms. However, most of B is absorbed in the form of H_3BO_3 (Oertli and Grgurevic 1975). As B uptake does not depend on temperature and is not affected by respiration inhibitors, we infer that it is a passive process.

B in the form of B $(OH)_3$ seems to be the only nutrient with high permeability that overcomes membranes (cytoplasm and tonoplasts) passively, without the need for an intermediate process mediated by a protein (Welch 1995). B passive uptake occurs through diffusion, as it is transformed into compounds that are complexed in the cell wall or cytoplasm soon after it enters the cell, becoming non-exchangeable, decreasing its internal concentration in cells, and favoring its diffusion gradient from the external to the internal medium.

The location of boron application can also affect the amount absorbed. Boaretto (2006), who studied application of 1 kg ha^{-1} of B in the soil and in the citrus leaf in production, found that the amount absorbed was 65 and 17 g ha^{-1} of B for application to soil and leaf, respectively. The authors concluded that B uptake efficiency by the roots is approximately 3.5 times higher than uptake efficiency by leaves.

The presence of B affects the activity of specific components of the membrane and can increase the capacity of the root to absorb P, Cl, and K (Malavolta 1980).

Transport

Once absorbed, boron in H_3BO_3 form is transported unidirectionally through the transpiratory current.

Redistribution

Boron is practically not redistributed in plants in general, as it is immobile in the phloem. B having a quite limited redistribution in most crops has the following practical implications:

(a) Deficiency symptoms appear in new leaves or growth regions;
(b) The plant needs a continuous supply to live;
(c) B must be applied in the soil to reach the plant root system, consequently supplying its absorption and transport to the shoot.

Recent research indicates that B can be mobile in certain plants that produce polyols (simple sugar—primary photosynthetic metabolite) that bind with B, forming a sugar–boron complex that is mobile in the plant (Brown and Hu 1998). Plants that do not have these polyols in the phloem but have sucrose do not have mobile boron in the phloem. According to Marschner (1997), sucrose does not from stable complexes with B for not having cis-diol bonds. Thus, some plants have polyols in the phloem sap of species of the genus *Pyrus, Malus,* and *Prunus*, according to Hu and Brown (1997) and Hu et al. (1996), resulting in B mobility in plants. B was also considered mobile in other species (apple, plum, and cherry) (Brown and Hu 1998); (broccoli) (Shelp 1988).

Studies on boron mobility in plants can be performed using isotopic technique. Research with *Vigna unguiculata L.* with isotope [10]B applied exclusively via leaves ([10]B-L), via leaf and substrate ([10]B-L+10B-S), and exclusively via substrate ([10]B-S) observed that part of the [10]B was redistributed to the new leaf in 62%, 59%, and 21%, respectively. Thus, foliar spraying was effective to enrich leaves that emerged after application of the treatments, indicating mobility in the phloem (Fig. 10.2).

Nutritional status may affect boron mobility. Boaretto (2006) observed in citrus seedlings that B redistribution was close to 20% in deficient plants. In plants with adequate nutrient content, the redistribution doubled (i.e., approximately 40%). Prado et al. (2013), using the same technique, found that B is near immobile in beetroot and tomato plants.

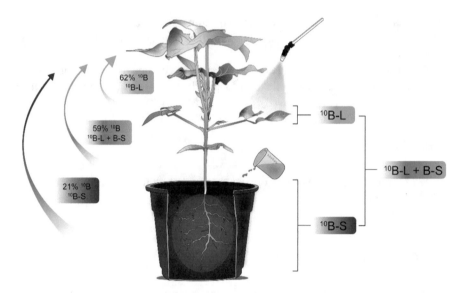

Fig. 10.2 Percentage of ^{10}B applied exclusively via leaves (^{10}B-L), via leaf and substrate (^{10}B-L+10B-S), and exclusively via substrate (^{10}B-S) redistributed to the new leaf that emerged after application of treatments

Finally, we note that boron mobility also depends, at least in part, on application and, possibly, on species. B accumulated in the leaf and coming from the soil would be immobile, while B stored in the leaf from foliar applications would be mobile (Malavolta 2006).

10.3   Participation in Plant Metabolism

Boron is the only nutrient that had its essentiality determined by indirect method (Marschner 1997). However, it is currently accepted that B also satisfies the direct criterion, as it activates several enzymes and acts as a constituent of the cell wall. According to Epstein and Bloom (2006), the main functions of B are related to cell wall structure and pectic substances associated with it, especially to the middle lamella. Therefore, we will discuss the various roles that B plays in plant life.

Cell Wall Synthesis and Cell Elongation

Boron deficiency changes the synthesis of compounds forming the cell wall (pectin, hemicelluloses, and lignin precursors) (carbohydrate metabolism), being found in cis-borate ester complexes. In B-deficient plants, there are irregularities in the

Fig. 10.3 Cambium cells
of rubber plant without
boron, indicating detached
cells without the presence
of middle lamella
constituents (**a**) and with
boron, indicating cells with
the presence of middle
lamella constituents (**b**)

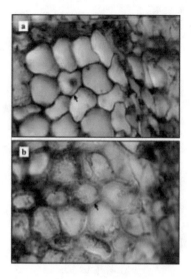

deposition of cementing substances in cambium cells, which can collapse the stem
of plants such as rubber tree (Moraes et al. 2002, Fig. 10.3).

B acts in the biosynthesis of the cell wall, helping Ca in the deposition and formation of pectates that are part of these structures. Plants that required more B are
those that contain more B content complexed in the cell wall.

General aspects of deficiency occur initially with reduced development (elongation) of meristematic tissues (extremities of root and branches), which become disorganized, causing tissue death. However, Cohen and Lepper (1977) considered that
cessation of pumpkin root elongation submitted to boron omission was caused by
the lack of division of meristematic cells and not by lack of cell elongation, suggesting that this micronutrient acts as cell division regulator. We believe that B influences cell division processes, changing AIA level through activation of enzymes
that oxidize this hormone.

B participates in the synthesis of the uracil nitrogenous base, which for being a
RNA component, decreases RNA synthesis and, consequently, protein synthesis. As
syntheses of RNA, ribose, and protein are the most important processes in meristematic tissues, cell division and differentiation are seriously impaired, affecting the
growth of young plant parts (Mengel and Kirkby 1987). Even root growth can be
paralyzed after 48 h of B omission (Amberger 1988). B deficiency promotes rapid
hardening of the cell wall, as it forms complexes with carbohydrates and controls
the disposition of cellulose micelles, not allowing the cell to increase volume above
normal (Malavolta 1980). Plant elongation is also hampered, as B directly affects
the formation of xylemic vessels (growth and differentiation).

In legumes, lack of B must affect cell wall synthesis of root nodules, allowing
O_2 flow and reducing biological N fixation (Blevins and Lukaszewski, 1998). In
Vigna unguiculata (L) plants, B deficiency decreased root development (Fig. 10.4).

Fig. 10.4 Root system of *Vigna unguiculata* (L) plants grown in nutrient solution, where the left side was supplied with B and the right side is B-deficient

B concentrations above normal protect root growth in situations where high Al levels would normally be inhibitors (Lenoble et al. 2000). Ruiz et al. (2006) add that Al toxicity can be minimized by the effect of B on enzymes related to glutathione (GSH) metabolism, which is considered an important mechanism of plants to mitigate environmental stress. B stimulates GSH biosynthesis in leaves and GSH is transported to the roots, reducing active oxygen action, normally produced by Al toxicity.

Sotiropoulos et al. (2006) also add that increased B concentrations in the culture medium increased enzyme activity in apple leaves (peroxidase, catalase, and superoxide dismutase).

Membrane Integrity

B effects are restricted only to plasmalema and not to the tonoplast, forming plasmalema constituents with cis-diol borate complexes. B deficiency decreases K and P uptake, ATPase activity of plasmalema, RNAase activity, and content of

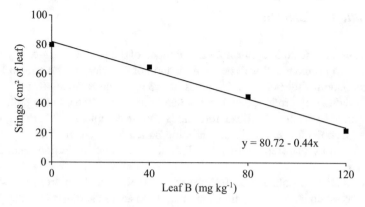

Fig. 10.5 Relationship of foliar B and attack intensity of red mite in palm oil seedlings 20 days after infestation

phospholipids and galactolipids; decreases protein content in the membranes; damages plasmalemma formation and stability; decreases the function of protein channels; and increases solute efflux. B-deficient plants accumulate phenolic compounds (deviation from glucose metabolism), leading to polyphenol oxidase enzyme action, which produces toxic quinones responsible for the formation of reactive oxygen species. In addition, glutathione reductase which could promote detoxification is inhibited, peroxidizing lipids and affecting membrane integrity.

There is a relationship between B and pest incidence (Rajaratnam and Hock 1975). Correlation between B doses and mite incidence is linearly negative (Fig. 10.5). This is explained by the fact that B plants have a plasmalema with increased efflux of free amino acids, which may serve as food for eventual parasites, while plants with adequate B content have increased plasmalema integrity, less available food, and decreased pest infestation.

Carbohydrate Transport

B facilitates sugar transport as it combines with carbohydrates, giving membranes an ionizable borate–sugar complex, which is more soluble (Gauch and Dugger Jr 1953). B has a beneficial effect on maintaining the structure and functioning of the conducting vessels. B deficiency reduces sucrose transport from leaves to other plant parts due to increased callose production (polysaccharide similar to cellulose), causing phloem obstruction which is the main transport pathway for sucrose (Loué 1993). B-deficient plants decrease carbohydrate transport, accumulating in leaves, as growth points (drains) have decreased requirement or metabolic activity.

We note that adequate B content in plants increases sugar excretion, facilitating mycorrhization.

Reproductive Growth

Germination of the pollen grain and development of the pollen tube are very dependent on B to promote cell wall deposition, and not elongation. The critical B level for germination of the pollen grain ranges from 3 μg^{-1} (maize) to 50–60 μg^{-1} (grapevine). B deficiency produces unviable pollen grains (Fig. 10.6). In the pollen tube, B is required to inactivate callose, forming B-callose complexes. Otherwise, phytoalexins (phenols) are synthesized, inhibiting pollen tube growth. Thus, B requirement by plants is often more critical in the reproductive period than in the vegetative period.

In *Petunia*, the pollen tube follows the natural B gradient from the stigma, through the stylo to the avian, indicating that it can act as chemostatic agent (Blevis and Lukaszewski 1998).

Lima Filho (1991) found in coffee plant that the boron supply increased the number of viable buds and plagiotropic branches, reducing the number of undeveloped buds. In plants of *Vigna unguiculata* L, B deficiency darkens pods and decreases pod and grain numbers, impairing their quality compared to a plant with B sufficiency (Fig. 10.7).

In mango tree, boron also has a positive effect on the reproductive part, increasing the number of fruits per panicle, besides fruit dry matter (Singh and Dhillon 1987). Wojcik (2006) studied foliar boron application (1.2 kg ha^{-1}) in post-harvesting of apple (3–4 weeks before leaf abscission), associated or not with urea. They observed that boron did not affect fruit quality (firmness and soluble solids), although improving plant reproductive growth without the addition of urea.

Silva et al. (1995) observed beneficial relationship between foliar B content and production in cotton plant, which was one of the most responsive crops to B application. Prado et al. (2006) also observed direct relationship between B content in the shoot and root with dry matter production in passion fruit seedlings.

B also affects such ascorbate metabolism, as it stimulates the activity of an enzyme (NADH oxidase) that acts on ascorbate production. B also participates in auxin metabolism, although results are contradictory. For example, high AIA levels

Fig. 10.6 Viable pollen grain, with B supply, and non-viable, with boron deficiency

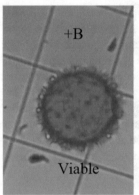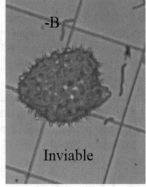

Fig. 10.7 Aspects of pods and grains of *Vigna unguiculata* (L) grown with nutritive solutions with boron deficiency and sufficiency

are only associated with B deficiency in plants where the scarcity of this element stimulates the production of certain phenolic compounds (caffeic acid) that inhibit AIA oxidase. Otherwise, this deficiency decreases AIA level (Marschner 1995). We highlight that the binding of B with the phenolic compound caffeic acid, besides blocking the formation of toxic quinine, favors the synthesis of phenolic alcohols that are lignin precursors.

Boron can affect the biological N fixation. The presence of oxygen can completely inhibit nitrogenase activity. Thus, a physiological mechanism determines the oxygen amount responsible for the process of cellular respiration without compromising enzyme activity. This mechanism may be related to B action in the cell wall, whose importance lies in the relationship between the membranes of the legume and the rhizobia (Bolaños et al. 1996). According to Cakmak and Römheld (1997), B contributes to strengthen the barrier in relation to the diffusion of O_2 by interacting with OH groups of glycopeptides in the polysaccharide layer of the envelope, protector of nitrogenase.

B is also involved in N metabolism, as B deficiency accumulates N amino acids in relation to N protein due to B action in the synthesis of nucleic acids. B-deficient plants have increased RNAase activity, hydrolyzing RNA, and decreased uracil synthesis (nitrogenous base) from orotic acid, which is a precursor of thiamine and cytosine.

10.4   Crop Nutritional Requirements

B and Zn are the micronutrients that most limit crop production in tropical regions.

Boron concentration in tissues of monocots ranges from 6 to 18 mg kg^{-1}, while ranging from 20 to 60 mg kg^{-1}in dicots. B deficiency in the monocot group has decreased incidence in the field. Cotton, brassica, sunflower, soy, peanut, and alfalfa are, in general, more susceptible to boron deficiencies than grasses. In addition, these are the plants that form tuberous roots (carrots, beetroots, and potatoes) and latex producers (euphorbiae) (Marschner 1995). Therefore, there is a direct relationship between B content in leaves and crop production, such as in cotton (Silva et al. 1995). The authors found a narrow range between optimal and toxic foliar B content. Another study on cotton plants shows that B concentration in the nutrient solution for deficiency, sufficiency, and toxicity is relatively narrow (Fig. 10.8).

The total extraction of boron varies with the species and the amount is relatively small (Malavolta et al. 1997) (Table 10.1).

Sugarcane and cotton are the crops that most exported boron by total extraction by area: 300 and 165 g ha^{-1}. However, cotton was the crop that most exported B for each ton produced (130 g t^{-1}), indicating that this crop has a high B requirement. Wheat was the crop that most exported B with harvest (133 g t^{-1}). Thus, replacing boron in areas cultivated with this crop is important.

Typically, crop requirements differ according to the ability of plants to accumulate boron in cell walls. In this sense, Hu et al. (1996), who studied 14 species, observed significant correlation between species that most require B and cell wall pectin level. The authors add that, in general, more B is required by dicots compared to monocots, presumably due to increased pectin content in the first plant group.

The ability of plants to retain B in the cell wall can even hamper knowledge on plant nutritional status. Boaretto et al. (1997) warn that often non-correlation between B levels in leaves and yield can be explained by difficulty in removing the boron retained in the leaf cuticle or bound in the cell wall pectic layer, performing its metabolic function, overestimating foliar B content.

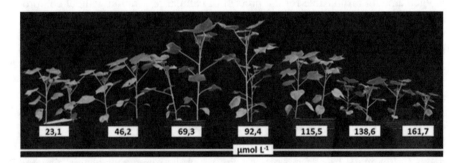

Fig. 10.8 B concentration in the nutrient solution in the growth of cotton plants inducing micronutrient deficiency, sufficiency, and toxicity

Table 10.1 Boron requirements of some crops

Crop	Plant part	Dry matter produced t ha^{-1}	B accumulated Plant part g ha^{-1}	B accumulated Total [3]	B required to produce 1 t of grains [4] g t^{-1}
Annual					
Cotton	Reproductive (cotton/cottonseed)	1.3	43 (33)[2]	165	130
	Vegetative (stem/branch/leaf)	1.7	117		
	Root	0.5	5		
Soybean[1]	Grains (pods)	3	–	100	33
	Stem/branch/leaf	6	–		
Maize[1]	Grains	6.4	20 (3.1)	80	12.5
	Crop residues	–	60		
Rice	Grains	3	6 (2.0)	107	35.6
	Stems	2	24		
	Leaves	2	34		
	Husk	1	13		
	Root	1	30		
Wheat	Grains	3	400 (133)	–	–
	Straw	3.7	–		
Perennial/semi-perennial					
Sugarcane	Stems	100	200 (2.0)	300	3.0
	Leaves	25	100		

[1] Malavolta (1980); [2] Nutrient export through the grains produced (kg t^{-1}): B accumulated in the grains/grain dry matter; [3] Suggests the total nutritional requirement of the crop for the respective yield level; [4] Suggests the relative nutritional requirement of B of the crop for production of one ton of the commercial product (grains/stems), obtained by the following formula: B accumulated in the plant (vegetative + reproductive part)/dry matter of the commercial product

Nutrient Uptake Rate

We note that boron uptake in soybean is relatively slow in the first 30 days (0.2 g/ha/day). From this period, uptake rate is high, reaching the maximum at 60–90 days (1.8 g/ha/day). From this point on, comprising the period of 90–120 days, B uptake rate sharply decreases (Bataglia and Mascarenhas 1977). In general, plants have increased B requirement in the reproductive phase.

Ross et al. (2006) observed that application of 0.3–1.11 kg B ha^{-1} during vegetative (V2) or reproductive (R2) growth was adequate for soybean production.

10.5   Symptoms of Nutritional Deficiencies and Excesses

Deficiency

As B is immobile in the phloem in several species, deficiency symptoms occur in new organs, leaves, or roots (Fig. 10.9). However, in species where B is mobile, symptoms occur in the new leaf, such as in *Vigna unguiculata* L.

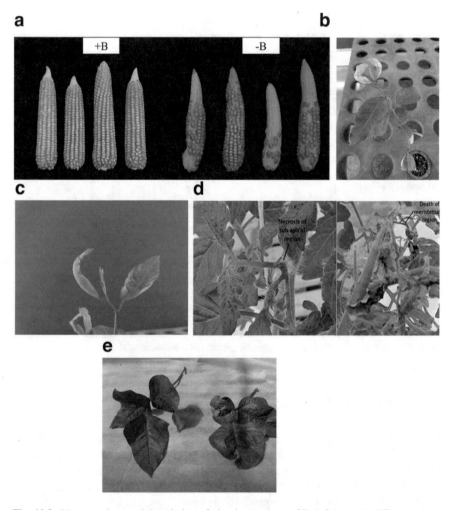

Fig. 10.9 Photos and general description of visual symptoms of B deficiency in different crops. Maize: Malformed ears with flaws in the rows and deformed grains (Lordkaew et al. 2011 Fig. 10.8a); Eucalyptus: Elongated, aqueous, or transparent streaks that subsequently turn white or dry on new leaves (Fig. 10.8b). The apical meristem deforms and dies (Fig. 10.8c); Tomato: Necrosis and death of the meristematic and subapical region. Distorted leaves with chlorosis in the leaf blade, especially at the margins (Milagres et al. 2019 Fig. 10.8d); Green bean: Plant with adequate B supply and another with new leaves deformed (Fig. 10.9e)

B deficiency initially causes relatively quickly metabolic changes in plants, including reducing uracil synthesis, which alters protein synthesis.

B roles in plant life help explain the following deficiency symptoms:

- Inhibition of shoot and root growth and even death of the terminal buds (possibly stimulating lateral sprouts);
- Shortening of internodes and small and deformed leaves/fruits;
- Thickened (carbohydrate accumulation), hard, and even brittle leaves;
- Decreased seed production;
- Necrotic leaves: excessive accumulation of phenols and AIA (AIA oxidase inhibition) (Coke and Whittington 1968).
- Stem is wrinkled, cracked, and brittle, often with spots or streaks.
- In fruits, symptoms similar to those of the stem may appear, and substances may accumulate in the albedo, as conduction of sugars to the roots is reduced.
- The presence of protruding veins may be caused by increased lignin. In B-deficient plants, there is decreased complexation of borate ions with phenols, particularly caffeic acid, increasing production of phenolic alcohols, which are part of the lignin structure (Pilbeam and Kirkby 1983).

Furthermore, B-deficient plants can affect Ca and K uptake and translocation (Ramon et al. 1990).

Toxicity

In the literature, it is common to find reports of the narrow limit between adequate and toxic B dose. Therefore, toxicity risks increase in the field, especially in sandy soils. However, toxicity symptoms can occur with increased B contents, depending on species, as they have transport speeds to the shoot, a fact that defines toxicity.

Typically, symptoms are characterized by chlorotic mottling (200 mg kg^{-1}) and necrotic spots (> 1500 mg kg^{-1}) on the edges of older leaves (regions of B accumulation), due to increased respiration rate in these sites, evolving to the leaf center (Fig. 10.10). Critical toxicity levels can range according to species, from 100 mg B kg^{-1} in maize to 1000 mg B kg^{-1} in pumpkin (Marschner 1995) and 444 mg B kg^{-1} in (old) leaves of citrus cv. Navelina (Papadakis et al. 2004). We note that B toxicity can be confounded with K deficiency, and according to Pereira et al. (2005), it is also similar to Ca deficiency, with burning of leaf mardings (in lettuce), besides inducing Zn deficiency.

Fageria (2000) evaluated the effects of B in the soil and in the shoot of several annual crops, confirming that range of adequate and toxic doses are narrow, 0.4–4.7 and 3–8.7 mg kg^{-1} of B, respectively. Toxic contents of B in the shoot are between 20 and 153 mg kg^{-1} of B for the five crops evaluated.

Fig. 10.10 Evolution of severity of B toxicity symptom in soybean plants

References

Amberger A. Pflanzenernährung: Ökologische und phisiologische Grundlagen Dynamik und Stoffwechsel der Nährelemente. Stuttgart: Eugen Ulmer; 1988.

Bataglia OC, Mascarenhas HAA. Absorção de nutrientes pela soja. Campinas: Instituto Agronômico; 1977.

Blevins DG, Lukaszewski KM. Boron in plant structure and function. Annu Rev Plant Physiol Plant Mol Biol. 1998;49:481–500. https://doi.org/10.1146/annurev.arplant.49.1.481.

Boaretto RM. Boro (10B) em laranjeira: absorção e mobilidade. Piracicaba, 120p. Tese, Centro de Energia Nucelar na Agricultura; 2006.

Boaretto AE, Tiritan CS, Muraoka T. Effects of foliar applications of boron on citrus fruit and on foliage and soil boron concentration. In: Bell RW, Rerkasem B, editors. Boron in soils and plants. Dordrecht: Kluwer Academic Publishers; 1997. p. 121–3.

Bolaños L, Brewin NJ, Bonilla I. Effects of boron on rhizobium-legume cell-surface interactions and nodule development. Plant Physiol. 1996;110:1249–56. https://doi.org/10.1104/pp.110.4.1249.

Brown PH, Hu H. Manejo do boro de acordo com sua mobilidade nas diferentes culturas. Informações Agronômicas. 1998;84:13.

Cakmak I, Römheld V. Boron deficiency-induced impairments of cellular functions in plants. Plant Soil. 1997;193:71–83. https://doi.org/10.1023/A:1004259808322.

Cohen MS, Lepper R. Effects of boron on cell elongation and division in squash roots. Plant Physiol. 1977;59:884–7. https://doi.org/10.1104/pp.59.5.884.

Coke L, Whittington WJ. The role of boron in plant growth. IV. Interrelationships between boron and Indol-3yl-acetic acid in the metabolism of bean radicles. J Exp Bot. 1968;19:295–308.

Epstein E, Bloom A. Nutrição mineral de plantas: princípios e perspectivas. Português editor: Maria Edna Tenório Nunes. Londrina: Planta; 2006.

Fageria NK. Adequate and toxic levels of boron for rice, common bean, corn, soybean and wheat production in cerrado soil. Rev Bras Eng Agríc Ambient. 2000;4:57–62. https://doi.org/10.1590/S1415-43662000000100011.

Gauch HG, Dugger WM Jr. The role of boron in the translocation of sucrose. Plant Physiol. 1953;28:457–66.

Hu H, Brown PH. Absorption of boron by plants roots. In: Dell B, Brown PH, Bell RW, editors. Boron in soils and plants: reviews. Dordrecht: Kluwer Academic; 1997. p. 49–58.

Hu H, Brown PH, Labavitch JM. Species variability in boron requirement is correlated with cell wall pectin. J Exp Bot. 1996;47:227–32.

Lenoble ME, Blevins DG, Miles RJ. Extra boron maintains root growth under toxic aluminum conditions. Informações Agronômicas. 2000;92:3–4.

Lima Filho OF. Calibração de boro e zinco para o cafeeiro (*Coffea arabica* L. cv. catuaí amarelo). Dissertação, Centro de Energia Nuclear na Agricultura; 1991.

Lordkaew S, Dell B, Jamjod S, et al. Boron deficiency in maize. Plant Soil. 2011;342:207–20. https://doi.org/10.1007/s11104-010-0685-7.

Loué A. Oligoelements en agriculture. Paris: SCPA Nathan; 1993. 577p

Malavolta E. Elementos de nutrição de plantas. São Paulo: Agronômica Ceres; 1980. 251p

Malavolta E. Manual de nutrição mineral de plantas. São Paulo: Agronômica Ceres; 2006.

Malavolta E, Vidal AA, Gheller AC, et al. Effects of the deficiencies of macronutrients in two soybean (*Glycine max* (L.) Merr.) varieties, Santa Rosa and UFV-1 grown in nutrient solution. An Esc Super Agric Luiz de Queiroz. 1980;37:473–84. https://doi.org/10.1590/S0071-12761980000100030.

Malavolta E, Vitti GC, Oliveira SA. Avaliação do estado nutricional das plantas: princípios e aplicações. Piracicaba: Associação Brasileira de Potassa e do Fósforo; 1997. 319p

Marschner H. Mineral nutrition of higher plants. London: Academic Press; 1995.

Marschner H. Functions of mineral: micronutrients. In: Marschner H, editor. Mineral nutrition of higher plants. San Diego: Academic Press; 1997. p. 313–404.

Mengel K, Kirkby EA. Principles of plant nutrition. Worblaufen-Bern: International Potash Institute; 1987.

Milagres CC, Maia JTLS, Ventrella MC, et al. Anatomical changes in cherry tomato plants caused by boron deficiency. Braz J Bot. 2019;42:319–28. https://doi.org/10.1007/s40415-019-00537-y.

Moraes LAC, Moraes VHF, Moreira A. Relação entre flexibilidade do caule de seringueira e a carência de boro. Pesq Agrop Brasileira. 2002;37:1431–6. https://doi.org/10.1590/S0100-204X2002001000011.

Oertli JJ, Grgurvic E. Effect of pH on the absorption of boron by excised barley roots. Agron J. 1975;67:278–80. https://doi.org/10.2134/agronj1975.00021962006700020028x.

Papadakis IE, Dimassi KN, Bosabalidis AM, et al. Effects of B excess on some physiological and anatomical parameters of 'Navelina' orange plants grafted on two rootstocks. Environ Exp Bot. 2004;51:247–57. https://doi.org/10.1016/j.envexpbot.2003.11.004.

Pilbeam DJ, Kirkby EA. The physiological role of boron in plants. J Plant Nutr. 1983;6:563–82. https://doi.org/10.1080/01904168309363126.

Prado RM, Natale W, Rozane D. Boron application the nutricional status and dry matter production of passion fruit cuttings. Rev Bras Frutic. 2006;28:305–9. https://doi.org/10.1590/S0100-29452006000200034.

Prado RM, et al. Applying boron to coconut palm plants: effects on the soil, on the plant nutritional status and on productivity boron to coconut palm trees. J Soil Sci Plant Nutr. 2013;13:79–85. https://doi.org/10.4067/S0718-95162013005000008.

Pereira C, Junqueira AMR, Oliveira SA. Nutrients balance and tip burn incidence in lettuce from hydroponic system – NFT. Hortic Bras. 2005;23:810–814. https://doi.org/10.1590/S0102-05362005000300024.

Raij BV, Cantarella H, Quaggio JA, et al. Recomendações de adubação e calagem para o Estado de São Paulo. Campinas: Instituto Agronômico & Fundação IAC; 1996.

Rajaratnam JA, Hock LI. Effect of boron nutrition on intensity of red spider mite attack on oil palm seedlings. Exp Agric. 1975;11:59–63.

Ramon AM, Carpena-Ruiz RO, Gárate A. The effects of short term deficiency of boron on potassium, calcium and magnesium distribution in leaves and roots of tomato (*Lycopersicon esculentum*) plants. Dev Plant Soil Sci. 1990;21:287–90.

Rosolem CA, Leite VM. Coffee leaf and stem anatomy under boron deficiency. Rev Bras Ciênc Solo. 2007;31:477–83. https://doi.org/10.1590/S0100-06832007000300007.

Ross JR, Slaton NA, Brye K, et al. Boron fertilization influences on soybean yield and leaf and seed boron concentrations. Agron J. 2006;98:198–205. https://doi.org/10.2134/agronj2005-0131.

Ruiz JM, Rivero RM, Romero L. Boron increases synthesis of glutathione in sunflower plants subjected to aluminum stress. Plant Soil. 2006;279:25–30. https://doi.org/10.1007/s11104-005-7931-4.

Shelp BJ. Boron mobility and nutrition in brocoli (Brassica oleracea var. italica). Ann Bot. 1988;61:83–91. https://doi.org/10.1093/oxfordjournals.aob.a087530.

Silva NM, Carvalho LH, Kondo JI, et al. Ten years of cotton fertilization with boron. Bragantia. 1995;54:177–85. https://doi.org/10.1590/S0006-87051995000100020.

Singh Z, Dhillon BS. Effect of foliar application of boron on vegetative and panicle growth, sex expression, fruit retention and physicochemical characters of fruits of mango (*Mangifera indica* L.) cv. Dusehri. Trop Agric. 1987;64:305–8.

Sotiropoulos TE, Molassiotis A, Almaliotis D, et al. Growth, nutritional status, chlorophyll content, and antioxidant responses of the apple rootstock mm 111 shoots cultured under high boron concentrations in vitro. J Plant Nutr. 2006;29:575–83. https://doi.org/10.1080/01904160500526956.

Welch RM. Micronutrient nutrition of plants. Crit Rev Plant Sci. 1995;14:49–82. https://doi.org/10.1080/07352689509701922.

Wojcik P. Effect of postharvest sprays of boron and urea on yield and fruit quality of apple trees. J Plant Nutr. 2006;29:441–50. https://doi.org/10.1080/01904160500524894.

Chapter 11
Zinc

Keywords Zn availability · Zinc uptake · Zinc transport · Zinc redistribution · Assimilatory zinc · Zinc deficient

Zinc is one of the most limiting micronutrients to crops due to its low availability in tropical soils. It is paramount in enzymatic processes vital to plant metabolism and high productivity. In this chapter, we will discuss initially (i) basic aspects of Zn in the soil; (ii) Zn uptake, transport, and redistribution; (iii) Zn metabolism; (iv) nutritional requirements for Zn in crops; (v) Zn extraction, export, and accumulation by the main crops; (vi) Zn deficiency and toxicity symptoms.

11.1 Introduction

Zn is a limiting micronutrient to most crops due to its low concentration in the soil, as it is often adsorbed to clay (such as hydrated Fe_2O_3), representing 30–60% of the total, and the other part is adsorbed to organic matter. In the soil solution, most of the zinc is in the form of soluble organic complexes. In addition to the issue of low Zn concentration in the soil, several factors greatly affect its availability. The most important is the soil pH: The higher it is, the lower Zn availability in the soil solution, especially in sandy soils that have received high doses of limestone. In addition to pH, high doses of phosphate fertilizers can induce Zn deficiency problems. Soils with a high organic matter content can adsorb Zn, or the microbiota can temporarily immobilize it. In addition, soils with a high moisture, associated with low temperatures, can also temporarily decrease Zn availability. In soils with an acidic subsurface layer and with a low Zn content, gypsum with Zn residues will have a beneficial effect on agricultural production (Alonso et al. 2006). Thus, it is important to know the factors that affect the availability of Zn in the soil as they affect plant absorption and nutrition.

The factors that affect the availability of Zn in the soil must be controlled in order to maintain the Zn concentration at levels appropriate to crops. Whenever the concentration is low (<0.7 mg dm^{-3}) or even medium (0.7–1.5 mg dm^{-3}) (DTPA extractor) (Raij et al. 1996), there is in general a potential response of plants to the

© Springer Nature Switzerland AG 2021
R. de Mello Prado, *Mineral nutrition of tropical plants*,
https://doi.org/10.1007/978-3-030-71262-4_11

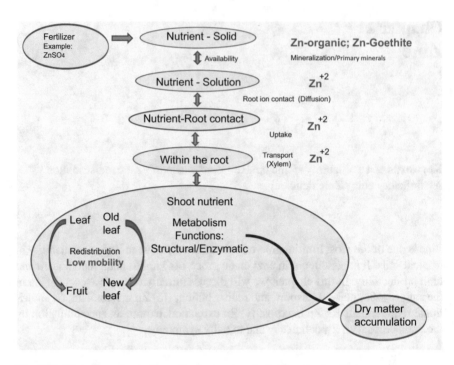

Fig. 11.1 Zinc dynamics in the soil–plant system indicating the nutrient passage processes through the different compartments of the plant

application of this micronutrient, which varies due to the crop's nutritional requirements.

In the study of zinc, it is important to know all the "compartments" that the nutrient travels in the soil and the plant system. The latter is an important factor in understanding its role in plant physiology and in the formation of crops (Fig. 11.1).

11.2   Uptake, Transport, and Redistribution of Zinc

Uptake

The Zn-root contact is explained especially by diffusion, which depends on the concentration gradient close to the root (Oliver and Barber 1966). In Oxisol, it was found that the movement of Zn depends on diffusion (Oliveira et al. 2010). Zinc is absorbed by roots as Zn^{2+} actively by a carrier, such as ZNT1, identified by Pence et al. (2000). ZIPn genes not only act on the absorption of Zn, especially in its deficiency but also allow the entry of Mns and Cu, but not in Fe. However, at a high pH value, it can be absorbed in the monovalent form $ZnOH^+$. Other nutrients (cations) at high concentrations in the medium can competitively inhibit the absorption of Zn.

Also, the accompanying ion can increase the absorption of Zn obeying the following decreasing order when applied to the root: chelate (lignosulfonate) > nitrate = sulfate > chloride, or to the leaf: chloride > nitrate = chelate > sulfate.

Zinc absorption can be affected by the cultivar. Yagi et al. (2006) found in sorghum a greater absorption of Zn by cv.BR 304 compared to cv. BR 310.

Transport

Zinc, in the same form as the absorbed Zn^{2+}, undergoes passive and active radial transport. The Zn complexes that are transported are citric and malic acid (Malavolta 2006). Then, there is the long-distance transport, from the roots to shoots, by the xylem.

Redistribution

Due to the low stability because of organic chelators, zinc is practically not found in the phloem, as its redistribution in the plant is very limited. It is thus considered little mobile. However, in plants generally well supplied with Zn, its mobility in the phloem can increase. Franco et al. (2005) verified a greater mobility of Zn in plants well supplied with this nutrient in beans; however, in coffee, this did not occur. The Zn in the phloem is complexed to organic compounds with molecular weights between 1000 and 5000, called phytochelatins (short chain peptides containing repeated units of glutamine and cysteines) (Malavolta 2006). The author adds that Zn, as Fe, exists in two fractions in plants, active and inactive, and if the proportion of the former is greater, the nutrient will have a greater redistribution. At the seedling phase, there may be a greater redistribution of Zn, but this needs to be better understood. In leaf applications, part of Zn can be complexed with cuticle components, thus reducing its redistribution.

11.3 Participation in Plant Metabolism

Zinc has no defined structural function. However, it is an activator of several enzymes, although it can be also part of the constitution of some of them. The zinc in the plant (Zn^{2+}) is neither oxidized nor reduced (i.e., it is not subject to valence changes and has no redox activity), unlike the other micronutrients. However, it has a tendency to form tetrahedral complexes. The most well-known functions of zinc are:

Synthesis of Indoleacetic Acid (AIA)

The lack of Zn in the plant can degrade the indoleacetic acid (increase in the activity of indoleacetic acid oxidase) or reduce its synthesis. In the case of synthesis of indoleacetic acid, although there is discussion about it, it is common to indicate that Zn is required for the synthesis of the amino acid tryptophan, a precursor to the biosynthesis of indoleacetic acid. The tryptophan synthase enzyme requires Zn for its activity (Fig. 11.2).

Zn deficiency results in decreased cell volume and less apical growth due to disturbance in the metabolism of auxins, such as indoleacetic acid (reduced synthesis or degradation itself). Therefore, Zn is more important for maintaining auxin in its active state than it is relevant to its synthesis (Skoog 1940). In plants submitted to Zn deficiency, there is a drastic decrease in the concentration of auxin even before the appearance of visual symptoms.

The auxin synthesis sites are the meristematic tissues of different organs (leaves, roots, flowers, etc.). The main physiological effect is cell elongation from the breaking of (noncovalent) bonds between hemicelluloses and celluloses in the cell wall, allowing the influx of water and causing pressure on the cell wall, thus increasing its plasticity. In fact, this mechanism of elongation of the cell wall is provided by acidification. Initially, auxin induces the cytoplasm to secrete H^+ ions into the adjacent primary wall through an ATPase (it uses ATP hydrolysis energy to generate protons) located in the plasma membrane, causing a reduction in pH, which in turn activates certain enzymes (endo-transglycosylase and β-glucan synthase), which are normally inactive at a higher pH, and hydrolyzes the cell wall, causing the breakdown of cell wall bonds and a fast growth response (Castro et al. 2005). After the cellular extension is complete, these bonds are reformed (by enzymatic action), and this process is irreversible.

Protein Synthesis (RNA) and Nitrate Reduction

Another factor that may inhibit growth is that plants deficient in Zn show a great decrease in the level of RNA, which results in less protein synthesis and difficulty in cell division. This is explained by the fact that Zn inhibits RNAase (RNA

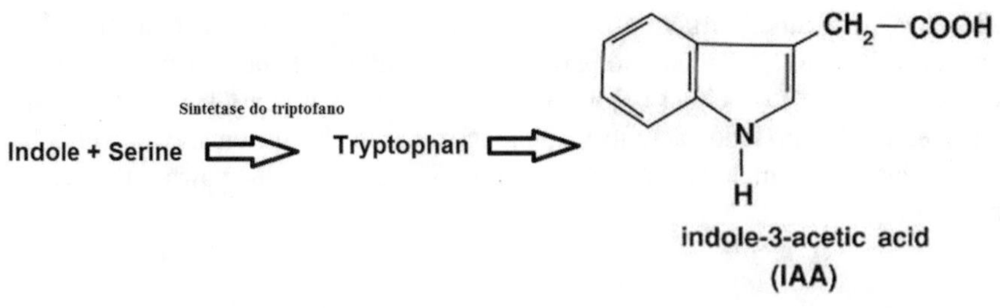

Indole + Serine ⟹ Tryptophan ⟹ indole-3-acetic acid (IAA)

Sintetase do triptofano

Fig. 11.2 Summary of metabolic reaction for indoleacetic acid synthesis

disintegrator) and is also part of the RNA polymerase, which synthesizes RNA. Zn is part of active proteins involved in DNA transcription (Takatsuji 1999). Furthermore, Zn is part of ribosomes (site of protein synthesis), and its deficiency leads to disintegration (Marschner 1986). However, with its replacement, the process is reversed (Fig. 11.3).

In plants deficient in Zn, protein and indoleacetic acid synthesis decreases (Cakmak et al. 1989). This is because there is an accumulation of amino acids associated with a decrease in indoleacetic acid, which thus decreases the production of dry matter (Table 11.1). There are indications that the main role of Zn is not the synthesis of tryptophan because, in plants deficient in Zn, there were high concentrations of tryptophan (Table 11.1). Thus, the role of Zn is in the metabolic pathway of tryptophan to auxin. The auxin acts in the initial event of cell expansion, known as the theory of acid growth (Rayle and Cleland 1992), as mentioned earlier.

Zn is also related to the metabolism of N in plants supplied with nitrate, and its deficiency leads to the accumulation of $N\text{-}NO_3$, which may decrease the synthesis of amino acids.

Enzyme Structure and Enzyme Activity

Zn, together with Cu, is part of the structure of the superoxide dismutase enzyme and others. The maintenance of superoxide dismutase activity is important because it decomposes oxidative radicals (O_2^-) produced from oxygen, protecting the cell from its harmful effects (degradation of AIA, oxidation of lipids), which in turn can affect the integrity of membranes. In addition, hydrogen peroxide may be produced, which catalase can detoxify ($H_2O_2 => H_2O + \frac{1}{2}O_2$) activated by Zn. Zn also participates in the composition of the RNA polymerase enzyme. In addition, the enzyme alcohol dehydrogenase catalyzes the reduction of acetaldehyde to ethanol. Another

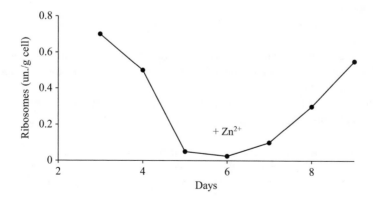

Fig. 11.3 Effects of omission (for six days) and reapplication of Zn on the number of ribosomes in Euglena

Table 11.1 Effects of zinc on the production of bean dry matter and on the content of organic components in the apical part of the plant

Zn applied	Dry matter	Apical part of the plant (leaves + branches)				
		Zn	Amino acids	Protein	Tryptophan	Indole acetic acid
	g for 3 plants	μ g^{-1} dry matter	μ Mol g^{-1} dry matter	mg g^{-1} fresh matter	μ Mol g^{-1} dry matter	ng g^{-1} fresh matter
+Zn (1 μM)	8.2	52	82	28	0.37	239
−Zn	3.7	13	533	14	1.32	118
−Zn + Zn[a]	4.5	141	118	30	0.27	198

[a]Zn was supplied for three days (3 μM)

enzyme is carbonic anhydrase, which catalyzes the dissolution of CO_2 (CO_2 + H_2O => HCO_3^- + H^+) as a step prior to its assimilation.

Finally, Zn acts on enzymatic activity, either to complement the enzyme/substrate bond or in the effects of the conformation of molecules. Zn activates important enzymes, such as ribulose 1,5-diphosphate carboxylase (present in chloroplasts), and therefore affects the photosynthetic rate of plants. Respiration activity decreases because the key enzyme aldolase is inhibited by the lack of Zn, and ATP synthesis consequently decreases at the level of glycolysis and of terminal electronic transport (Malavolta 2006).

In addition, Zn may play some role in the formation of chlorophyll, also affecting photosynthesis.

Zn deficiency promotes loss of membrane integrity due to the high level of reactive oxygen species (low dismutase activity) and structural protein destabilization due to the breakdown of Zn bonds with sulfhydryl groups (-SH). This contributes to increase the plant's susceptibility to diseases and inhibits the redistribution of P from shoots to roots.

The main functions of Zn are in the constitution of enzymes and in metabolic processes (Table 11.2).

In the literature, the beneficial effects of Zn applied to the soil surface (20 kg ha^{-1}) have been verified in the production of annual crops, such as soybeans, through the residual effect (Ritchey 1978).

In fruit trees, Zn doses increase the shoot and root dry matter of guava seedlings (Natale et al. 2002, Fig. 11.4), passion fruit (Natale et al. 2004), and papaya (Corrêa et al. 2005).

Thus, the beneficial effects of Zn at moderate doses are established due to the known function of this metal in the synthesis of auxin, which stimulates the development and elongation of young parts of plants (Malavolta et al. 1997).

Table 11.2 Functions of zinc in plants

Enzymatic constituent	Processes
Carbonic anhydrase	Hormonal control (AIA)
Phosphomannose isomerase	Protein synthesis
Lactic dehydrogenase	
Alcoholic dehydrogenase	
Aldolase	
Glutamic dehydrogenase	
Pyruvic carboxylase	
Tryptophan synthetase	
Ribonuclease	

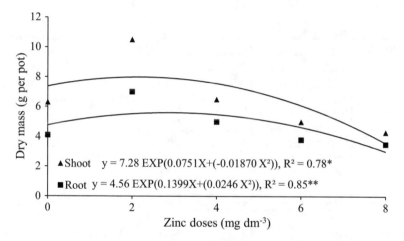

Fig. 11.4 Effects of zinc application in a substrate of Red-Yellow Argisol on the production of shoot and root dry matter of guava seedlings at 135 days after transplanting

11.4   Nutritional Requirements of Crops

Extraction and Export of Nutrients

Normally, total crop extraction is greater for grasses, such as sugarcane (720 g ha^{-1}) and corn (544 g ha^{-1}), than to legumes such as soybeans (145 g ha^{-1}) (Table 11.3).

The crop requirement of zinc does not reach 1 kg ha^{-1}. However, its presence in plants is fundamental because it has several functions in the metabolism (discussed above), so that the cost/benefit reduction of this micronutrient (and also others) in the production systems is worthwhile.

As for the requirement for zinc per unit of product harvested, the rice crop is the most demanding (122 g t^{-1}) and has the lowest nutritional efficiency. Therefore, it is important to supply this nutrient to meet its requirements and maintain a high

Table 11.3 Zinc requirements of major crops

Crop	Plant part	Dry matter produced	Accumulated Zn		Zn required for production of 1 t of grain[4]
			Plant part	Total [3]	
		t ha^{-1}	g ha^{-1}		g t^{-1}
Annual					
Cotton	Reproductive (cotton/seed)	1.3	16 (5)[2]	60	46
	Vegetative (stem/branch/leaf)	1.7 (DM)	42		
	Root	0.5 (DM)	2		
Soybean	Grains (pods)	2.4	102 (42)	145	60
	Stem/branch/leaf	5.6	43		
Beans	Grains	0.9	0.03	–	–
	Stem	0.4			
	Leaves	1.2			
	Root	0.1			
Corn [1]	Grains	6.4	178 (28)	544	85
	Crop remains	–	366		
Rice	Grains	3	30 (10)	366	122
	Stems	2	101		
	Leaves	2	38		
	Husk	1	147		
	Root	1	50		
Wheat	Grains	3	40 (13)	90	30
	Straw	3.7	50		
Semi-perennial/perennial					
Sugarcane	Stems	100	500 (5)	720	7.2
	Leaves	25	220		
Coffee tree[1]	Grains (nut)	2	80 (40)	–	–
	Trunk, branches, and leaves	–	–		

[1]Malavolta (1980); [2] Relative export of nutrients by produced grains (g t^{-1}): Zn accumulated in the grains/dry matter of grains; [3] Suggests a nutritional requirement (total) by crop area for the respective level of productivity; [4] Suggests a relative nutritional requirement of Zn of the crop to produce one ton of commercial product (grains/stalks); obtained by the formula: Zn accumulated in the plant (vegetative + reproductive part)/dry matter of the commercial product

production. The greater efficiency in the use of Zn by rice genotypes is determined by the greater efficiency in absorption, which is related to a larger root surface area (Gao et al. 2005). Soybean and coffee are the crops that most export Zn during harvest, reaching 42 and 40 g t^{-1} of grains, respectively.

New studies have emerged on the agronomic biofortification of crops to enrich the Zn content in grains/fruits (food), thus improving human nutrition by the leaf spraying of the nutrient on reproductive organs, such as in rice crops (Alvarez et al. 2019). This aspect is important because in the last decades, there has been a decrease in the levels of micronutrients in food, which fosters research in biofortification.

Nutrient Absorption Curve

The accumulation of Zn by soy is slow during the first 30 days (0.4 g ha^{-1} per day) and high after 30 days of sowing (2.0 g ha^{-1} per day). It reaches the maximum speed in the period of 60–90 days (2.1 g ha^{-1} per day); after that, at 90–120 days (1.2 g ha^{-1} per day), there is a decrease in the absorption speed at the final phase of the plant cycle (Bataglia and Mascarenhas 1977). Despite this, due to low soil mobility, its application is indicated before or even during crop sowing.

11.5 Symptoms of Nutritional Deficiencies and Excesses

Deficiencies

Normally, Zn deficiency in different species results in the shortening of internodes and the leaf becomes small. However, there are yellowish (or white) bands between the rib and the edges of leaves (Fig. 11.5). Due to the difficulty of the internodes to lengthen, successive knots form, and the leaves get closer, thus appearing the rosette symptom.

Plants deficient in Zn may lead to chlorosis induced by Fe deficiency. However, the symptoms of small leaves characterize Zn deficiency.

Plants deficient in Zn may have a high P content due to the breakdown of membranes, and in parallel, there is inhibition of the redistribution of P from the plant parts to the root. This can cause a very superficial rooting and low production.

Zn deficiency may be related to plant diseases, which could be explained by the following factors: less phenols and lignin, more sugars and free amino acids, substrate for pathogens, and disorganized walls (Beretta et al. 1986).

Excess

The toxicity of Zn is manifested by the decrease in leaf area followed by chlorosis, and a reddish-brown pigment may appear in the plant, perhaps a phenol. Furthermore, it decreases the absorption of K. In the xylem of some plants intoxicated by Zn, plug

Fig. 11.5 Photos and general description of the visual symptoms of Zn deficiency in different crops. Vine: The leaves become small and asymmetrical with internerval yellowing, the margin of leaves can become necrotic in severe deficiency (Fig. 11.5a); the grapes are small, seedless, and yellow in color (Bavaresco et al. 2010, Fig. 11.5b); Corn: Symptoms start on youngest leaves with white or yellowish bands between the main vein and the edges (Fig. 11.5c); there may be necrosis and purple tones; new leaves unfolding in the growing region are whitish or pale yellow in color and deformed; the plants have short internodes (Fig. 11.5d). The ears are small and without grains at the tip (Fig. 11.5e); Soybean: The leaflets with zinc deficiency become smaller with chlorotic areas between the ribs; these symptoms are more severe in basal leaves. Deficient chlorotic leaves are yellow-brown in color and die prematurely (Fig. 11.5f). Zn-deficient soybeans are yellow-brown in color when observed from a distance. Ripening delays and few pods are produced; Banana: Plants take on aspects of dwarfism with short internodes; the leaves have reddish-yellow spots with necrosis on the margins (Fig. 11.5g); Sugarcane: It starts with chlorotic streaks on the leaf, and then, a wide band of chlorotic tissue forms on each side of the central rib, but does not extend to the leaf margin, except in severe cases of deficiency; internerval tissues remain green initially, but soon the entire leaf blade may become chlorotic, extending to the base; noticeably short and wide leaves in the middle part and asymmetrical; leaf necrosis when the deficiency is severe; reduced tillering, shorter internodes, and thin stems (Fig. 11.5h)

buffer containing Zn accumulates, which hinders the rise of the raw sap (Malavolta et al. 1997).

In addition, excess Zn can cause symptoms similar to Fe deficiency, as there is a decrease in its absorption, in addition to P. There are plants with a high tolerance to Zn that may reach a content of 20 g of Zn kg^{-1} (Küpper et al. 1999).

Fageria (2000) studied the effects of Zn in the soil–plant system for the production of dry matter. Unlike B, adequate and toxic dose ranges of Zn are broader: from 1–10 to 40–110 mg Zn kg^{-1} of soil, respectively, while the adequate levels of zinc in the plant range from 18 to 67 mg kg^{-1} of shoot dry matter, and the toxic doses range from 100 to 673 mg kg^{-1}, depending on the crop.

The toxicity of Zn reached a content in millet shoots of 451 mg kg^{-1} (Silva et al. 2010) and for black oats 494 mg kg^{-1} (Abranches et al. 2009). In sunflower, Zn deficiency and toxicity occur with a content of 20 and 240 mg kg^{-1}, respectively

(Khurana and Chatterjee 2001). According to this last author, the excess of Zn caused a significant decrease, in addition to biomass, in the concentration of chlorophyll (a, b) and soluble proteins in plants.

In peanuts, there is Zn toxicity when the Ca:Zn ratio is equal to or lower than 50 (Parker et al. 1990).

References

Abranches JL, Batista GS, Ramos SB, et al. Response of oats to the application of zinc in Oxisol. Agrária. 2009;4:278–82. https://doi.org/10.5039/agraria.v4i3a8.

Alonso FP, Arias S, Ordónez Fernándes R, et al. Agronomic implications of the supply of lime and gypsum by-products to palexerults from western Spain. Soil Sci. 2006;171:65–81. https://doi.org/10.1097/01.ss.0000200557.253069.50.

Alvarez RCF, Souza JP, Prado RM, et al. Effects of foliar spraying with new zinc sources on rice seed enrichment, nutrition, and productivity. Acta Agric Scand Sect B Soil Plant Sci. 2019;69:511–5. https://doi.org/10.1080/09064710.2019.1612939.

Bataglia OC, Mascarenhas HAA. Absorção de nutrientes pela soja. Campinas: Instituto Agronômico; 1977.

Bavaresco L, Gatti M, Fregoni M. Nutritional deficiencies. In: Delrot S, Medrano H, Or E, Bavaresco L, Grando S, editors. Methodologies and results in grapevine research. Dordrecht: Springer; 2010. p. 165–91.

Beretta MJG, Fogaça M, Moraes WBC. Experimental induction of amorpiious-like pluos in citrus and their possible association with factors. Pesq Agrop Brasileira. 1986;21:1261–5.

Cakmak I, Marschner H, Bangerth F. Effect of zinc nutritional status on growth, protein metabolism and levels of indole-3-acetic acid and other phytohormones in bean (*Phaseolus vulgaris* L.). J Exp Bot. 1989;40:405–12. https://doi.org/10.1093/jxb/40.3.405.

Castro PRC, Kluge RA, Peres LEP. Manual de fisiologia vegetal: teoria e prática. Piracicaba: Agronômica Ceres; 2005.

Corrêa MCM, Natale W, Prado RM, et al. Adubação com zinco na formação de mudas de mamoeiro. Caatinga. 2005;18:245–50.

Fageria NK. Adequate and toxic levels of zinc for rice, common bean, corn, soybean and wheat production in cerrado soil. Rev Bras Eng Agríc Ambient. 2000;4:390–5. https://doi.org/10.1590/S1415-43662000000300014.

Franco IAL, Martinez HEP, Zabini AV et al. Translocation and compartmentation of zinc by ZnSO4 e ZnEDTA applied on coffee and bean seedlings leaves. Cienc Rural. 2005;35:332–339. https://doi.org/10.1590/S0103-84782005000200013.

Gao X, Zou C, Zhang, et al. Tolerance to zinc deficiency in rice correlates with zinc uptake and translocation. Plant Soil. 2005;278:253–61. https://doi.org/10.1007/s11104-005-8674-y.

Khurana N, Chatterjee C. Influence of variable zinc on yield, oil content, and physiology of sunflower. Commun Soil Sci Plant Anal. 2001;32:3023–30. https://doi.org/10.1081/CSS-120001104.

Kupper H, Zhao FJ, McGrath SP. Cellular compartmentation of zinc in leaves of the hyperaccumulator Thlaspi caersulescens. Plant Physiol. 1999;119:305–11.

Malavolta E. Elementos de nutrição de plantas. São Paulo: Agronômica Ceres; 1980. 251p

Malavolta E. Manual de nutrição mineral de plantas. São Paulo: Agronômica Ceres; 2006.

Malavolta E, Vitti GC, Oliveira SA. Avaliação do estado nutricional das plantas: princípios e aplicações. Piracicaba: Associação Brasileira de Potassa e do Fósforo; 1997. 319p

Marschner H. Mineral nutrition of higher plants. London: Academic Press; 1986.

Natale W, Prado RM, Corrêa MCM, et al. Response of guava to zinc application. Rev Bras Frutic. 2002;24:770–3. https://doi.org/10.1590/S0100-29452002000300052.

Natale W, Prado RM, Leal RM, et al. Effects of the zinc application on the development, nutritional status and dry matter production of passion fruit cuttings. Rev Bras Frutic. 2004;26:310–4. https://doi.org/10.1590/S0100-29452004000200031.

Oliveira EMM, Ruiz HA, Alvarez-V VH, et al. Nutrient supply by mass flow and diffusion to maize plants in response to soil aggregate size and water potential. Rev Bras Cienc Solo. 2010;34:317–27. https://doi.org/10.1590/S0100-06832010000200005.

Oliver S, Barber SA. Mechanisms for the movement of Mn, Fe, B, Cu, Zn, Al and Sr from the soil to the soil to the surface of soybean roots. Soil Sci Soc Am Proc. 1966;30:468–70. https://doi.org/10.2136/sssaj1966.03615995003000040021x.

Parker MB, Gaines TP, Walker ME, et al. Soil zinc and pH effects on leaf zinc and the interaction of the leaf calcium and zinc on zinc toxicity of peanuts. Commun Soil Sci Plant Anal. 1990;21:2319–32. https://doi.org/10.1080/00103629009368383.

Pence NS, Larsen PB, Ebbs SD, et al. The molecular physiology of heavy metal transport in the Zn/Cd hyperaccumulator *Thlaspi caerulescens*. Proc Natl Acad Sci U S A. 2000;97:4956–60. https://doi.org/10.1073/pnas.97.9.4956.

Raij BV, Cantarella H, Quaggio JA et al. Recomendações de adubação e calagem para o Estado de São Paulo. Instituto Agronômico e Fundação IAC, Campinas; 1996.

Rayle DL, Cleland RE. The acid growth theory of auxin-induced cell elongation is alive and well. Plant Physiol. 1992;99:1271–4. https://doi.org/10.1104/pp.99.4.1271.

Ritchey KD. Residual zinc effects – agronomic-economic research on tropical soils: annual report for 1976–1977. Raleigh: North Carolina State University; 1978.

Silva TMR, Prado RM, Vale DW, et al. Zinc toxicity in millet grown in a red dystrophic Oxisol. Agrária. 2010;5:336–40. https://doi.org/10.5039/agraria.v5i3a652.

Skoog F. Relationships between zinc and auxin in the growth of higher plants. Am J Bot. 1940;27:939–51. https://doi.org/10.2307/2436564.

Takatsuji H. Zinc finger proteins: the classic zinc finger emerges in contemporany plant science. Plant Mol Biol. 1999;39:1073–8. https://doi.org/10.1023/A:1006184519697.

Yagi RM, Simili FF, Araújo JC, et al. Zinc application in seeds and its effect on germination, nutrition and initial development of sorghum. Pesq Agrop Brasileira. 2006;41:655–60. https://doi.org/10.1590/S0100-204X2006000400016.

Chapter 12
Manganese

Keywords Mn availability · Manganese uptake · Manganese transport ·
Manganese redistribution · Assimilatory manganese · Manganese deficient

Manganese is one of the most absorbed micronutrients by plants, but it may have low availability in alkaline soils. It is paramount in enzymatic processes vital to the plant, providing a high productivity. In this chapter, we discuss initially (i) basic aspects of Mn in the soil; (ii) Mn uptake, transport, and redistribution; (iii) Mn metabolism; (iv) nutritional requirements for Mn in crops; (v) Mn extraction, export, and accumulation by the main crops; and (vi) Mn deficiency and toxicity symptoms.

12.1 Introduction

Unlike B and Zn, Mn is the second most abundant micronutrient in tropical soils, second only to Fe. The availability of manganese in the soil depends mainly on the pH, the redox potential, the organic matter, and the balance with other cations (Fe, Ca, and Mg). In the weathering processes of the matrix rock, the release of manganese occurs, and it is quickly transformed into the oxides Mn^{3+} (Mn_2O_3.n H_2O) or Mn^{4+} (MnO_2.n H_2O) through chemical and biological reactions. Only a small portion remains in the soil solution as the ion Mn^{2+}, which is the form available to plants (Bartlett 1988). Thus, the results of these chemical reactions, and especially the biological ones, may lead to a greater or lesser concentration of forms available to plants. Therefore, the knowledge of these reactions of Mn in the soil system allows a more adequate management of this nutrient, benefiting the nutrition of plants.

The high concentration of Mn in acidic soils, which can cause toxicity in plants, while soils receive high doses of corrective materials to correct acidity, can induce low availability of Mn, causing deficiencies in plants, especially in originally poor soils. Thus, it is important to maintain adequate concentrations of Mn in the soil. It is more advantageous to apply Mn in soils with low (<1.5 mg dm^{-3}) or medium Mn concentrations (1.5–5.0 mg dm^{-3}) (DTPA extractor) (Raij et al. 1996), in which the responses of the crops are greater. Usually, plant responses to manganese are most

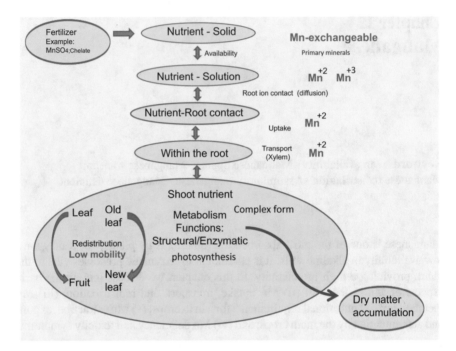

Fig. 12.1 Manganese dynamics in the soil–plant system indicating the nutrient passage processes through the different compartments of the plant

likely to occur in soils with high levels of organic matter, high pH, and limy or sandy soils (Wiese 1993) receiving high rainfall volumes.

In the study of manganese in the plant system, it is important to know all the "compartments" that Mn travels from the soil solution, from roots to shoots (leaves/ fruits), that is, from the soil until its incorporation into an organic compound or as an enzymatic activator, where it will perform vital functions to enable the maximum accumulation of dry matter in the final agricultural product (grains, fruits, etc.) (Fig. 12.1).

12.2   Uptake, Transport, and Redistribution of Manganese

Uptake

The path of manganese from the soil to the roots occurs through diffusion mechanisms, therefore requiring applications of the nutrient close to the root system of plants.

Manganese is actively absorbed by the roots as Mn^{2+}. In general, heavy metals, such as Mn, can be transported by four different families: ATPases of the type CPx,

Nramp, facilitators of diffusion of cations, and ZIP (Williams et al. 2000). In plants with Fe deficiency, IRT1 is important for the absorption of Fe^{2+} and also Mn^{2+} and Zn^{2+}. In addition, ZIPn contributes to Zn, Mn^{2+}, and Cu^{2+} absorption in deficient plants. As Mn^{2+}, it has similar chemical properties (ionic radius) as the nutrients Ca^{2+}, Fe^{2+}, Zn^{2+}, and especially Mg^{2+}. Their presence can inhibit its absorption and even transport. It should be noted that the absorption of Cu^{2+}, Zn^{2+}, and Fe^{2+} can double with Mn deficiency (Yu and Rengel 1999). However, the opposite is also true, with an emphasis on Fe^{2+}.

Finally, it should be noted that the efficiency of Mn absorption by plants submitted to low nutrient concentration is controlled genetically (Foy et al. 1988).

In crops of transgenic plants, the application of glyphosate increases oxidizing microorganisms transforming Mn^{2+} into Mn^{4+} (not absorbed), thus decreasing the availability of micronutrients in the soil. However, more research is needed to verify whether these changes are important in the absorption of Mn by plants.

Transport

The transport in the xylem is done in the cationic form Mn^{2+}, since the chelate of Mn with citric acid has a low stability constant and dissociates easily, being able to be displaced and subjected to precipitation with anions such as $H_2PO_4^-$. However, there may be chelates of Mn associated with low-molecular-weight nitrogen compounds.

Redistribution

The transport of Mn is done unidirectionally by the xylem from the roots to shoots. The opposite is very seldom because its concentration in the phloem is very low, and therefore the redistribution is limited.

The redistribution of Mn is complex in plants, as variation may occur depending on the genotype and the environment. The Mn accumulated in the leaves is not mobilized, although it can be removed by washing. However, the nutrient contained in the roots and stems can be redistributed, but its importance as a supplier of Mn varies with the species (Malavolta 2006).

12.3   Participation in Plant Metabolism

Within the cell, Mn^{2+} forms weak bonds with organic ligands and can be quickly oxidized to Mn^{3+}, Mn^{4+}, and Mn^{6+}. Because of this relative ease of changing in the oxidation state, Mn plays an important role in the oxidation processes in the plant, such as electron transport in photosynthesis and detoxification of reactive oxygen species.

Manganese is involved with plant enzyme systems either as a constituent (cofactor) or as an enzyme activator.

Mn participates directly in the chemical composition of two enzymes (S enzyme and peroxide dismutase), which perform the following functions in plants:

The S enzyme contains four Mn atoms, which performs the most well-known function of this nutrient in plants, together with chlorine, that is, the photochemical breakdown of the water molecule (Hill reaction) $2H_2O => 4H^+ + 4e + O_2$, as Mn has a high capacity for successive oxidations because its electrons are transferred to the photosystem II ($4e => 4Mn^{3+} => 4Mn^{2+} => 4e$). The reduction of photosynthesis can also be affected more by the destruction of chloroplasts, decreasing the concentration of chlorophyll (Mn is bound in the structures of the membranes of thylakoids).

When the electron flow is impaired, there is a corresponding negative effect on subsequent reactions, such as photophosphorylation, fixation of CO_2, and reduction of nitrite and sulfate.

Inhibition of the Hill reaction may form reactive oxygen species. To eliminate this harmful effect (necrosis), the action of the peroxide dismutase enzyme is essential. It catalyzes the reaction $O_2^- + 2H^+ => H_2O_2 + O_2$. Subsequently, the H_2O_2 under the action of catalase or other peroxidase turns into water.

In addition to the enzymes cited earlier, Mn acts as a cofactor by activating more than 30 enzymes. As Mg, they function as bridges between ATP and group transfer enzymes (phosphokinases). Most reactions of enzymes activated by Mn are related to the Krebs cycle (respiration). They are redox, decarboxylation, or hydrolytic reactions. An important enzyme activated by Mn is nitrite reductase, as it enables the reduction of N and later its incorporation to carbon skeletons.

Thus, Mn participates in several processes in plants, which can be summarized as follows: Mn deficiency decreases cell elongation and may reduce radical growth, indicating inhibition of lipid or gibberellic acid metabolism or a less flow of carbohydrates to roots.

As manganese activates RNA polymerase, it has an indirect effect on protein synthesis and cell multiplication, although, in the case of Mn deficiency, enzyme activity is more affected than protein synthesis. Mg can also activate this enzyme. However, Mn is much more efficient. For this same activity, the enzyme requires a concentration of Mg ten times that of Mn (Marschner 1986).

In plants deficient in Mn, there is normally an accumulation of amino acids (N-soluble) due to the decrease in photosynthesis, with reflexes in reduced protein synthesis and in the carbohydrate content (Table 12.1).

Given the role of Mn in protein synthesis (Mann et al. 2002), studies have indicated that an increase in Mn may lead to an increase in the protein and oil contents in soybean seeds (Fig. 12.2). It is worth mentioning the importance of adequate levels of protein and oil in seeds, as these elements are responsible for determining the quality and quantity of the final products, such as soybean bran and oil.

Table 12.1 Effects of manganese deficiency on the growth and organic components of bean plants

Parameter	Leaves		Root	
	+Mn	−Mn	+Mn	−Mn
Dry matter (g per plant)	0.6	0.5	0.2	0.1
N-protein (mg g^{-1} dry matter)	52.7	51.2	27.0	25.6
N-soluble (mg g^{-1} dry matter)	6.8	11.9	17.2	21.7
Soluble carbohydrates (mg g^{-1} dry matter)	17.5	4.0	7.6	0.9

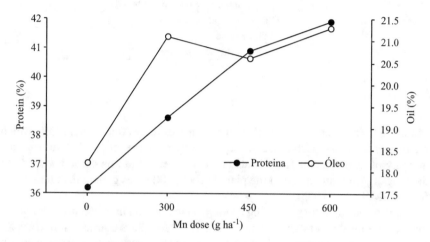

Fig. 12.2 Effects of manganese on protein and oil contents of soybean seeds (cv. Garimpo) (split leaf application at stages V4 and V8)

In hormonal control, Mn functions as a cofactor of the system and oxidation of AIA, and in plants with a nutrient deficiency, it presents a high AIA oxidase activity and no activity of the inhibitor. In addition, Mn is an important reaction factor derived in secondary compounds, such as anthranilic acid, which is a precursor to tryptophan and forms AIA. In N metabolism, nitrite and hydroxylamine reductase, and glutamine synthetase are activated by Mn, although the latter is less efficient. The nodulation of legumes requires a high AIA activity, which is affected by Mn (Malavolta 2006).

Mn is one of the main micronutrients and plays an important role in reducing damage by plant diseases. This is because Mn activates a series of enzymes in the biosynthesis of secondary metabolites. Its deficiency results in a decrease in the concentration of numerous substances, such as aromatic amino acids, phenolic compounds, coumarins, lignins, flavonoids, and indoleacetic acid, which may inhibit or hinder the proliferation of pathogens (of the soil or not). In addition to the involvement of Mn in lignification, constituting a physical barrier to the pathogen, there are other effects of this nutrient that affect the incidence of diseases in plants.

Table 12.2 Enzymes and biological processes influenced by Mn

Enzyme activator	Processes
Glutathione synthetase	Ionic absorption
Activation of methionine	Photosynthesis
ATPase	Respiration
Pyruvic kinase	Hormonal control
Enolase	Protein synthesis
Isocitric dehydrogenase	Resistance to diseases
Pyruvic decarboxylase	
Pyrophosphorylase (starch synthesis)	
Glutamine synthetase	
Malic enzyme	
Indolyl acetic acid oxidase	

Mn inhibits enzymes such as aminopeptidases (decreases free amino acids, which are fungus food) and pectin methylesterase (enzyme in the fungus that destroys the integrity of membranes). There is also a direct effect of Mn on causing toxicity to the fungus (Mn requirement in the plant is about 100 times greater than that of the fungus) (Malavolta 2006).

Mn is responsible for promoting an enzymatic function in metabolism, influencing the biological processes of plants (Table 12.2). Mn plays a structural role as part of proteins (nonenzymatic) such as manganin and concanavalin A (Malavolta 2006).

Crops present different probabilities of response to the application of Mn. Cultures have different probability of response to the application of Mn, and those with the highest response are: beans, cucumber, lettuce, barley, peas, soybeans, sorghum, spinach, beets, wheat, citrus, apple, peach, grapevine, rose, strawberry, and cassava (Lucas and Knezek 1973, cited by Marinho 1988).

Mascagni Júnior and Cox (1985) verified the response of soybean to the application of Mn (via leaf) (1.5 kg ha^{-1}).

12.4   Nutritional Requirements of Crops

As stated earlier, the study of the crop's nutritional requirement must reflect the total extraction of the nutrient from the soil and respect the extraction at each stage of crop development to satisfy the nutritional needs of crops aiming a maximum economic production.

The total content of Mn in the plant can vary from 1% to 2% (10–20 g kg^{-1}). These values may vary depending on the crop and other factors, which is the topic of Chap. 19.

Thus, for a proper discussion of the nutritional requirements of crops, two factors are equally important: total extraction/export of the nutrient and rate of absorption of this nutrient throughout the cultivation.

Extraction and Export of Nutrients

The total extraction of Mn depends on its contents in the plant and on the amount or the production of accumulated dry matter, which basically varies depending on the genotype (species or cultivar) and the soil.

As for the vegetable species, there is a variation in the quantity required depending on the crop (Table 12.3).

Based on the results of different crops, the total extraction of Mn varies from 130 (cotton) to 5700 g ha^{-1} (sugarcane). However, in relation to the relative value of manganese extraction, in g per ton produced, there is a higher demand for rice (154 g kg^{-1}), while soybeans are the most exported Mn, with 43 g t^{-1} of grains produced.

Thus, most of the Mn is contained in the vegetative part, especially in leaves, as in rice (Table 12.3). However, Mn can accumulate in flowers; Malavolta (2006) reported that Mn was the micronutrient with the greatest concentration in that organ.

Absorption Curve

The study of the manganese absorption curve (nutrient accumulated depending on the length of cultivation) is important because it allows determining the times when the element is most required and correcting the deficiencies that may occur during the crop development. By analyzing the absorption curve of soybeans, there is a small absorption in the initial period (0–30 days); in the following period (30–60 days), the absorption speed increases more than ten times, 3.2 g/ha/day of Mn, indicating a high demand at this phase, and reaches a maximum in the period of 60–90 days, with 5.7 g/ha/day of Mn (Bataglia and Mascarenhas 1977).

Normally, the application of Mn in planting furrows meets the plant's requirement throughout the cycle. In case of initial symptoms of deficiency during the initial crop development, the information on Mn regarding plant's nutrition may assist in the management of this correction via leaf for both the season and the number of applications.

As for the timing of application, the information on the absorption curve is valuable. In the case of soybeans, the initial period of greatest absorption speed occurs 30 days after sowing; therefore, the application must coincide with that time.

Table 12.3 Manganese requirements of the main crops

Crop	Plant part	Dry matter produced	Accumulated Mn		Mn required for production of 1 t of grain[4]
			Plant part	Total [3]	
Annual					
Cotton	Reproductive (cotton/seed)	1.3	19 (15)[2]	130	100
	Vegetative (stem/branch/leaf)	1.7	106		
	Root	0.5	5		
Soybean	Grains (pods)	2.4	102 (43)	312	130
	Stem/branch/leaf	5.6	210		
Corn [1]	Grains	6.4	53 (8.3)	767	120
	Crop remains	–	714		
Rice	Grains	3	52	461	154
	Stems	2 (DM)	96		
	Leaves	2 (DM)	226		
	Husk	1	57		
	Root	1 (DM)	30		
Wheat	Grains	3	90 (30)	250	83
	Straw	3.7	160		
Semiperennial/perennial					
Sugarcane	Stems	100	1200 (12)	5700	57
	Leaves	25	4500		
Coffee tree[1]	Grains (nut)	2	40 (20)	–	–
	Trunk, branches, and leaves	–	–		

[1] Malavolta (1980). [2] Relative export of nutrients by produced grains (g t^{-1}): Mn accumulated in the grains/dry matter of grains. [3] Suggests a nutritional requirement (total) by crop area for the respective level of productivity. [4] Suggests a relative nutritional requirement of Mn of the crop to produce one ton of commercial product (grains/stalks); obtained by the formula: Mn accumulated in the plant (vegetative + reproductive part)/dry matter of the commercial product

As for the number of applications, it is necessary to know the mobility of Mn in the plant. As Mn is not very mobile, more than one application will be needed. According to the results of application of Mn in corn, Mascagni Jr and Cox (1985) observed that two applications (fourth and eighth completely expanded leaf) provided a greater production compared to only one application.

Fig. 12.3 Pictures and general description of the visual symptoms of Mn deficiency in different crops. Corn: Internerval chlorosis of the youngest leaves (thick reticulate) when the deficiency is moderate; in more severe cases, long, white streaks appear on the tissue, and the tissue in the middle of the chlorotic area may die and become detached; thin stems (Fig. 12.3a). Vine: Mn deficiency causes chlorosis between the veins of younger leaves. Except for the veins, vine leaves turn pale green and then pale yellow (Bavaresco et al. 2010 Fig. 12.3b). Coffee: Chlorosis starts in the youngest leaves of the branches. At first, many white spots appear on newer leaves, which then come together to turn a yellowish color, almost an egg yolk color (Fig. 12.3c). Soybean: Chlorotic areas between the veins, which evolve to necrosis with a brown coloration that develop on the leaves as the deficiency becomes severe. Mn deficiency differs from that of Fe in that the ribs remain green and protruding (Fig. 12.3d). Sorghum: Chlorosis between the youngest leaf vein in sorghum plants (Fig. 12.3e)

12.5   Symptoms of Nutritional Deficiencies and Excesses of Manganese

Deficiency

In general, Mn deficiency is characterized by chlorosis (yellowing) of the surface of young leaves, which may progress to between the ribs, known by a thick reticulate (the ribs form a thick green network on a yellow background) (Fig. 12.3), although it may vary from one species to another. It should be noted, however, that the symptoms of deficiency may vary depending on the species.

Mn deficiency decreases the germination of pollen grains and the development of corn plant seeds (Sharma et al. 1991).

Toxicity

Toxicity appears, initially, also in young leaves. It is characterized by marginal chlorosis, brown spots that evolve to necrotic ones on the surface of the leaf, and leaf wrinkling, especially in legumes. It is at these necrotic points that Mn accumulates at high concentrations. Plants may show a reduced growth, with a stunted aspect.

In addition to brown spots, there is an irregular distribution of chlorophyll, in addition to induction of auxin and Fe deficiency (Souza and Carvalho 1985).

One way to relieve toxicity by Mn^{2+} (corn leaf content >350 mg kg^{-1}) is to increase the availability of Mg in the soil, which reduces the absorption of Mn (Mengel and Kirkby 1987).

It should be noted that the genetic control of plants for tolerance to toxicity by Mn is driven by a small number of genes (Foy et al. 1988), which makes the selection of these plants difficult.

Some tropical species have a high tolerance to Mn toxicity, such as Brachiaria, which has an element content in the first cut of 440 mg kg^{-1} (Puga et al. 2011) up to 9967 mg kg^{-1} (Guirra et al. 2011). Puga et al. (2011) add that in the second cut of forage, the levels reach 1315 mg kg^{-1}, exceeding the maximum tolerable to cattle (1000 mg kg^{-1}), according to NRC (2001), with risk of intoxication.

References

Bartlett RJ. Manganese redox reactions and organic interaction in soils. In: Graham RD, Hannam RJ, Uren NC, editors. Manganese in soils and plants. Dordrecht: Kluwer Academic Publisher; 1988. p. 59–73.

Bataglia OC, Mascarenhas HAA. Absorção de nutrientes pela soja. Campinas: Instituto Agronômico; 1977.

Bavaresco L, Gatti M, Fregoni M. Nutritional deficiencies. In: Delrot S, Medrano H, Or E, Bavaresco L, Grando S, editors. Methodologies and results in grapevine research. Dordrecht: Springer; 2010. p. 165–91.

Foy CD, Scoot BJ, Fisher JA. Genetic differences in plant tolerance to manganese toxicity. In: Graham RD, Hannam RJ, Uren NC, editors. Manganese in soil and plants. Dordrecht: Kluwer Academic; 1988. p. 293–307.

Guirra APPM, Fiorentin CF, Prado RM, et al. Tolerance of marandu grass to manganese doses. Biosci J. 2011;27:413–9.

Malavolta E. Elementos de nutrição de plantas. São Paulo: Agronômica Ceres; 1980. 251p

Malavolta E. Manual de nutrição mineral de plantas. São Paulo: Agronômica Ceres; 2006.

Mann EN, Resende PM, Mann R, et al. Effect of manganese application on yield and seed quality of soybean. Pesq Agrop Brasileira. 2002;37:1757–64. https://doi.org/10.1590/S0100-204X2002001200012.

Marinho ML. Respostas das culturas aos micronutrientes ferro, manganês e cobre. In: Borket CM, Lantmann AF, editors. Enxofre e micronutrientes na agricultura brasileira. Londrina: Embrapa; 1988. p. 239–64.

Marschner H. Mineral nutrition of higher plants. Academic Press, London; 1986.

Mascagni HJ Jr, Cox FR. Evaluation of inorganic and organic manganese fertilizer sources. Soil Sci Soc Am J. 1985;49:458–61. https://doi.org/10.2136/sssaj1985.03615995004900020037x.

Mengel K, Kirkby EA. Principles of plant nutrition. Worblaufen-Bern: International Potash Institute; 1987.

National Research Council – NRC. Nutrient requirements of dairy cattle. Washington: National academy of science; 2001.

Puga AP, Prado RM, Melo DM, et al. Effects of manganese on growth, nutrition and dry matter production of plants of *Brachiaria brizantha* (cv. MG4) in greenhouse conditions. Rev Ceres. 2011;58:811–6. https://doi.org/10.1590/S0034-737X2011000600019.

Raij BV, Cantarella H, Quaggio JA, et al. Recomendações de adubação e calagem para o Estado de São Paulo. Campinas: Instituto Agronômico & Fundação IAC; 1996.

Sharma CP, Sharma PN, Chatterjee C, et al. Manganese deficiency in maize affects pollen viability. Plant Soil. 1991;138:139–42. https://doi.org/10.1007/BF00011816.

Souza DMG, Carvalho LJCB. Nutrição mineral de plantas. In: Goedert WJ, editor. Solos dos cerrados: tecnologias e estratégias de manejo. Planaltina: Embrapa; 1985. p. 75–98.

Wiese MV. Wheat and other small grains. In: Bennett WF, editor. Nutrient deficiencies and toxicities in crop plants. Saint Paul: APS Press; 1993. p. 27–33.

Williams LE, Pittman JK, Hall JL. Emerging mechanisms for heavy metal transport in plants. Biochim Biophys Acta Biomembr. 2000;1465:104–26. https://doi.org/10.1016/S0005-2736(00)00133-4.

Yu Q, Rengel Z. Micronutrient deficiency influences plant growth and activities of superoxide dismutases in narrow-leaf edlupins. Ann Bot. 1999;183:175–82. https://doi.org/10.1006/anbo.1998.0811.

Chapter 13
Iron

Keywords Fe availability · Iron uptake · Iron transport · Iron redistribution · Assimilatory iron · Iron deficient

Iron is one of the most absorbed micronutrients by crops, but it may have low availability in alkaline soils. It is paramount in enzymatic processes vital for crops to obtain a high productivity. In this chapter, we discuss initially (i) basic aspects of Fe in the soil; (ii) Fe uptake, transport, and redistribution; (iii) Fe metabolism; (iv) nutritional requirements for Fe in crops; (v) Fe extraction, export, and accumulation by the main crops; and (vi) Fe deficiency and toxicity symptoms.

13.1 Introduction

There are several factors that can affect the iron available in the soil: the imbalance in relation to other metals (Mo, Cu, and Mn), the excess of P, the effects of high pH (excessive liming), soil soaking, and low temperatures, among others.

It is important to maintain adequate concentrations of Fe in the soil. It is more advantageous to apply Fe in soils with low (<5 mg dm^{-3}) or medium Fe concentrations (5–12 mg dm^{-3}) (DTPA extractor) (Raij et al. 1996), in which the responses of the crops are greater. Normally, an adequate supply of iron to plants depends much more on pH, moisture, and aeration conditions than on the total content present in the soil, which is normally abundant. The availability of iron is higher in tropical soils with acidic pH (<6.0) and with a decrease in the potential for redox (Eh) of the soil. In some situations of very aerated tropical soils, the ionic concentration of Fe^{2+} may be low.

In alkaline soils, iron deficiency is common in several regions of the world, such as in Europe and Asia. In these regions, Fe is the micronutrient that most limits productivity. Studies on nutrition and plant breeding aiming genotypes with a high nutritional efficiency for Fe are common, such as those with fruit trees (Prado and Alcântara-Vara 2011).

In the study of Fe in the plant system, it is important to know all the "compartments" that Mn travels from the soil solution, from roots to shoots (leaves/fruits),

© Springer Nature Switzerland AG 2021
R. de Mello Prado, *Mineral nutrition of tropical plants*,
https://doi.org/10.1007/978-3-030-71262-4_13

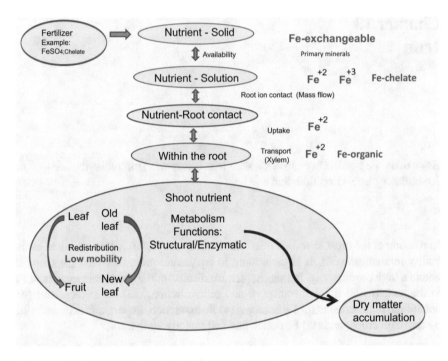

Fig. 13.1 Iron dynamics in the soil–plant system indicating the nutrient passage processes through the different compartments of the plant

that is, from the soil until its incorporation into an organic compound or as an enzymatic activator, where it will perform vital functions to enable the maximum accumulation of dry matter in the final agricultural product (grains, fruits, etc.) (Fig. 13.1).

13.2   Uptake, Transport, and Redistribution of Iron

Uptake

The Fe-root contact receives greater contribution from the mass flow. However, the process of diffusion and root interception are also important. Oliver and Barber (1966) indicated that the mechanism of movement of Fe in the soil varies with environmental conditions. In conditions of low transpiration, the diffusion must always dominate the transport of the nutrient in the soil.

There are important factors that decrease the diffusive flow of Fe in soils, such as low moisture, increases in the pH value, and the concentration of P (Nunes et al. 2004), which can decrease the movement of the element in the soil and therefore the contact with roots, leading to Fe deficiency.

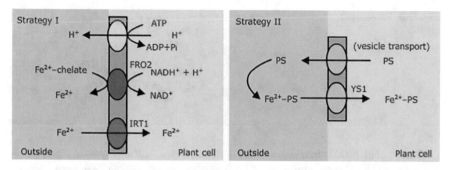

Fig. 13.2 Acquisition of iron by two different strategies in higher plants. Strategy I involves transporters and enzymes, while strategy II involves phytosiderophores

The preferred form absorbed by the roots is Fe^{2+}. However, it can be absorbed as Fe^{3+} and Fe-chelated. The Fe transport in the membranes is conducted by proteins belonging to the Nramp, ZIP, and other families (Curie and Briat 2003).

The activity of H-ATPases present in membranes can induce effluxes of phenolic compounds (chelates) and H^+ that can solubilize Fe (hydroxide) and chelate it until it reaches the root surface (or in ELA) and de-chelating it by the action of reductases of releasing Fe^{2+} for absorption. Therefore, plants that are more capable of absorbing iron present strategies I and II (Kathpalia Bhatla 2018, Fig. 13.2). In strategy I, they can reduce Fe^{3+} in the rhizosphere to Fe^{2+} by proton efflux (Diem et al. 2000). In addition, the amount of Fe^{3+} is low in cultivated soils (pH ~6.5) $[Fe^{3+} + 3OH^- \Leftrightarrow Fe(OH)_3]$. Another form of Fe absorption is via siderophore (chelate), especially in grasses (strategy II), where this Fe chelate is absorbed without reduction. Furthermore, absorption of Fe-chelate can occur regardless of soil pH.

The use of a nutrient solution with an ammonium N source decreases the pH value and increases the solubility of Fe in the rhizosphere even in the cellular apoplast, thereby increasing the soluble content of the micronutrient in leaves (Prado and Alcântara-Vara 2011).

It should be noted that high concentrations of other ions in the soil solution (P, Mn, and Zn) may inhibit Fe absorption by ionic competition. However, Mo can increase Fe absorption. The absorption of Fe is also governed by the genetic factor. Sorghum is one of the most inefficient species in Fe absorption, especially at the initial phase (up to four weeks after germination) (Malavolta 1980).

Transport

Iron is transported through the xylem via the respiratory chain, predominantly as citric acid chelate. Chelate means "tweezers," in which the metal is surrounded by an organic compound (Fig. 13.3), preventing reactions of the metal with other substances in the medium.

Fig. 13.3 Scheme of a chelate

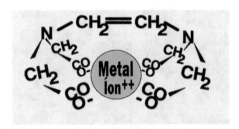

Soares et al. (2001) found in studies with eucalyptus testing doses of Zn in nutrient solution that the translocation of Fe from the roots to shoots decreased, regardless of the species, from 21% in the control to only 2% with 1600 μM of Zn. This indicates a strong relationship between the decrease in dry matter production and the occurrence of induced Fe deficiency in plants.

Redistribution

Iron is poorly redistributed in the plant. Therefore, deficiency symptoms appear on the youngest leaves. A high leaf content can occur in plants deficient in iron, as Fe may be present in the plant tissue in the form of complexes (Fe^{3+}) or as precipitates (Fe-P).

13.3   Participation in Plant Metabolism

Most of the iron in plants (~80%) is located in chloroplasts, as phytoferritin (reserve protein), therefore affecting photosynthesis. In general, iron is important in chlorophyll biosynthesis and is part of protein classes, in enzymatic constituents that transport electrons, and also in the activation of enzymes. It is essential for protein synthesis and helps to form some enzymatic respiratory systems.

Chlorophyll and Protein Biosynthesis

Fe participates in chlorophyll biosynthesis and is important in the reaction that results in the formation of ALA (delta aminolevulinic acid), which is the precursor to porphyrin, a component of this pigment. In protein synthesis, especially that of chloroplasts, Fe participates in a protein of ribosomes (site of synthesis) and also reflects in the increase in RNA content.

Cytochrome and Ferredoxin Compounds and Enzyme Activation

Fe participates in two main groups of proteins in the plant: hemoproteins and proteins with Fe-S groups (Fe joins the cysteine thiol group). They participate in the plant's redox system. In the first group, there are cytochromes c that have prosthetic groups with two Fe atoms of copper, present in the mitochondria, which transport electrons during respiration processes, among others (cytochrome oxidase, catalase, peroxidase, and leg-hemoglobin). The importance of Fe is attested in the electron transport chain of oxidative phosphorylation that generates the largest part of ATP. In the other group, ferredoxin (contains Fe) and the cytochrome b6f complex (contains Fe and S) present in the thylakoid membrane of chloroplasts, which transports electrons resulting from the breakage of the water molecule. They act on the metabolic processes of photosynthesis, resulting in chemical energy (NADPH and ATP), reduction of N_2 (biological fixation), nitrite, sulfate, and even incorporation of N (GOGAT). Fe participates in the heme group, which is a prosthetic group of several proteins, including leg-hemoglobin.

During the metabolism of plants, in the photochemical phase of photosynthesis, an excited electron is transferred to oxygen and produces superoxide (O_2^-), which can also be produced when the plant is subjected to stress. By the action of superoxide dismutase (SOD), this O_2^- radical is transformed into H_2O_2 (energetic oxidizer). The two products (O_2^- and H_2O_2) being harmful to cells. Thus, the superoxide dismutase enzyme (SOD) inactivates O_2^- in H_2O_2. Then, catalase (photorespiration process) and also peroxidase (heme protein), when activated by Fe, inactivates H_2O_2 producing H_2O and O_2 (Fig. 13.4).

Fe is also involved in the synthesis of lipoxygenase (hemoenzyme) that catalyzes the oxidation of linoleic and linolenic acids into several other compounds (including traumatin and jasmonic acid). Therefore, its deficiency can affect growth, senescence, and resistance to diseases.

The functions of Fe in plants can be summarized as constituents of compounds and also as an enzyme activator with reflexes in vital physiological processes in the life of plants (Table 13.1).

Iron also has a structural function. It participates in chelated compounds (with di- and tricarboxylic acids), phytoferritin (with P), and Leg H6 (Malavolta 2006).

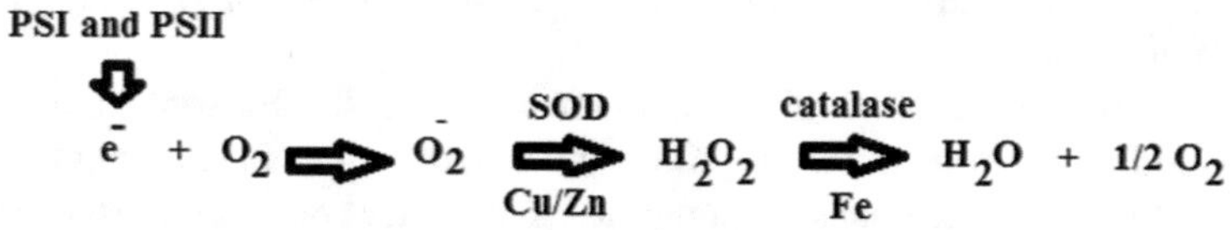

PSI and PSII

$$\Downarrow$$

$$\bar{e} + O_2 \Rightarrow O_2^- \overset{SOD}{\underset{Cu/Zn}{\Rightarrow}} H_2O_2 \overset{catalase}{\underset{Fe}{\Rightarrow}} H_2O + 1/2\, O_2$$

Fig. 13.4 Effects of enzymatic systems on inactivation of reactive oxygen species

Table 13.1 Summary of functions of iron in plants

Enzyme constituent/activator	Process
Heme-peroxidase	Photosynthesis
Catalase	Respiration
Cytochrome oxidase	Biological N fixation
Leg-hemoglobin	Assimilation of N and S
Sulfite reductase	
Sulfite oxidase	
Ferredoxin	
Dehydrogenase	
Nitrogenase	
Nitrite reductase	
Nitrate reductase	
Hydrogenase	
Aconitase	

13.4 Mineral Requirements of the Main Crops

The demand for Fe in crops varies from 6442.6 g ha^{-1} (coffee) to 8890 g ha^{-1} (cane-plant), while the export by harvest varies from 12.7 g t^{-1} (ratoon cane) to 408 g t^{-1} (lettuce) (Table 13.2). It is pertinent to point out that in plants of direct consumption by humans, studies aiming to obtain genotypes with a greater Fe accumulation, as a public health measure, have disseminated. According to Grusak and Dellapenna (1999), there are about 2 billion people with anemia due to iron deficiency.

Typically, crops are likely to respond differently to iron application (Lucas and Knezek 1972). Low probability of response: mint and wheat; medium response: alfalfa, asparagus, cabbage, corn, oats, and coffee; and high response: beans, barley, broccoli, cauliflower, soy, sorghum, spinach, beetroot, rice, tomato, citrus, apple, peach, pear, vine, rose, strawberry, and pineapple.

The iron absorption curve in soybean crops occurs slowly at the beginning of plant growth (0–30 days) (5.7 g ha^{-1} per day), reaching a maximum in the period of 60–90 days (15.5 g ha^{-1} per day), and thereafter there is a marked decrease until the end of the production cycle (2.6 g ha^{-1} per day) (Bataglia and Mascarenhas 1977).

Agronomic biofortification is important to enrich the Fe content in grains/fruits (food), improving human nutrition and reducing anemia resulting from leaf spray on reproductive organs, although it has no effect on productivity, only on quality.

Table 13.2 Iron requirements of some crops

Crop[1]	Plant part	Dry matter produced	Accumulated Fe		Fe required for production of 1 t of grain[4]
			Plant part	Total [3]	
		t ha^{-1}	g ha^{-1}		g t^{-1}
Annual					
Lettuce [5]	Shoot	1.21	646 (646)[2]	–	–
Beans	Grains	1	262 (262)	–	–
Citrus	Fruits	1	66 (66)	–	–
Coffee tree (6 years)	Stem	1.97	901.5	6442.6	
	Branches	4.07	794.2		1049.3
	Leaves	3.56	3082.9		
	Fruits	6.14	1663.9 (271)		
	Plant cane				
Sugarcane	Stems	100	2378 (23.8)	8890	88.9
	Leaves		6512		
	Ratoon cane				
	Stems	100	1207 (12.7)	5745	57.5
	Leaves		4538		

[1] Malavolta (1980). [2] Relative export of nutrients by produced grains (g t^{-1}): Fe accumulated in the grains/dry matter of grains. [3] Suggests a nutritional requirement (total) by crop area for the respective level of productivity. [4] Suggests a relative nutritional requirement of Fe of the crop to produce one ton of commercial product (grains/stalks); obtained by the formula: Fe accumulated in the plant (vegetative + reproductive part)/dry matter of the commercial product. [5] Garcia et al. (1982)

13.5   Symptoms of Nutritional Deficiencies and Excess of Iron

Deficiency

The start of Fe deficiency reflects a reduction in the size of chloroplasts, protein synthesis, and chlorophyll content. Thus, the symptoms appear initially in the young plant parts as a chlorosis (leaves turn yellow) due to the lower synthesis of chlorophyll, while only the ribs may stay green for some time, standing out as a fine reticulate (thin green network of ribs on the yellow background) and may evolve to "bleaching" (Fig. 13.5). However, with the evolution of symptoms, even the veins become chlorotic. Therefore, Fe accumulates in older leaves as insoluble oxides or inorganic compounds (Fe-P). This decreases Fe entry into the phloem, causing deficiency symptoms in new leaves.

The iron-deficient plant causes a reduction in the size of chloroplasts and in protein synthesis (Castro et al. 2005), and physiological changes (decrease in chlorophyll and photosynthetic rate) and biochemical changes (decrease in peroxidase and catalase activity and increase in concentration de H_2O_2) (Molassiotis et al. 2006).

Fig. 13.5 Pictures and general description of the visual symptoms of Fe deficiency in different crops. Apple tree: It starts with internerval chlorosis from the tip to the base of the youngest leaves (Valentinuzzi et al. 2019, Fig. 13.5a). Vine: There is an internerval chlorosis along the entire length of the leaf blade with some necrotic spots, with only the veins remaining (fine reticulate of veins) in the younger leaves (Fig. 13.5b), which progress to old leaves, reaching the entire plant (Bavaresco et al. 2010, Fig. 13.5c). Corn: The characteristic symptoms are internerval chlorosis (fine reticulate), which occurs in the youngest leaves. Along the evolution of symptoms, the green color is completely lost, including on the main veins (Fig. 13.5d). Pear tree: The youngest leaves turn yellow, the veins remain green; as the symptoms progress, the veins turn yellow and the leaf margins become necrotic and dry (Fig. 13.5e). Sunflower: At the onset of symptoms, the youngest leaves become chlorotic following the gradient from the leaf's basal region to the center (image on the left); as symptoms progress, new leaves become necrotic and dry, and chlorosis extends to older leaves (image on the right) (Fig. 13.5f)

Thus, several factors may cause Fe deficiency in plants, such as low levels of Fe in the soil, high levels of P in the soil, extreme temperatures, genetic differences, low organic matter content in acidic soils, and free $CaCO_3$ (Lucas and Knezek 1972).

At a cellular level, "chlorotic" leaves deficient in Fe may have a higher Fe content than other leaves (green). This occurs due to the accumulation of iron in the apoplast due to the high pH caused by the increase in HCO_3^- or even by the nitrate reduction process ($NO_3^- + 8H^+ + 8e \Rightarrow NH_3 + 2H_2O + OH^-$), which blocks Fe^{3+} reductase and causes a decrease in the active iron content (Fe^{2+}) (HCl N extractor or chelating agents) (Romheld 1987). This effect of cellular pH in decreasing chlorosis occurs because spraying Fe^{2+} or diluted acid corrects this nutritional disorder (Malavolta 2006).

Toxicity

Iron toxicity can occur in periods of excessive rain or in flooded soils, such as flooded rice (reduction of $Fe^{3+} \Rightarrow Fe^{2+}$), and in some plants, the Fe content may reach 50 mg kg^{-1}.

The toxic action of Fe, on the other hand, is usually evident in plant shoots. The roots seem insensitive to high concentrations of Fe and seem affected only indirectly as a result of the inhibition in the growth of shoots (Foy 1976).

In sorghum, excess Fe makes the leaves lighter, with lesions ranging from blackish to a straw color in the borders. In soybean, the toxicity of Fe is similar as that of Mn, except in the case of excess Fe, when leaves are less brittle than with excess Mn.

It is worth mentioning that in some situations, excess Fe inhibits the absorption of Mn and, thus, the symptoms may be similar to Mn deficiency. It may also be similar to K deficiency.

Foy et al. (1978) add that the toxicity of manganese is difficult to be studied in isolation because of the interactions existing between it and other elements, such as phosphorus, calcium, iron, aluminum, and silicon. The authors point out that such interactions may be responsible for the diversity of symptoms in plants and for the reductions in the growth due to the excess of manganese in different species.

References

Bataglia OC, Mascarenhas HAA. Absorção de nutrientes pela soja. Campinas: Instituto Agronômico; 1977.

Bavaresco L, Gatti M, Fregoni M. Nutritional deficiencies. In: Delrot S, Medrano H, Or E, Bavaresco L, Grando S, editors. Methodologies and results in grapevine research. Dordrecht: Springer; 2010. p. 165–91.

Castro PRC, Kluge RA, Peres LEP. Manual de fisiologia vegetal: teoria e prática. Piracicaba: Agronômica Ceres; 2005.

Curie C, Briat JF. Iron transport and signaling in plants. Annu Rev Plant Biol. 2003;54:183–206. https://doi.org/10.1146/annurev.arplant.54.031902.135018.

Diem HG, Duhoux E, Zaid H, et al. Cluster roots in Casuarinaceae: role and relationship to soil nutrient factors. Ann Bot. 2000;85:929–36. https://doi.org/10.1006/anbo.2000.1127.

Foy CD. Differential aluminium and manganese tolerances of plant species and varieties in acid soils. Ciênc Cult. 1976;28:150–5.

Foy CD, Chaney RL, White MC. The physiology of metal toxicity in plants. Annu Rev Plant Biol. 1978;29:511–66. https://doi.org/10.1146/annurev.pp.29.060178.002455.

Garcia LLC, Haag PE, Minami K, et al. Nutrição mineral de hortaliças. XL. Concentração e acúmulo de micronutrientes em alface (*Lactuca sativa* L.) cv. Brasil 48 e Clause's Aurélia. An Esc Super Agric Luiz de Queiroz. 1982;39:485–504. https://doi.org/10.1590/S0071-12761982000100028.

Grusak MA, Dellapenna D. Improving the nutrient composition of plants to enhance human nutrition and health. Annu Rev Plant Physiol Plant Mol Biol. 1999;50:133–61. https://doi.org/10.1146/annurev.arplant.50.1.133.

Kathpalia R, Bhatla SC. Plant mineral nutrition. In: Bhatla SC, Lal MA, editors. Plant physiology, development and metabolism. Singapore: Springer; 2018. p. 37–81. https://doi.org/10.1007/978-981-13-2023-1_2.

Lucas RE, Knezek BD. Climatic and soil conditions promoting micronutrient deficiencies in plants. In: Mortvedt JJ, Giordano PM, Lindsay WL, editors. Micronutrients in agriculture. Madison: Soil Science Society of America; 1972. p. 265–88.

Malavolta E. Elementos de nutrição de plantas. São Paulo: Agronômica Ceres; 1980. 251p

Malavolta E. Manual de nutrição mineral de plantas. São Paulo: Agronômica Ceres; 2006.

Molassiotis A, Tanou G, Diamantidis G, et al. Effects of 4-month Fe deficiency exposure on Fe reduction mechanism, photosynthetic gas exchange, chlorophyll fluorescence and antioxidant defense in two peach rootstocks differing in Fe deficiency tolerance. J Plant Physiol. 2006;163:176–85. https://doi.org/10.1016/j.jplph.2004.11.016.

Nunes FN, Barros NF, Albuquerque AW, et al. Diffusive flux of iron in soils influenced by phosphorus rates and levels of acidity and moisture. Rev Bras Ciênc Solo. 2004;28:423–9. https://doi.org/10.1590/S0100-06832004000300003.

Oliver S, Barber SA. Mechanisms for the moviment of Mn, Fe, B, Cu, Zn, Al and Sr from the soil to the soil to the surface of soybean roots. Soil Sci Soc Am Proc. 1966;30:468–70. https://doi.org/10.2136/sssaj1966.03615995003000040021x.

Prado RM, Alcântara-Vara E. Tolerance to iron chlorosis in non grafted quince seedlings and in pear grafted onto quince plants. J Soil Sci Plant Nutr. 2011;11:119–28. https://doi.org/10.4067/S0718-95162011000400009.

Raij BV, Cantarella H, Quaggio JA, et al. Recomendações de adubação e calagem para o Estado de São Paulo. Campinas: Instituto Agronômico & Fundação IAC; 1996.

Romheld V. Different strategies for iron acquisition in higher plants. Physiol Plant. 1987;70:231–4. https://doi.org/10.1111/j.1399-3054.1987.tb06137.x.

Soares CRFS, Grazziotti PH, Siqueira JO, et al. Zinc toxicity on growth and nutrition of *Eucalyptus maculata* and *Eucalyptus urophylla* in nutrient solution. Pesq Agrop Brasileira. 2001;36:339–48. https://doi.org/10.1590/S0100-204X2001000200018.

Valentinuzzi F, Venuti S, Pii Y, et al. Common and specific responses to iron and phosphorus deficiencies in roots of apple tree (*Malus × domestica*). Plant Mol Biol. 2019;101:129–148. https://doi.org/10.1007/s11103-019-00896-w.

Chapter 14
Copper

Keywords Cu availability · Copper uptake · Copper transport · Copper redistribution · Assimilatory copper · Copper deficient

Copper is not one of the most absorbed micronutrients by crops, but it is a key factor in enzymatic processes to ensure high crop yields. In this chapter, we discuss initially (i) basic aspects of Cu in the soil; (ii) Cu uptake, transport, and redistribution; (iii) Cu metabolism; (iv) nutritional requirements for Cu in crops; (v) Cu extraction, export, and accumulation by the main crops; and (vi) Cu deficiency and toxicity symptoms.

14.1 Introduction

Copper is present in the soil as Cu^{2+}. It is strongly bound to organo-mineral colloids. The proportion of copper complexed by organic compounds in the soil solution may reach 98%. Thus, the organic form plays an important role in regulating its mobility and availability in the soil solution. Therefore, it can be inferred that the higher the content of organic matter, the lower the availability of copper in plants. The availability of this element is strongly related to soil pH.

It is important to maintain adequate concentrations of Cu in the soil. It is more advantageous to apply Cu in soils with low (<0.3 mg dm^{-3}) or medium Cu concentrations (0.3–1.0 mg dm^{-3}) (DTPA extractor) (Raij et al. 1996), in which the responses of the crops are greater. Typically, the crops most susceptible to copper deficiency are cereals (wheat, corn, rice, oats, and barley). However, Cu deficiency may occur in other crops (vegetables and fruits) (Gupta 1997).

In the study of copper, it is important to know all the "compartments" that the nutrient travels from the soil solution to shoots, its physiological function in plants, and its role in plant production (Fig. 14.1).

© Springer Nature Switzerland AG 2021
R. de Mello Prado, *Mineral nutrition of tropical plants*,
https://doi.org/10.1007/978-3-030-71262-4_14

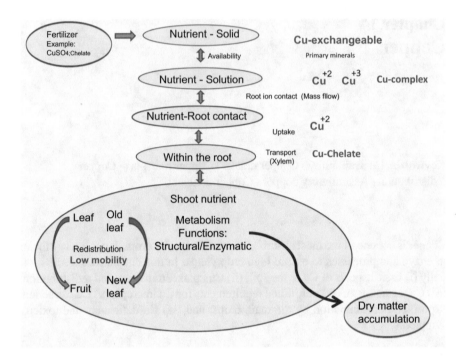

Fig. 14.1 Copper dynamics in the soil–plant system indicating the nutrient passage processes through the different compartments of the plant

14.2   Uptake, Transport, and Redistribution of Copper

Uptake

The path of copper from the soil to the absorption site occurs due to the mass flow (95%) (Oliver and Barber 1966). In the case of Cu, it is more effective in shorter distances; in longer distances, it can move by leaching complexed to organic radicals.

Copper can be absorbed as Cu^{2+} (preferably) and Cu complexed to chelates (citric, tartaric, malic, oxalic acids, metallophores, and phenols, among others). In the active absorption process, there is a competition between P and Zn for the same sites as the carrier, and Mn does not interfere with it. In general, heavy metals, such as Cu, can be transported by four different families: ATPases of the type CPx, Nramp, facilitators of diffusion of cations, and ZIP (Williams et al. 2000). The presence of vesicular-arbuscular mycorrhizae may increase Cu absorption.

There may be competitive inhibitions between copper and zinc, among others ($H_2PO_4^-$, K^+, Ca^{2+}, and NH_4^+) (Malavolta 2006). The ability of copper to move ions, mainly Fe and Zn, from exchange sites has been identified as the main cause of this inhibition (Mengel and Kirkby 1987).

Transport

After its absorption, copper is transported via xylem in the form of Cu-chelated (amino acids) through the process of transpiration to shoots. In the xylem, almost all Cu (>99%) is in the form of a complex (Marschner 1995), such as nitrogen compounds of low molecular weight.

Redistribution

Copper is considered not very mobile in the phloem. Therefore, deficiency symptoms occur in the youngest leaves. However, in situations where copper is high in the middle of the plant, there may be redistribution (Cu-chelated) because it has the ability to form stable leaf chelates for fruits. This does not occur when copper is deficient in the medium. Thus, as Cu has a high affinity with N (amino group), it can form soluble compounds and therefore be transported in plants (Loneragan 1981).

14.3 Participation in Plant Metabolism

It is an iron-like transition element, easily transporting electrons. It is therefore very relevant for the physiological processes of redox.

One of the main functions of copper is as an activator or constituent of enzymes. Most of the Cu is present in leaf chloroplasts, and more than half is bound to plastocyanin (electron donor for the photosystem I), which acts especially in electronic transport with valence changes, although other enzymes also perform this electron transport and are activated by Cu (laccase, ascorbic acid oxidase, and the cytochrome oxidase complex). This occurs due to the ease of Cu to change valence (Fig. 14.2).

The electronic transport mediated by copper can, however, occur without change of valence (amine oxidase, tyrosinase, and galactose oxidase).

In this role of $\bar{e}$ transporter, copper acts in several metabolic processes in vegetables. One of the vital processes affected by copper deficiency is photosynthesis, which performs the electronic transport between the photosystems (I and II) through plastocyanin and other enzymatic systems. Also in photosynthesis (dark phase), Cu activates ribulose diphosphate carboxylase, responsible for the entry of CO_2 in

Fig. 14.2 Valence change of copper

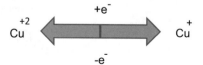

organic compounds. In respiration, Cu and Fe are constituents of the enzyme cytochrome oxidase (two atoms of Cu and two atoms of Fe), acting on electron transport of oxygen in mitochondria, thus affecting oxidative phosphorylation. Cu constitutes the enzyme ascorbate oxidase, which acts in the oxidation of ascorbic acid into dehydroascorbic acid.

In addition, Cu also participates in the prosthetic group of superoxide dismutase, which protects the plant from the deleterious effects of superoxide radicals (O_2^-), producing H_2O_2, which in turn is reduced to H_2O and O_2 by catalase. Cu also activates enzymes (polyphenol oxidase, diamine oxidase) that catalyze the oxidation of phenolic compounds into ketones, precursors of lignin (precursor + H_2O_2➔ phenyl propanol radical + H_2O, a reaction mediated by peroxidase, which follows the pathway until forming lignin). Therefore, its deficiency causes an accumulation of phenolic compounds and a decrease in lignin synthesis, and tissues such as the xylem may collapse.

In legumes, copper is required by nodes and, thus, there is an increase in nodulation and consequently the fixation of N. This can be explained because Cu deficiency causes a reduction in the supply of carbohydrates for nodulation. In addition, polyphenol oxidase polyphenoloxidase activity may be inhibited, accumulating diphenols, producing less melanin, and inhibiting rhizobia. Finally, Cu can retain more Fe in roots and thereby increase the production of leg-hemoglobin.

Copper, in coffee, promotes a more uniform grain ripening and has a tonic effect, which is explained by an inhibition in the production of ethylene, which in turn is involved in the senescence process, maintaining the leaves active for longer (Malavolta et al. 1997, 2006).

The functions of copper are thus its enzymatic role and its reflexes in several vital physiological processes for plants (Table 14.1).

Typically, crops are likely to show different responses to the application of copper. A high response occurs in alfalfa, barley, carrots, lettuce, oats, spinach, wheat, citrus, sugar cane, and coffee (Marinho 1988).

Table 14.1 Enzymes and biological processes affected by Cu in plants

Enzyme constituent/activator	Process
Ascorbate oxidase	Photosynthesis
Polyphenol oxidase, cresolase, catecholase, or tyrosinase	Respiration
Laccase	Hormonal relations
Plastocyanin	N fixation
Diamine oxidase	(Indirect)
Cytochrome oxidase	Metabolism secondary compounds
Ribulose diphosphate carboxylase	Secondary compounds

14.4   Nutritional Requirements of Crops

The study of the crop's nutritional requirement must reflect the total extraction of the nutrient from the soil and respect the extraction at each stage of plant development to satisfy the nutritional needs of crops aiming a maximum economic production.

It should be noted that the total Cu content in the plant may vary from 1 to 5 g kg^{-1}; however, it may reach 100 g kg^{-1} in old leaves. These values may vary depending on the crop and other factors, which is the topic of Chap. 19.

Thus, for a proper discussion of the nutritional requirements of crops, two factors are equally important: total extraction/export of the nutrient and rate of absorption of this nutrient throughout the cultivation.

Extraction and Export of Nutrients

The total extraction of copper depends on its contents in the plant and the amount of dry matter accumulated. Therefore, it varies according to production, which, in turn, depends on the species, variety/hybrid, availability of Cu in the soil, crop management, among others.

Sugarcane and corn are the crops that most extract copper by area: 270 and 181 g ha^{-1}, respectively (Table 14.2). The cotton and rice crops are the most demanding: 45 and 38 g t^{-1} of harvested product. The crops that most export copper in the harvest are coffee, soybeans, and wheat (15, 14, and 10 g t^{-1}, respectively).

Lisuma et al. (2006) verified the response of corn to the application of copper in pot (20 mg kg^{-1}) and field conditions (10 kg ha^{-1}) using volcanic soils.

Absorption Curve

In the evaluation of the copper absorption curve in cotton, there is a low absorption speed in the period of 0–30 days (0.2 g/ha/day), increasing more than four times in the period of 30–60 days (reaching 0.9 g/ha/day). In the subsequent period, the maximum absorption speed occurs (1.2 g/ha/day); in the period of the end of the cycle (90–120 days), there is stabilization with low absorption (0.02 g/ha/day) (Mendes 1965).

It thus seems that, at the reproductive phase, the plant's requirement is the highest. Thus, Cu deficiency problems tend to occur more frequently during the reproductive period of plants compared to the vegetative period.

In a long-term study with application of Cu to wheat, Brennan (2006) observed that the dose of 1.38 kg ha^{-1} of Cu was sufficient to satisfy the adequate nutrition of this crop (Cu leaf = 1.4 mg kg^{-1}) for 28 years. In this period, the crop removed only 2–3% of the applied Cu.

Table 14.2 Copper requirements for major crops

Crop	Plant part	Dry matter produced	Accumulated Cu		Cu required for production of 1 t of grain[d]
			Plant part	Total[c]	
Annual					
Cotton	Reproductive (cotton/seed)	1.3	2 (1.5)[b]	59	45
	Vegetative (stem/branch/leaf)	1.7	44		
	Root	0.5	13		
Soybean[a]	Grains (pods)	2.4	34 (14)	64	27
	Stem/branch/leaf	5.6	30		
Corn[a]	Grains	6.4	25 (3.9)	181	28.3
	Crop remains	–	156		
Rice	Grains	3	10 (3.3)	114	38
	Stems	2	6		
	Leaves	2	5		
	Husk	1	18		
	Root	1	75		
Wheat	Grains	3	30 (10)	40	13
	Straw	3.7	10		
Semiperennial/perennial					
Sugarcane	Stems	100	180 (1.8)	270	2.7
	Leaves	25	90		
Coffee tree[a]	Grains (nut)	2	30 (15)	–	–
	Trunk, branches, and leaves	–	–		

[a]Malavolta (1980)

[b]Relative export of nutrients by produced grains (g t^{-1}): Cu accumulated in the grains/dry matter of grains

[c]Suggests a nutritional requirement (total) by crop area for the respective level of productivity

[d]Suggests a relative nutritional requirement of Cu of the crop to produce one ton of commercial product (grains/stalks); obtained by the formula: Cu accumulated in the plant (vegetative + reproductive part)/dry matter of the commercial product

14.5　Symptoms of Nutritional Deficiencies and Excess of Copper

Deficiency

Symptoms vary by species and may not be as easy to identify as those of other micronutrients. A moderate deficiency sometimes causes only less growth and a reduced harvest without characteristic symptoms, while more severe deficiencies may cause yellowing (or a bluish-green color) of leaves. They can become withered or have the edges curled up or, still, may be larger than normal; even death in the growth regions of the branches may occur. It should be noted that the deformation/curvature of plant tissues (leaves) is because of the role of this nutrient in lignifying the cell wall. This event may even cause plants to tip over. These symptoms appear in the younger plant organs, since Cu is not redistributed (Fig. 14.3). Deficient plants show weak stems or culms and a tendency to wilt even when there is enough moisture.

In perennial crops, the peels may become rough and covered in bubbles, and a gum may exude from fusion in the "exanthem" of the peel.

In annual plants, at the early stages of development, severe deficiency may lead to seedling death. At later stages, the leaves may become twisted, and there may be even a greater sterility of pollen grains (excessive accumulation of auxin), reducing grain production, especially in cereals.

Cu deficiency can increase disease incidence in plants. Therefore, Cu plays an important fungistatic role due to the following factors: Cu increases lignification (physical barrier); Cu deficiency may present less active O_2 harmful to the pathogen, less wall proteins, less induction of alexines, disorganization of the cell wall and membranes, causing oxidation of lipids by not dissipated reactive oxygen species; and lack of tonic effect (Malavolta 2006).

Fig. 14.3 Pictures and general description of the visual symptoms of Cu deficiency in different crops. Soybean: Chlorosis of the youngest leaves (**a**). Sugarcane: Leaves eventually discolored that become paper-thin and curled up when the deficiency is severe; stems and meristems lose their turgidity ("fallen top" disease), the leaves bend, and the stub appears as it were crushed and reduced tillering (**b**). Corn: Yellowing of new leaves as soon as they start to unfold, then the tips curl and show necrosis; the leaves are yellow and show strips similar as those caused by lack of iron; the margins are necrotic (**c**); chlorosis in the form of stretch marks that start at the base of leaf insertion; the stem is soft and folds (**d**). Rice: The newer leaves are blue-green, becoming chlorotic at the tips. Chlorosis develops downward, along the main vein on both sides, followed by dark brown necrosis of the tips. The leaves curl, maintaining the appearance of needles along their entire length or, occasionally, in the middle of the leaf, and the last base develops normally (**e**). Wheat: In young plants, the leaves may curve downward from the base. The blade becomes deformed with chlorosis (**f**)

Toxicity

Most of the excess copper in plants occurs predominantly complexed and bound to phytochelatins. Cu toxicity is not common; however, during the early stages, the reduction in growth is evident. The reduction in root length is a good indicator of copper toxicity. Corn grown in a 15.7 µM Cu nutritive solution inhibits root growth (Ali et al. 2002). Thus, in the root system, Cu toxicity causes reduced branching, thickening, and less root growth (damage to membrane permeability) and may induce Fe deficiency and appearance of necrotic spots. Therefore, the general symptoms of toxicity, in general, start at the roots (death) and advance to older leaves, then the intermediate ones, and lastly the newer ones (large watery spots appear, which are then blackened, as they were burnt) (Malavolta 2006).

In sorghum, Cu toxicity makes the internerval tissue lighter in color, similar as the Fe deficiency, with red strips along the margins (Clark 1993). This is because Cu has the ability to remove Fe from its positions, causing its deficiency (Mengel and Kirkby 1987). The excess of Cu can decrease the Mn content in cauliflower plants (Chartterjee and Chartterjee 2000). The toxicity in corn manifests with leaf contents of Cu > 70 mg kg^{-1} (Mengel and Kirkby 1987). However, there are species (*Commelina communis*, *Rumex acetosa*) with a high tolerance to Cu. They present a high content of this element (500–1000 mg kg^{-1} DM) without causing symptoms (Tang et al. 1999). Other plants, conversely, are sensitive to Cu, such as rosacea (apple, plum, etc.). Excessive Cu inhibits plant growth and prevents important cellular processes, such as electron transport in photosynthesis (Yruela 2005) and exerts a destructive effect on the integrity of chloroplast membranes, also decreasing photosynthesis (Mocquot et al. 1996).

References

Ali NA, Bernal MP, Mohammed A. Tolerance and bioaccumulation of copper in *Phragmites australis* and *Zea mays*. Plant Soil. 2002;239:103–11. https://doi.org/10.1023/A:1014995321560.

Brennan RF. Long-term residual value of copper fertiliser for production of wheat grain. Aust J Exp Agric. 2006;46:77–83. https://doi.org/10.1071/EA04271.

Clark RB. Sorghum. In: Bennet WF, editor. Nutrient deficiencies and toxicities in crop plants. Saint Paul: APS Press/The American Phytopathological Society; 1993. p. 21–6.

Chatterjee J, Chatterjee J. Phytotoxicity of cobalt, chromium and copper in cauliflower. Environ Pollut. 2000;109:69–74. https://doi.org/10.1016/S0269-7491(99)00238-9.

Gupta UC. Copper in crop and plant nutrition. In: Richardson WH, editor. Handbook of copper compounds and applications. New York: Marcel Dekker; 1997. p. 203–29.

Lisuma JB, Semoka JMR, Semu E. Maize yield response and nutrient uptake after micronutrient application on a volcanic soil. Agron J. 2006;98:402–6. https://doi.org/10.2134/agronj2005.0191.

Loneragan JF. Distribution and movement of cooper in plants. In: Lon Eragan JR, Robson AD, Grahan RD, editors. Copper in soil and plants. London: Academic; 1981. p. 165–88.

Malavolta E. Elementos de nutrição de plantas. São Paulo: Agronômica Ceres; 1980. 251p.

Malavolta E. Manual de nutrição mineral de plantas. São Paulo: Agronômica Ceres; 2006.

Malavolta E, Vitti GC, Oliveira SA. Avaliação do estado nutricional das plantas: princípios e aplicações. Piracicaba: Associação Brasileira de Potassa e do Fósforo; 1997.

Marinho ML. Respostas das culturas aos micronutrientes ferro, manganês e cobre. In: Borket CM, Lantmann AF, editors. Enxofre e micronutrientes na agricultura brasileira. Londrina: Embrapa; 1988. p. 239–64.

Marschner H. Mineral nutrition of higher plants. London: Academic; 1995.

Mendes HC. Cultura e adubação do algodoeiro. São Paulo: Instituto Brasileiro de Potassa; 1965.

Mengel K, Kirkby EA. Principles of plant nutrition. Worblaufen-Bern: International Potash Institute; 1987.

Mocquot B, Vangronsveld J, Clijsters H, et al. Copper toxicity in young maize (*Zea mays* L.) plants: effects on growth, mineral and chlorophyll contents, and enzyme activities. Plant Soil. 1996;182:287–300. https://doi.org/10.1007/BF00029060.

Oliver S, Barber SA. Mechanisms for the movement of Mn, Fe, B, Cu, Zn, Al and Sr from the soil to the soil to the surface of soybean roots. Soil Sci Soc Am Proc. 1966;30:468–70. https://doi.org/10.2136/sssaj1966.03615995003000040021x.

Raij BV, Cantarella H, Quaggio JA, et al. Recomendações de adubação e calagem para o Estado de São Paulo. Campinas: Instituto Agronômico & Fundação IAC; 1996.

Tang S, Wilke BM, Huang C. The uptake of copper by plants dominantly growing on copper mining soils along Yangtze river, the people's republic of china. Plant Soil. 1999;209:225–32. https://doi.org/10.1023/A:1004599715411.

Williams LE, Pittman JK, Hall JL. Emerging mechanisms for heavy metal transport in plants. Biochim Biophys Acta Biomembr. 2000;1465:104–26. https://doi.org/10.1016/S0005-2736(00)00133-4.

Yruela I. Copper in plants. Braz J Plant Physiol. 2005;17:145–56. https://doi.org/10.1590/S1677-04202005000100012.

Chapter 15
Molybdenum

Keywords Mo availability · Molybdenum uptake · Molybdenum transport ·
Molybdenum redistribution · Assimilatory molybdenum · Molybdenum deficient

Molybdenum is absorbed in low amounts by plants. In some crops, Mo contents in seeds meet the plant's requirements. It is a nutrient that performs biological functions involved with N. Therefore, it ensures a high crop productivity. In this chapter, we will discuss initially (i) basic aspects of Mo in the soil; (ii) Mo uptake, transport, and redistribution; (iii) Mo metabolism; (iv) nutritional requirements for Mo in crops; (v) Mo extraction, export, and accumulation by the main crops; (vi) Mo deficiency and toxicity symptoms.

15.1 Introduction

Molybdenum is a metal found in the soil, but as an oxyanion in the form of molybdate MoO_4^{2-} in its highest valence form. Its properties are similar as those of nonmetals and other divalent inorganic anions. Thus, in acidic soils, phosphate and molybdate have a similar behavior in relation to the strong adsorption of hydrated iron oxides and, on absorption, molybdate competes with sulfate. Molybdate is a weak acid. Dissociation decreases with a pH decrease from 6.5 to below 4.5, which favors the formation of polyanions. The solubility of MoO_4^{2-} can be estimated as follows: Soil-Mo + 2 $OH^- \Leftrightarrow MoO_4^{2-}$ + soil-$2OH^-$

The effects of liming on acidic soils to increase availability of Mo are evident, and it is estimated that it is increased a hundredfold by increasing each pH unit.

Due to the effects of pH on the availability of Mo, the expression ratio of limestone substitution for Mo emerged. To maintain the same crop production (soybean), Quaggio (1997) reported that the application of lime close to 70%, to maintain base saturation [(K+Ca+Mg)×100/(K+Ca+Mg+H+Al)], requires 25 g ha^{-1} of Mo, while in a soil base saturation of 60%, the requirement for Mo increases to 50 g ha^{-1}.

In the study of molybdenum in the plant system, it is important to know all the "compartments" that Mn travels from the soil solution, from roots to shoots (leaves/fruits), that is, from the soil until its incorporation into an organic compound or as

© Springer Nature Switzerland AG 2021
R. de Mello Prado, *Mineral nutrition of tropical plants*,
https://doi.org/10.1007/978-3-030-71262-4_15

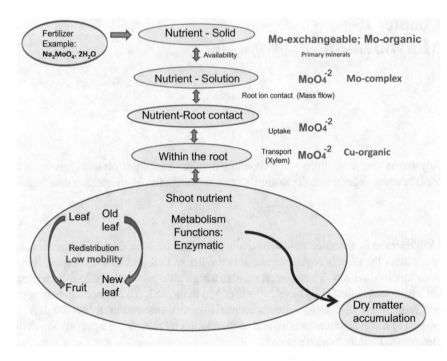

Fig. 15.1 Molybdenum dynamics in the soil–plant system indicating the nutrient passage processes through the different compartments of the plant

an enzymatic activator, where it will perform vital functions to enable the maximum accumulation of dry matter in the final agricultural product (grains, fruits etc.; Fig. 15.1).

15.2 Uptake, Transport, and Redistribution of Molybdenum

Uptake

The Mo-root contact is basically by mass flow, especially at higher soil concentrations. Typically, molybdenum is absorbed as MoO_4^{2-} when the pH of the medium is equal to or greater than 5.0, and as $HMoO_4^{-}$ when the pH is lower than 5.0.

The absorption process is via symporter and the presence of other ions can affect the absorption of Mo. The presence of $H_2PO_4^{-}$ has a synergistic effect on the absorption of Mo, whereas other nutrients inhibit its absorption as SO_4^{2-} and others (Cl^-, Cu^{2+}, Mn^{2+}, and Zn^{2+}).

Transport

In the xylem, molybdenum can be transported as MoO_4^{-2} (higher oxidation state, Mo^{6+}) complexed with the SH groups of amino acids or with the OH groups of carbohydrates, in addition to other polycarboxylate compound groups, such as organic acids.

Redistribution

In general, Mo is not very mobile in the phloem of most species. However, Jongruaysup et al. (1994) verified in *Vigna mungo* that the mobility of Mo may vary depending on the level of supply: low mobility occurs in deficient plants and high mobility occurs in plants well supplied with Mo.

15.3 Participation in Plant Metabolism

Normally, molybdenum deficiency can be strongly associated with nitrogen metabolism due to the requirement of Mo for nitrogenase activity and N fixation. In addition to legumes, crucifers are particularly demanding in Mo. In plants, a given nutrient may undergo changes in oxidation state between the Mo^{6+} (oxidized) and Mo^{5+} (reduced) forms, therefore participating in the redox system.

Thus, molybdenum participates as a constituent of several enzymes, especially those that act in the metabolism of nitrogen and sulfur, which are related to the transfer of electrons. Regarding nitrogen metabolism, Mo affects biological fixation (nitrogenase) and reduction of nitrate and nitrite (reductases). Regarding nitrate reductase, it contains three subunits: FAD (flavin), heme (cytochrome), and the unit with Mo. Thus, during nitrate reduction, electrons are transferred directly from Mo to the nitrate. Mo is also involved with the metabolism of sulfur (sulfite reductase) in redox reactions. In addition, it has a significant effect on pollen formation.

As for nitrogenase, this enzyme system catalyzes the reduction of atmospheric N_2 into NH_3, a reaction by which the *Rhizobium* of root nodules supplies nitrogen to the host plant. For this reason, Mo-deficient legumes often show symptoms of nitrogen deficiency. Nitrogenase, as stated in the chapter on nitrogen, is formed by the Fe-protein and Fe-Mo protein complexes, the latter of which contains molybdenum and iron ions as cofactors, both necessary for enzyme activation.

Molybdenum is necessary for plants when N is absorbed as NO_3^- because it is a component of the nitrate reductase enzyme. This enzyme catalyzes the biological reduction of NO_3^- to NO_2^-, which is the first step toward the incorporation of N as NH_2 in proteins.

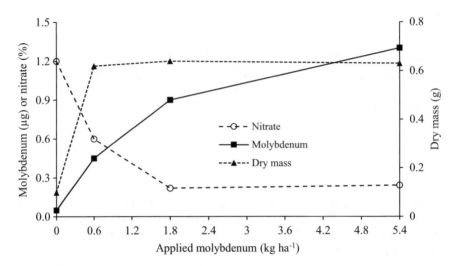

Fig. 15.2 Production of dry matter and contents of Mo and nitric N in corn plants grown under different levels of Mo in the soil, 18 days after germination

Molybdenum is essential to the enzyme xanthine dehydrogenase, which acts in the conversion of xanthine into uric acid during the catabolism of purines (adenine and guanine). Also, it is part of the oxidase responsible for the synthesis of abscisic acid (Marenco and Lopes 2005).

In this sense, Brown and Clark (1974) verified the beneficial effects of Mo in decreasing the accumulation of nitrate and the positive effects on the production of dry matter in corn plants (Fig. 15.2).

The greatest efficiency of the symbiotic fixation of N_2 by peanuts occurs with the application of molybdenum (Hafner et al. 1992).

15.4   Nutritional Requirements of Crops

The optimal level of Mo in plants is very low, close to 1 mg kg^{-1}. This value may vary depending on the crop and other factors, which is the topic of Chap. 19.

The critical level of Mo in the nodules of legumes is about ten times higher than that of leaves due to its high demand.

Extraction and Export of Nutrients

The content of molybdenum in plant tissues is very low. It is the lowest among all nutrients, therefore the requirement of plants is also low. As for extraction by area, soybeans extract the most Mo (13 g ha^{-1}), while the most demanding requirement per ton of harvested products is that of soy and cotton (5.4 and 1.1 g t^{-1}, respectively) (Table 15.1).

Soybean is also the crop that exports molybdenum the most (4.5 g t^{-1}).

As the requirement of plants is small, the recommendations of application of Mo in legumes, such as soybean, indicate applications via seed (30–60 g ha^{-1}) or leaves (1–2 kg ha^{-1}) for this crop.

As for the application via seed, research data have indicated that an adequate nutrition of Mo in the mother plant in seed fields leads to the greatest benefits in nodulation and production by applying lower Mo doses (close to 20 g ha^{-1}; Table 15.2).

Table 15.1 Molybdenum requirements of major crops

Crop	Plant part	Dry matter produced t ha^{-1}	Accumulated Mo Plant part g ha^{-1}	Total[c]	Mo required for production of 1 t of grain[d] g t^{-1}
Annual					
Cotton	Reproductive (cotton/seed)	1.3	0.2 (0.15)[b]	1.4	1.1
	Vegetative (stem/branch/leaf)	1.7	1.0		
	Root	0.5	0.2		
Soybean	Grains (pods)	2.4	11 (4.5)	13	5.4
	Stem/branch/leaf	5.6	2		
Corn[a]	Grains	6.4	2.5 (0.39)	4.0	0.6
	Crop remains	–	1.5		
Rice	Grains	3	0.3 (0.1)	1.4	0.5
	Stems	2	0.1		
	Leaves	2	0.3		
	Husk	1	0.4		
	Root	1	0.3		
Semiperennial/Perennial					
Sugarcane[a]	Stems	100	2 (0.02)	–	–
	Leaves	25	–		
Coffee tree[a]	Grains (nut)	2	0.5 (0.25)	–	–
	Trunk, branches, and leaves	–	–		

[a]Malavolta (1980)
[b]Relative export of nutrients by produced grains (g t^{-1}): Mo accumulated in the grains/dry matter of grains
[c]Suggests a nutritional requirement (total) by crop area for the respective level of productivity
[d]Suggests a relative nutritional requirement of Mo of the crop to produce one ton of commercial product (grains/stalks); obtained by the formula: Mo accumulated in the plant (vegetative + reproductive part)/dry matter of the commercial product

Table 15.2 Effects of the use of soybean seeds with different Mo contents and Mo doses applied to the seed, number and mass of nodules, and production of grains

Dose of Mo	Nodulation (10 plants)		Production
	Number	Dry matter (mg)	kg ha^{-1}
Rich seed Mo + 0 g ha^{-1}	230	350	3378
Rich seed Mo + 10 g ha^{-1}	240	430	3508
Rich seed Mo + 20 g ha^{-1}	210	330	3641
Rich seed Mo + 40 g ha^{-1}	200	350	3102
Mean seed rich in Mo	220	365	3407
Medium seed Mo + 0 g ha^{-1}	190	280	3049
Medium seed Mo + 10 g ha^{-1}	180	230	3217
Medium seed Mo + 20 g ha^{-1}	190	240	3045
Medium seed Mo + 40 g ha^{-1}	180	310	3306
Mean seed with average Mo	185	265	3154
Poor seed Mo + 0 g ha^{-1}	200	260	2766
Poor seed Mo + 10 g ha^{-1}	200	280	3075
Poor seed Mo + 20 g ha^{-1}	170	230	3020
Poor seed Mo + 40 g ha^{-1}	180	290	3129
Mean seed poor in Mo	**188**	**265**	**2998**
DMS 5%	25.3	58.3	262

Absorption Curve

Normally in soybeans, the molybdenum absorption curve starts slowly until it reaches 30 days (0.01 g ha^{-1} per day), and becomes high after this period, with a peak at 60–90 days (0.24 g ha^{-1} per day). However, unlike other micronutrients, the absorption of Mo at the reproductive (final) phase (90–120 days; 0.23 g ha^{-1} per day) remains high, probably to meet the high activity of the nitrogenase system at this stage of the plant's life cycle.

Another option for supplying Mo is via leaf. The leaf application of Mo in common beans, at a dose of 80 g ha^{-1}, increases the leaf contents of Mo and also of N without the need to split or fraction the applied dose (Pires et al. 2005).

15.5   Symptoms of Nutritional Deficiencies and Excess of Molybdenum

Deficiency

Due to its restricted mobility in plants, the symptoms of Mo deficiency described for some species occur in new leaves and in other species in old leaves. In general, internerval chlorosis occurs, similar as with Mn deficiency, in which the leaf margins tend to curve upwards or downwards (Fig. 15.3).

Fig. 15.3 Pictures and general description of the visual symptoms of Mo deficiency in different crops. Sugarcane: Disappearance of chlorophyll in spots randomly distributed in the blade; the spots develop brown centers with yellow or orange halos, which may coalesce or overlap (**a**); Beet leaves: The leaf blade is deformed by the appearance of wrinkles (**b**); Cauliflower: In vegetables, new leaves are deformed, and the leaf faces inwards (**c**)

In legumes, the characteristic symptom is N deficiency (uniform chlorosis in old leaves, which may progress to necrosis). There may be wilting of the margins of new leaves and upward (tomato) or downward curving of the blade (coffee tree). In Brassica, the new leaves grow almost without blades, presenting only the main vein.

Molybdenum deficiency is more likely to occur in plants that received N in the form of nitrate and not ammonium, since the nitrate in plants needs to be reduced (by enzymes activated by Mo) and ammonium does not.

Toxicity

Molybdenum toxicity in crops is not common and is found only when there are very high levels of Mo. However, Gris et al. (2005) found that high concentrations of ammonium molybdate (160 g ha^{-1}) applied via leaf in soybeans grown in a no-tillage system may have caused a toxic effect on plants, thus presenting less production than control plants.

Molybdenum toxicity can result in internerval chlorosis of leaves, similar as Fe deficiency, and new leaves may be distorted.

In sorghum, symptoms appear as a dark violet color on the entire blade, and can be distinguished from P deficiency symptoms, which result in dark green leaves with overlapping dark red spots (Clark 1993).

Plants may show a greater tolerance than animals to excess Mo. Forages with a high Mo content (5 to 10 mg kg^{-1}) may cause toxicity (molybdenosis) to ruminants.

References

Bataglia OC, Mascarenhas HAA. Absorção de nutrientes pela soja. Campinas: Instituto Agronômico; 1977.

Brown JC, Clark RB. Differential response of two maize inbreds to molybdenum stress. Soil Sci Soc Am Proc. 1974;38:331–3.

Clark RB. Sorghum. In: Bennet WF, editor. Nutrient deficiencies and toxicities in crop plants. Saint Paul: APS Press/The American Phytopathological Society; 1993. p. 21–6.

Gris EP, Castro AMC, Oliveira FF. Soybean yield in response to molybdenum and *Bradyrhizobium japonicum* inoculation. Rev Bras Ciênc Solo. 2005;29:151–5. https://doi.org/10.1590/S0100-06832005000100017.

Hafner H, Ndunguru BJ, Bationo A, et al. Effects of nitrogen, phosphorus and molybdenum application on growth and symbiotic N_2 fixation of groundnut in an acid sandy soil in Niger. Fertil Sci Res. 1992;31:69–77.

Jongruaysup S, Dell B, Bell RW. Distribution and redistribution of molybdenum in black gram (*Vigna mungo* L. Hepper) in relation to molybdenum supply. Ann Bot. 1994;73:161–7.

Malavolta E. Elementos de nutrição de plantas. São Paulo: Agronômica Ceres; 1980. 251p.

Malavolta E, Vitti GC, Oliveira SA. Avaliação do estado nutricional das plantas: princípios e aplicações. Piracicaba: Associação Brasileira de Potassa e do Fósforo; 1997. 319p

Marenco RA, Lopes NF. Fisiologia vegetal: fotossíntese, respiração, relações hídricas e nutrição mineral. Viçosa: UFV; 2005.

Pires AA, Araujo GAA, Leite UT, et al. Molybdenum partitioning and foliar application time on common bean leaves mineral composition. Acta Sci. 2005;27:25–31.

Quaggio JA, Raij, B Van, Piza Júnior CT (1997) Frutíferas. In: Recomendações de adubação e calagem para o Estado de São Paulo. Raij, B Van, Cantarella H, Quaggio JA et al (ed). Instituto Agronômico, pp. 121-125

Chapter 16
Chlorine

Keywords Cl availability · Chlorine uptake · Chlorine transport · Chlorine redistribution · Assimilatory chlorine · Chlorine deficient

Chlorine is one of the most absorbed micronutrients by crops, but the plant's biological requirement is relatively low. In this chapter, we will discuss initially (i) basic aspects of Cl in the soil; (ii) Cl uptake, transport, and redistribution; (iii) Cl metabolism; (iv) nutritional requirements for Cl in crops; (v) Cl extraction, export, and accumulation by the main crops; (vi) Cl deficiency and toxicity symptoms.

16.1 Introduction

Chlorine is not fixed by soil organic matter or clays and is easily leached. It is one of the first elements removed from minerals by weathering processes (accumulating in the seas).

Chlorine, along with Mo, increases its soil availability with the increase in pH.

In Brazil, normally, there is no chlorine deficiency, since the commonly used source of potassium (potassium chloride; lowest cost source) contains chlorine. Thus, chlorine is added to production systems in significant quantities via fertilizer and rains and has soil reserves but is highly leached into the soil due to its high solubility. Therefore, the dynamics of chlorine in the production systems is high, and there is rarely a deficiency. Chlorine toxicity may occur in certain situations, however rarely, due to the high tolerance of plants to high levels of Cl in tissues.

High levels of chlorine in the soil are associated with sodium, alkaline, or saline soils found in arid regions such as in the northeast of Brazil, among others.

Upon studying chlorine in the plant system, it is important to know all the "compartments" that the nutrient travels from the soil solution to roots and shoots (leaves/fruits; Fig. 16.1).

© Springer Nature Switzerland AG 2021
R. de Mello Prado, *Mineral nutrition of tropical plants*,
https://doi.org/10.1007/978-3-030-71262-4_16

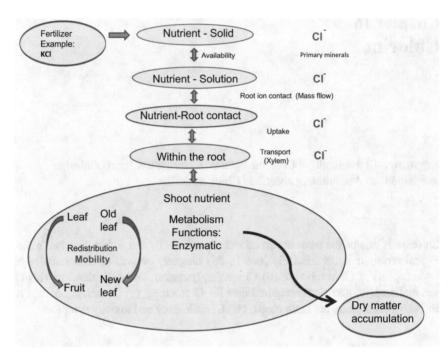

Fig. 16.1 Chlorine dynamics in the soil–plant system indicating the nutrient passage processes through the different compartments of the plant

16.2   Uptake, Transport and Redistribution of Chlorine

Uptake

Plants actively absorb chlorine in the Cl⁻ form. Its absorption can be competitively inhibited by NO_3^- and SO_4^{2-}.

Recently, new genes encoding transporters mediating Cl− influx (ZmNPF6.4 and ZmNPF6.6), Cl− efflux (AtSLAH3 and AtSLAH1), and Cl− compartmentalization (AtDTX33, AtDTX35, AtALMT4, and GsCLC2) have been identified and characterized. These transporters have proven to be highly relevant for nutrition, long-distance transport and compartmentalization of Cl−, as well as for cell turgor regulation and stress tolerance in plants (Colmenero-Flores et al. 2019).

Transport

Chlorine has a high mobility in the xylem in the same absorbed form (Cl⁻).

Redistribution

Its mobility in the plant is considered in general high, and there is a redistribution from mature leaves to the most demanding parts.

16.3 Participation in Plant Metabolism

Among the micronutrients, chlorine is the most required. Its contents in plants are high. In general, most of the chlorine in plants is not linked to the constitution of organic compounds. However, there are some compounds that contain Cl, such as chloroindole acetic acid (Engvild 1986). Thus, its main function is enzymatic as an enzyme cofactor that acts in water photolysis (photosynthesis) and in other enzymes such as ATPases of the tonoplast. It is known that Cl affects this ATPase, although monovalent cations do not affect it. It promotes, using the energy of ATP, the pumping of H^+ from the cytosol, allowing the absorption of nutrients; for example, the addition of Cl (as KCl) increases the absorption of Zn.

There are also osmotic effects of Cl on the mechanism of opening and closing of stomata and on the balance of electrical charges, although the need for this nutrient by the plant is greater to exert an osmotic effect. In the vacuole, the Cl content is about three times higher than in the cytoplasm.

The main function of Cl in plants is its participation in the Hill reaction, in the membranes of thylakoids (in chloroplasts), not as an electron transporter (single valence anion) but as a cofactor of the enzyme–Mn complex (i.e., Cl is a bridge between Mn atoms and the enzyme) by breaking the water molecule and releasing electrons to the photosystem II (Fig. 16.2).

Browyer and Leegood (1997) state that the specific role of Cl in photosynthesis is that it can regulate the access of water in the water oxidation site in the photosystem II.

Chlorine acts to regulate the osmotic pressure of the cell. The accumulation of this element inside the cell decreases the water potential, becoming lower than the external environment. Thus, there is a potential gradient of water that favors its entry into the cell, and therefore there is hydration of tissues (turgor) and opening of stomata. In addition to Cl^-, the opening and closing of stomata are regulated by the flow of K, accompanied by malate. Therefore, Cl acts as a counteranion to K^+, thus performing osmoregulation. The effect of Cl^- on the opening and closing of stomata may again affect photosynthesis due to the flow of CO_2. Cl can also minimize the effects of water deficit due to the improvement in water savings, as it can increase water potential and reduce transpiration.

Fig. 16.2 Scheme illustrating the role of Cl as an enzymatic cofactor

There is a need for further studies on Cl's participation in metabolic processes. Recent studies have shown participation of Cl in the following processes:

- stimulus in ATPase (maintains cytoplasmic pH>7) in the tonoplast. ATPase, when pumping protons from the cytoplasm to vacuums, generates H^+ electro-chemical gradient, which drives the transport of ions and organic compounds (Kirsch et al. 1996);
- inhibition of protein synthesis or degradation, since in plants deficient in Cl there is an increase in amino acids and amides;
- stimulus in the synthesis of asparagine (asparagine synthetase) and activation of other enzymes (ATPases; pyrophosphatases act in the metabolism of N);
- stimulates cell division (Harling et al. 1997).

Finally, chlorine has also been associated with mitigating certain plant diseases (root rot).

16.4   Nutritional Requirements of Crops

The chlorine concentration in plants is high (more or less equivalent to that of macronutrients), reaching up to 20,000 mg kg^{-1}. However, it is considered a micronutrient because the plant requirement is quite lower. An excellent plant growth occurs with 340–1000 mg kg^{-1}. It is noted, therefore, that unlike boron, the levels of chlorine deficiency and toxicity vary greatly.

Thus, for a proper discussion of the nutritional requirements of crops, two factors are equally important: total extraction/export of the nutrient and rate of absorption of this nutrient throughout the cultivation.

Extraction and Export of Nutrients

Research data on chlorine requirements are scarce. There are studies only for a few cultures, as well as few studies on absorption curves.

The extraction of chlorine by crops varies from 1.2 to 36 kg ha^{-1}. Corn has the highest extraction. Therefore, corn has the highest demand of Cl (5.6 kg t^{-1}) by produced grains and exports 313 g t^{-1} of grains (Malavolta et al. 1997) (Table 16.1).

When assessing the chlorine accumulation curve by soybean, there is a very slow absorption in the first 30 days (2.2 g ha^{-1} per day). In the period of 30–60 days (17.4 g ha^{-1} per day), the highest absorption rate occurs, remaining at this level until the end of the crop cycle (Bataglia and Mascarenhas 1977).

Palm tree has a high demand for Cl, especially in the period from 4 to 6 years (Viégas et al. 2020, Table 16.2).

The palm fruit has a high Cl content compared to that of other species (Viégas et al. 2020, Fig. 16.3).

Table 16.1 Chlorine requirements of major crops

Crop	Plant part	Dry matter produced	Accumulated Cl		Cl required for production of 1 t of grain[d]
			Plant part	Total[c]	
		t ha^{-1}	g ha^{-1}		g t^{-1}
Annual					
Soybean[a]	Grains (pods)	2.4	568 (237)[b]	1197	499
	Stem/branch/leaf	5.6	629		
Corn[a]	Grains	6.4	2000 (313)	36,000	5625
	Crop remains	–	34,000		
Rice	Grains	3	0.4 (0.13)	1512	504
	Stems	2	8		
	Leaves	2	3		
	Husk	1	0.5		
	Root	1	1500		

[a]Malavolta (1980)
[b]Relative export of nutrients by produced grains (g t^{-1}): Cl accumulated in the grains/dry matter of grains
[c]Suggests a nutritional requirement (total) by crop area for the respective level of productivity
[d]Suggests a relative nutritional requirement of Cl of the crop to produce one ton of commercial product (grains/stalks); obtained by the formula: Cl accumulated in the plant (vegetative + reproductive part)/dry matter of the commercial product

Table 16.2 Cumulative absorption of chlorine in oil palm trees as a function of plant age

Age (years)	Total accumulation	Absorbed Cl	Cl absorption increase
	kg ha^{-1}	g ha^{-1} per day	
2	8.6	11.8	–
3	24.6	22.5	10.7
4	63.8	43.7	21.2
5	114.3	62.6	18.9
6	217.8	99.5	36.9
7	265.5	103.9	4.4
8	320.3	109.7	5.8

16.5   Symptoms of Nutritional Deficiencies and Excess of Chlorine

Deficiency

Depending on the plant, deficiency symptoms appear first in older leaves (tomato, lettuce, cabbage, beet) or in newer leaves (corn, zucchini): withering, chlorosis, bronzing and cupping. Bronzing, which is characteristic, may progress to leaf necrosis. The roots also develop less, becoming thick and without lateral nodes.

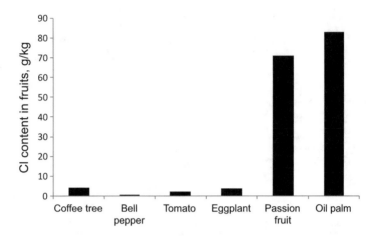

Fig. 16.3 Chlorine contents in fruits of different crops in comparison with those of the oil palm tree. * Average Cl content in palm oil fruits (2–8 years)

Under water deficit, the coconut palm deficient in Cl has a greater severity of wilting and may result in the early death of plants (Uexkull 1992).

In sugarcane, the appearance of short roots and few side branches may occur. In tobacco, the deficiency of Cl⁻ decreases leaf combustion.

It should be noted that such deficiencies of Cl have not occurred in plants in the field, since the source of potassium (KCl) contains Cl and even rainwater contains this element. Also, the critical level of Cl deficiency is low (1 mg kg⁻¹), and thus doses between 4 and 8 kg ha⁻¹ of Cl are enough to meet the nutritional needs of crops (Marschner 1995).

Toxicity

Most studies on chlorine have focused on toxicity problems caused by excessive absorption related to soil salinization problems (Castro et al. 2005).

Excesses of Cl⁻ (and also of Na⁺) in the protoplasm cause disturbances in ion balance, in addition to the specific effects of these ions on enzymes and cell membranes (Flores 1990).

The crops usually show a certain tolerance to high concentrations of Cl, showing great genotypic differences. Fageria (1985) states that a plant species can be considered tolerant to saline effect (NaCl) when the reduction in the dry matter production of shoots is less than 20%, and sensitive when this reduction is greater.

Marschner (1995) indicates that the toxicity of Cl can be reached with contents of 3500 mg kg⁻¹ (~10 mM in fresh leaf extract; most fruits, cotton, beans) or from 20,000 to 30,000 mg kg⁻¹ in barley, spinach, and lettuce. Since halophyte plants are known to have a high content of Cl in their tissues, they are plants that have developed different mechanisms of tolerance to salinity (formation of barriers to the transport of Cl and Na, elimination of salts through trichomes, leaf abscission with high salt content, and compartmentalization of salt in vacuoles). Thus, oftentimes plants can absorb dozens of times the amount of Cl normally required. This is thus a typical example of luxury consumption. However, when the concentration of Cl in the soil solution is further increased, the membranes lose selectivity, increasing the absorption of the ion by the crop. It should be noted that sensitive soy plants (Paraná) accumulate a large amount of Cl (30,000 mg kg^{-1}), while tolerant plants, with Cl exclusion mechanisms, present much lower leaf contents (1000–2000 mg Cl per kg of dry matter).

In coffee plants subjected to the application of potassium chloride (400 g per plant), there are high levels of Cl (5149 mg kg^{-1}) compared to control plants (433 mg kg^{-1} of Cl). Even so, plants do not show symptoms of toxicity (Catani et al. 1969). This also happens with banana, in which the contents of Cl reach on average 9200–9900 mg kg^{-1} (Gallo et al. 1972). Plants demanding in K, such as coffee and banana, accumulate high levels of Cl, since the most common source (lowest cost) is potassium chloride. Although there are some indications that some plants are more susceptible to Cl (tobacco, potatoes, sweet potatoes, and citrus), Cl as KCl must be replaced during application (Meurer 2006). This is also true for crops in nutrient solution, because, according to Marschner (1995), Cl reduces the absorption of NO_3^-. Abd El-Shamad and Shaddad (2000) found that the increase in the concentration of NaCl in the nutrient solution decreases the nitrate content in leaves and roots of corn.

Although crops do not show symptoms of toxicity that harm agricultural production, the excess of Cl may influence the quality of these products. Further research is needed to assess the effects of high levels of Cl on the quality of agricultural products.

Chlorine toxicity may cause the tips and the margins of leaves to burn, yellow, and fall prematurely. The symptoms of chlorine toxicity in sugarcane occur in a similar way as its deficiency, that is, the appearance of abnormal roots with few lateral roots.

The effects of Cl^- toxicity can sometimes be diminished by the presence of $CaSO_4.2H_2O$ in the substrate. Plants, when subjected to salinity, increase the concentration of Ca in the cytosol, which aggravates stress. This effect can be reduced by the effect of Ca in increasing the activity of Ca-ATPase (Niu et al. 1995), which removes Ca from the cytosol and eases stress. However, Silva et al. (2003) observed that supplemental calcium did not mitigate the inhibitory effects of NaCl-induced stress.

References

Abd El-Shamad HM, Shaddad MAK. Comparative effect of sodium carbonate, sodium sulfate, and sodium chloride on the growth and related metabolic activities of plants. J Plant Nutr. 2000;19:717–28. https://doi.org/10.1080/01904169609365155.
Bataglia OC, Mascarenhas HAA. Absorção de nutrientes pela soja. Campinas: Instituto Agronômico; 1977.

Browyer JR, Leegood RC. Photosynthesis. In: Dey PM, Harborne JB, Bonner JE, editors. Pant biochemistry. San Diego: Academic; 1997. p. 49–110.

Castro PRC, Kluge RA, Peres LEP. Manual de fisiologia vegetal: teoria e prática. Piracicaba: Agronômica Ceres; 2005.

Catani RA, Moraes FRP, Bergamim Filho H. The concentration of chlorine in coffee leaves. An Esc Sup Agr Luis Queiros. 1969;26:93–8.

Colmenero-Flores JM, Franco-Navarro JD, Cubero-Font P, et al. Chloride as a beneficial macronutrient in higher plants: new roles and regulation international. J Mol Sci. 2019;20:4686. https://doi.org/10.3390/ijms20194686.

Engvild KC. Chlorine-containing natural compounds in higher plants. Phytochemistry. 1986; 25:781–791. https://doi.org/10.1016/0031-9422(86)80002-4

Fageria NK. Salt tolerance of rice cultivars. Plant Soil. 1985;88:237–43. https://doi.org/10.1007/BF02182450.

Flores HE. Polyamines and plant stress. In: Lascher RG, Cumming JR, editors. Stress responses in plants: adaptation and acclimation mechanisms. New York: Wiley-liss; 1990. p. 217–39.

Gallo JR, Bataglia OC, Furlani PR, et al. Inorganic chemical composition of banana (*Musa acuminata*, Simmond), cultivar Nanicão. Ciên Cult. 1972;24:70–9.

Harling H, Czaja I, Schell J, et al. A plant cation-chloride co-transporter promoting auxin-independent tobacco protoplast division. Embo J. 1997;16:5855–66. https://doi.org/10.1093/emboj/16.19.5855.

Kirsch M, Zhigang A, Viereck R, et al. Salt stress induces as increased expression of V-type H+-ATPase in mature sugar beet leaves. Plant Mol Biol. 1996;32:543–7. https://doi.org/10.1007/BF00019107.

Malavolta E, Vidal AA, Gheller AC, et al. Effects of the deficiencies of macronutrients in two soybean (*Glycine max* L. Merr.) varieties, Santa Rosa and UFV- grown in nutrient solution. An Esc Super Agric Luiz de Queiroz. 1980;37:473–84. https://doi.org/10.1590/S0071-12761980000100030.

Malavolta E, Vitti GC, Oliveira SA. Avaliação do estado nutricional das plantas: princípios e aplicações. Piracicaba: Associação Brasileira de Potassa e do Fósforo; 1997, 319p.

Marschner H. Mineral nutrition of higher plants. London: Academic; 1995.

Meurer EJ. Potássio. In: Fernandes MS, editor. Nutrição mineral de plantas. Viçosa: Sociedade Brasileira de Ciência do Solo; 2006. p. 281–98.

Niu X, Bressan RA, Hasegawa PM, et al. Ion homeostasis in NaCl stress environments. Plant Physiol. 1995;109:735–42. https://doi.org/10.1104/pp.109.3.735.

Silva EB, Nogueira FD, Guimaraes PTG. Nutritional status of coffee tree evaluated by DRIS in response to potassium fertilization. Rev Bras Ciênc Solo. 2003;27:247–55. https://doi.org/10.1590/S0100-06832003000200005.

Uexkull HRV. Oil palm (*Elaeis guineensis* Jacq.). In: IFA, editor. World fertilizers use manual. Paris: IFA; 1992. p. 245–53.

Viégas IJM, Galvão JR, Silva AO. Chlorine nutrition of oil palm (*Elaeis guinq* jacq.) in eastern amazon. J Agric Stud. 2020;8:704–20. https://doi.org/10.5296/jas.v8i3.16243.

Chapter 17
Nickel

Keywords Ni availability · Nickel uptake · Nickel transport · Nickel redistribution · Assimilatory nickel · Nickel deficient

Nickel is a micronutrient associated with N in enzymatic processes; however, its improper use may cause phytotoxicity. In this chapter, we will discuss initially (i) basic aspects of Ni in the soil; (ii) Ni uptake, transport, and redistribution; (iii) Ni metabolism (reduction and assimilation); (iv) nutritional requirements for Ni in crops; (v) Ni extraction, export, and accumulation by the main crops; (vi) Ni deficiency and toxicity symptoms.

17.1 Introduction

Nickel (Ni) levels are usually higher in the topsoil, especially in clayey soils. This element is highly adsorbed in soils with high iron oxide and organic matter content (Mellis et al. 2004). Soil pH is the variable that most affects soil Ni availability, which decreases from pH 5.5 due to the formation of less soluble complexes (Temp 1991; Chen et al. 2009).

The total contents of this element in samples of uncontaminated soils in São Paulo State, Brazil, ranged from 14.8 to 50.2 mg kg^{-1} (Caridad Cancela 2002). Information on the adequate content of available soil Ni for crops is not yet known. The first basic studies that identified Ni as a nutrient were carried out in the 1980s (Eskew et al. 1983; Brown et al. 1987).

Nickel deficiency problems in field crops are rare, with more reports of toxicity problems in these contexts. The use of inadequate soil Ni rates can cause damage that is difficult to control, preventing the cultivation of many species. This is because the tolerance of crops to Ni varies widely among cultivated species, with some

Rafael Ferreira Barreto: Email – rafael.fb@outlook.com; São Paulo State University, Jaboticabal, Brazil

Flávio José Rodrigues Cruz: Email – fjrcpp@outlook.com; Rural Federal University of Pernambuco, Recife, Brazil

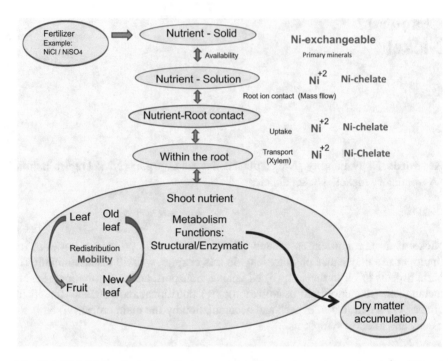

Fig. 17.1 Nickel dynamics in the soil–plant system, indicating the nutrient passage in the different plant compartments

plants being very sensitive to this element. In addition, excess soil Ni decreases microbial biomass, impairing soil life.

Nickel fertilization, either via soil, leaf, or seed treatment, must follow technical standards for physiological gains and crop growth. However, even with a beneficial effect on yield, this element may contaminate the edible product beyond the maximum allowed for human consumption (5 mg kg^{-1}), as indicated by ABIA (1985). Therefore, scientific knowledge on Ni nutrition in crops is essential to ensure balance for sustainable production with food security, avoiding soil contamination. In the study of Ni in the plant system, it is important to know all the compartments that the nutrient travels through, from the soil solution to roots and shoots (leaves/fruits; Fig. 17.1).

17.2   Uptake, Transport and Redistribution of Nickel

Uptake

The predominant form of contact between Ni and plant roots has not yet been described. The nutrient is absorbed predominantly as Ni^{+2}, and in the form of chelates. Absorption can be active, through nonspecific channels (Yusuf et al. 2011).

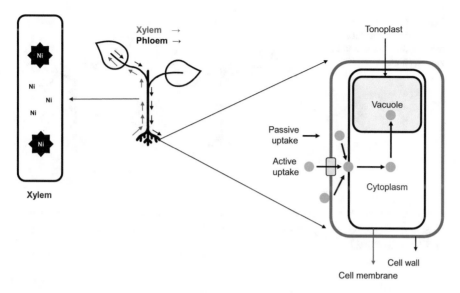

Fig. 17.2   Scheme illustrating nickel absorption and transport in plants

Nickel absorption by plant roots occurs mainly through passive diffusion and active transport (Fig. 17.2). The ratio between passive diffusion and active transport varies depending on the plant species, Ni form, and Ni content in the soil, or nutrient solution (Dan et al. 2002; Vogel-Mikus et al. 2005; Seregin and Kozhevnikova 2006).

In *Arabidopsis thaliana* plants under Fe deficiency, experimental evidence indicates that Ni is absorbed by iron transporters (AtIRT1) and accumulates (Nishida et al. 2011; Nishida et al. 2012). Soil Ni^{2+} can be absorbed via low-selective transporters from the (ZRT) (zinc)–(IRT) (iron)-like (ZIP) family, which is responsible for the absorption of Zn and Fe (Grotz et al. 1998; Milner et al. 2013; van der Pas and Ingle 2019).

In addition to root absorption, Ni can be foliarly absorbed. In a study applying a Ni radioisotope to the leaves of *Helianthus annuus*, 37% of the total Ni applied was translocated to other organs of the plant (Sajwan et al. 1996). Hirai et al. (1993) observed the same in eggplant, soybean, tomato, and oat plants.

Transport

Nickel transport in the xylem occurs in the same way as its absorption, being ionic and/or organic, in which the element is complexed with citrate, malate, and peptides (White 2012). The Ni transport pathway in plants covers from roots to shoots via transpiratory flow (Podar et al. 2004). However, organic acids and amino acids act as potential chelators that facilitate Ni transport in the xylem (Fig. 17.2), without which its transport would become slower (Cataldo et al. 1988; Briat and Lebrun 1999). Some studies have shown that Ni can be transported to the shoots via

Fig. 17.3 Illustration
indicating that nickel is
more mobile compared to
zinc in plants

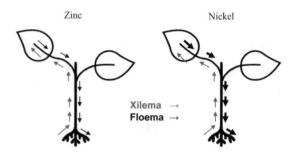

chelated xylem by organic acids such as citrate and malate (Homer et al. 1991;
Sagner et al. 1998) and by amino acids such as histidine (Persans et al. 1999; Kerkeb
and Kramer 2003).

Redistribution

Studies indicate that Ni is considered mobile in plants (Page and Feller 2005; Van
Bel 2003; Turgeon and Wolf 2009; Sharma and Dhiman 2013; Page and Feller
2015). Cataldo et al. (1978) experimentally demonstrated that 70% of the Ni present
in senescent soybean plants had been redistributed to the seeds. For example, in
comparison, nickel is more mobile than zinc in plants (Fig. 17.3). Nickel is supplied
to the meristematic parts of plants through its redistribution from old leaves to new
ones, and to seeds and fruits via phloem (Page et al. 2006).

17.3 Participation in Plant Metabolism

Nickel was effectively considered essential from a classical study: in barley cultivated in nutrient solution without Ni, the plants became unviable after three genera-
tions and did not germinate adequately (Brown et al. 1987). This same group of
researchers analyzed soybean grown in nutrient solution and observed necrosis at
the tip of the leaves (Eskew et al. 1983). According to these authors, the symptom
occurred from the accumulation of urea in toxic concentrations due to Ni deficiency.

The urease enzyme is composed of two Ni atoms, and catalyzes the conversion
of urea to CO_2 + NH_3. Ammonia (NH_3) is incorporated into carbon skeletons (glutamate), eliminating phytotoxicity.

A significant part of nitrogen is stored in different organic compounds, and in
some pathways it can originate urea, for example, (Polacco and Holland 1993): (a)
Ureide catabolism (allantoate and allantoin). Ureide-glycolate, a product of

allantoate degradation, is a precursor of urea. The ureides formed at the roots by biological N fixation by tropical legumes (soybeans, beans) are transported to the shoots, forming urea; (b) Ornithine cycle. In this cycle, the process starts from the glutamate semialdehyde that derives from the amino acids, ornithine, citrulline, argininosuccinate, and arginine. From arginine, under the action of the enzyme arginase, urea is formed.

Therefore, it is necessary to recycle urea N, formed in large quantities in the plant. This occurs from the action of urease (Ni), inducing an increase in the efficiency of use of the element, preventing its accumulation in the plant and, consequently, toxicity.

In addition to urease, Ni acts on other important enzymes. These enzymes participate in the photosynthetic pathway, including RuBisCO and aldolase, among others (Sheoran et al. 1990); in N metabolism, for example, nitrate reduction by malate dehydrogenase (Brown 2007); and in antioxidation, including dismutase superoxide (Schickler and Caspi 1999). In legumes, Ni may increase the efficiency of the nitrogenase-catalyzed reduction of N_2 to NH_3. This element is important for activating the hydrogenase that recycles part of the H_2 gas, which is vital for reducing atmospheric N (Evans et al. 1987).

Thus, the enzymatic action of Ni in plant metabolism would benefit different physiological processes, affecting crop growth and production in case the amount of the element in a plant does not meet its requirement. Finally, Ni participates in the synthesis of phytoalexins, which improve plant resistance to disease (Walker et al. 1985).

17.4 Crop Nutritional Requirements

Nickel demand is relatively very low, with few studies assessing the accumulation of this element throughout crop development. Seed Ni levels close to 0.1 mg kg^{-1} should ensure adequate germination (Brown et al. 1987). For example, low or high seed Ni content (0.04 or 8.32 mg kg^{-1}) did not affect soybean germination rate or shoot and root dry weight 28 days after germination (Kutman et al. 2013). Seed Ni content is very variable between species, reaching from 0.25 mg kg^{-1} in peas and beans to 8.0 mg kg^{-1} in oats (Kabata-Pendias and Pendias 2011).

A study by Malavolta et al. (2006) found that half of the Ni in citrus plants would be allocated to the flowers. This may indicate a high demand for the plant at this stage of crop development, which may infer the need for this element in plant metabolism. Therefore, it may be opportune to conduct research on foliar spraying of Ni at this plant stage, with a view to nourish the flower organ and induce a possible increase in flower setting. In general, the restricted information on adequate soil and leaf Ni levels, combined with the low demand of plants and possible toxicity risks, hinder the universal use of this element in crops.

However, in a study with 15 soybean genotypes, 4 had an increase in grain yield with soil supply of 0.5 mg kg^{-1} Ni (Freitas et al. 2018). Once soybean genotypes responsive to Ni fertilization have been identified, studies addressing doses can be performed. In a study with a soybean genotype responsive to Ni, the dose of highest grain yield was 3.35 mg kg^{-1}, the critical level in the leaf tissue was 1.8 to 2.2 mg kg^{-1} Ni, and the ideal grain Ni content ranged from 12.9 to 15.9 mg kg^{-1}. According to the authors, these values are well below the toxic levels expected for human ingestion of fresh grains (Freitas et al. 2019).

17.5   Deficiency and Toxicity Symptoms

Deficiency

The most well-known Ni deficiency symptom occurred in pecan, with rounded dark spots at the tip of new leaves and mouse ear-like distortion (Wood et al. 2006; Fig. 17.4). In leguminous species, such as soybean, Ni deficiency is manifested by the accumulation of urea, causing tissue necrosis (Kutman et al. 2013; Fig. 17.5). In another study with soybean plants, the increase in Ni doses increased the green color intensity of leaves, until a maximum was reached at the dose of 3.00 mg kg^{-1} Ni, after which symptoms of Ni toxicity were seen (Freitas et al. 2019; Fig. 17.6). There is a need for further studies on these symptoms in other species. Adequate foliar Ni contents range from 0.01 to 10 mg kg^{-1} (Brown et al. 1987). This wide variation also indicates the need for research on different species.

a   b

Fig. 17.4 Nickel deficiency in pecan; altered shape of blade to produce mouse ear-like or little leaf foliage (a), and blunting, crinkling, dark green zone, and necrosis of the apical portion of the leaf blade (b)

Fig. 17.5 Toxicity symptoms in youngest parts of 28-day-old urea-sprayed (2%) soybean (*Glycine max* cv. Nova) plants 3 days after foliar urea application. Plants were raised from seeds with low, medium, or high Ni concentrations with or without solution Ni supply under growth chamber conditions. †[1] Nickel concentrations of youngest parts. Values are means and standard deviations of three independent pot replicates, each containing three plants. †[2] n.d. not detectable (<1 µg kg^{-1})

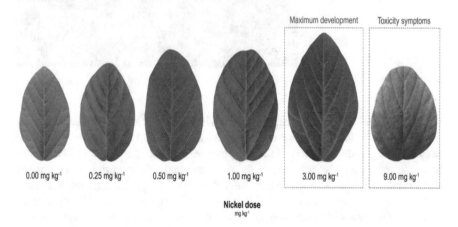

Fig. 17.6 Effect of six Ni doses applied via soil on soybean leaves (R1–R2). The increase in Ni doses increased the green color intensity of leaves, until a maximum was reached at the dose of 3.00 mg kg^{-1} Ni, after which symptoms of Ni toxicity were seen

Toxicity

At the molecular level, Ni toxicity stems from its action on the photosystem, disturbing the Calvin cycle and inhibiting electrical transport because of the excessive amounts of ATP and NADPH accumulated by the inefficiency of dark reactions

Fig. 17.7 Soybean leaves showing Ni toxicity symptoms. (A) Control, (B) 0.05 μmol L⁻¹, (C) 0.1 μmol L⁻¹, (D) 0.5 μmol L⁻¹, (E) 10 μmol L⁻¹, and (F) 20 μmol L⁻¹. Leaves were harvested after 7 days of exposure to Ni treatment

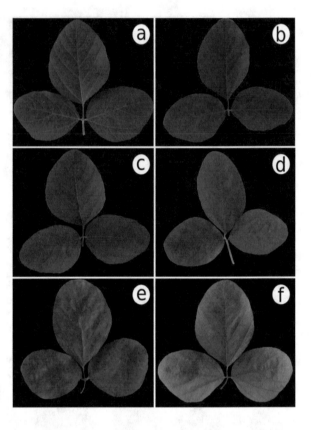

(Krupa et al. 1993). The Ni toxicity symptom is a chlorosis similar to that caused by micronutrient deficiency (Mn and Fe), decreasing root and shoot growth and possibly evolving to deformation of several plant parts (Mishra and Kar 1974). In soybean, the increase in Ni concentrations in the nutrient solution promoted chlorosis and a predominance of brown color (Reis et al. 2017; Fig. 17.7). In another study with soybean, plants treated with 9.0 mg kg⁻¹ Ni showed Ni toxicity symptoms characterized by the development of small black spots that became necrotic, surrounded by a yellow halo, mainly on the leaf margins (Freitas et al. 2019; Fig. 17.6). Nickel toxicity occurs when Ni content in the dry matter of plants is greater than 50 mg kg⁻¹, with the exception of accumulating and hyperaccumulating species (Adriano 1986). Lettuce, for example, is an accumulator; hyperaccumulators would be *Vellozia* spp., with more than 3000 mg kg⁻¹ in its leaves, and *Sebertia acuminate*, with 11,700 mg kg⁻¹ (Brooks et al. 1990). However, Marschner (1995) indicates that values greater than 10 mg kg⁻¹ Ni in the dry matter can cause toxicity for most crops. An important aspect is that although Ni is not at a toxic level for the plant, it can be at a critical level for human consumption.

References

ABIA – Associação Brasileira de Indústrias de Alimentação. Compêndio da Legislação de Alimentos, São Paulo. 1985.

Adriano DC. Trace elements in the terrestrial environmental. New York: Springer; 1986.

Boyd RS. Plant defense using toxic inorganic ions: conceptual models of the defensive enhancement and joint effects hypotheses. Plant Sci. 2012;195:88–95. https://doi.org/10.1016/j.plantsci.2012.06.012.

Briat JF, Lebrun M. Plant responses to metal toxicity. C R Acad Sci. 1999;322:43–54.

Brooks RR. Plants that hyperaccumulate heavy metals: Their role in phytoremediation, microbiology, archaeology, mineral exploration, and phytomining. Wallingford: CAB International; 1998. p. 380.

Brooks RR, Reeves RD, Baker AJM, et al. The Brazilian serpentine plant expedition (BRASPEX), 1988. Natl Geogr Res. 1990;6:205–19.

Brown PH, Welch RM, Cary EE. Nickel: a micronutrient essential for higher plants. Plant Physiol. 1987;85:801–3. https://doi.org/10.1104/pp.85.3.801.

Caridad Cancela R. Contenido de macro-micronutrientes, metales pesados y otros elementos en suelos naturales de São Paulo (Brasil) y Galicia (España). 573p. Tesis (Doctorado) – Universidad de A Coruña, A Coruña, España. (2002)

Cataldo DA, Garland TR, Wildung RE. Nickel in plants: II. Distribution and chemical form in soybean plants. Plant Physiol. 1978;62:566–70. https://www.jstor.org/stable/4265482

Cataldo DA, McFadden KM, Garland TR, et al. Organic constituents and complexation of nickel (II), iron (III), cadmium (II) and plutonium (IV) in soybean xylem exudates. Plant Physiol. 1988;86:734–9. https://doi.org/10.1104/pp.86.3.734.

Chen C, Huang D, Liu J. Functions and toxicity of nickel in plants: recent advances and future prospects. Clean Soil Air Water. 2009;37:304–13. https://doi.org/10.1002/clen.200800199.

Dan TV, Krishnaraj S, Saxena PK. Cadmium and nickel uptake and accumulation in scented Geranium (*Pelargonium* sp. 'Frensham'). Water Air Soil Pollut. 2002;137:355–64. https://doi.org/10.1023/A:1015590007901.

Eskew DL, Welch RM, Cary EE. Nickel: an essential micronutrient for legumes and possibly all higher-plants. Science. 1983;222:621–3. https://doi.org/10.1126/science.222.4624.621.

Evans, H.J, Harker AR, Papen H, et al. Physiology, biochemistry, and genetics of the uptake hydrogenase in rhizobia. Annu. Rev Microbiol. 1987;41:335–361. https://doi.org/10.1146/annurev.mi.41.100187.002003.

Freitas DS, Rodak BW, Reis AR, et al. Hidden nickel deficiency? Nickel fertilization via soil improves nitrogen metabolism and grain yield in soybean genotypes. Front Plant Sci. 2018;9:614. https://doi.org/10.3389/fpls.2018.00614.

Freitas DS, Rodak BW, Carneiro MAC, et al. How does Ni fertilization affect a responsive soybean genotype? A dose study. Plant Soil. 2019;441:567–86. https://doi.org/10.1007/s11104-019-04146-2.

Grotz N, Fox T, Connolly E, et al. Identification of a family of zinc transporter genes from Arabidopsis that respond to zinc deficiency. Proc Natl Acad Sci U S A. 1998;95:7220–4. https://doi.org/10.1073/pnas.95.12.7220.

Hirai M, Kawai-Hirai R, Hirai T, et al. Structural change of jack bean urease induced by addition of surfactants studied with synchrotron-radiation small-angle X-ray scattering. Eur J Biochem. 1993;215:55–61. https://doi.org/10.1111/j.1432-1033.1993.tb18006.x.

Homer FA, Reeves RD, Brooks RR, et al. Characterization of the nickel-rich extract from the nickel hyperaccumulator *Dichapetalum gelonioides*. Phytochemistry. 1991;30:2141–5. https://doi.org/10.1016/0031-9422(91)83602-H.

Kabata-Pendias A, Pendias, H. Trace Elements in Soils and Plants. 4th Edition, CRC Press, Boca Raton, 2011.

Kerkeb L, Kramer U. The role of free histidine in xylem loading of nickel in *Alyssum lesbiacum* and *Brassica juncea*. Plant Physiol. 2003;131:716–24. https://doi.org/10.1104/pp102.010686.

Kutman BY, Kutman UB, Cakmak I. Nickel-enriched seed and externally supplied nickel improve growth and alleviate foliar urea damage in soybean. Plant Soil. 2013;363:61–75. https://doi.org/10.1007/s11104-012-1284-6.

Krupa, Z, Siedlecka A, Maksymiec W, et al. In vitro responses of photosynthetic apparatus of Phaseolus vulgaris L. to nickel toxicity. Plant Physiol. 1993;142:664–66. https://doi.org/10.1016/S0176-1617(11)80899-0.

Malavolta E, de Moraes MF, Lavres Junior J et al. Micronutrientes em metais pesados - essencialidade e toxidez In: Paterniani, E. Ciência, agricultura e sociedade. Ed. Brasília, DF: EMBRAPA Informação Tecnológica; 2006. pp 117–154.

Marschner H. Mineral nutrition of higher plants. London: Academic; 1995.

Mellis EV, Cruz MCP, Casagrande JC. Nickel adsorption by soils in relation to pH, organic matter and iron oxides. Sci Agric. 2004;61:190–5. https://doi.org/10.1590/S0103-90162004000200011.

Milner MJ, Seamon J, Craft E, et al. Transport properties of members of the ZIP family in plants and their role in Zn and Mn homeostasis. J Exp Bot. 2013;64:369–81. https://doi.org/10.1093/jxb/ers315.

Mishra D, Kar M. Nickel in plant growth and metabolism. Bot Rev. 1974;40:395–452. https://doi.org/10.1007/BF02860020.

Nishida S, Tsuzuki C, Kato A, et al. AtIRT1, the primary iron uptake transporter in the root, mediates excess nickel accumulation in *Arabidopsis thaliana*. Plant Cell Physiol. 2011;52:1433–42. https://doi.org/10.1093/pcp/pcr089.

Nishida S, Aisu A, Mizuno T. Induction of IRT1 by the nickel-induced iron-deficient response in *Arabidopsis*. Plant Signal Behav. 2012;7:329–31. https://doi.org/10.4161/psb.19263.

Page V, Feller U. Selective transport of zinc, manganese, nickel, cobalt and cadmium in the root system and transfer to the leaves in young wheat. Ann Bot. 2005;96:425–34. https://doi.org/10.1093/aob/mci189.

Page V, Feller U. Heavy metals in crop plants: transport and redistribution processes on the whole plant level. Agronomy. 2015;5:447–63. https://doi.org/10.3390/agronomy5030447.

Page V, Weisskopf L, Feller U. Heavy metals in white lupin: uptake, root-to-shoot transfer and redistribution within the plant. New Phytol. 2006;171:329–41. https://doi.org/10.1111/j.1469-8137.2006.01756.x.

Persans MW, Xiange Y, Patnoe JMML, et al. Molecular dissection of histidine's role in nickel hyperaccumulation in *Thlaspi goesingense* (Hálácsy). Plant Physiol. 1999;121:1–10. https://doi.org/10.1104/pp.121.4.1117.

Podar D, Ramsey MH, Hutchings MJ. Effect of cadmium, zinc and substrate heterogeneity on yield, shoot metal concentration and metal uptake by *Brassica juncea*: implications for human health risk assessment and phytoremediation. New Phytol. 2004;163:313–24. https://doi.org/10.1111/j.1469-8137.2004.01122.x.

Polacco JC, Holland MA. Roles of urease in plant cells. Int Rev Cytol. 1993;145:65–103. https://doi.org/10.1016/S0074-7696(08)60425-8.

Reis AR, Barcelos JPQ, Osório CRWS, et al. A glimpse into the physiological, biochemical and nutritional status of soybean plants under Ni-stress conditions. Environ Exp Bot. 2017;144:76–87. https://doi.org/10.1016/j.envexpbot.2017.10.006.

Sagner S, Kneer R, Wanner G, et al. Hyperaccumulation, complexation and distribution of nickel in *Sebertia acuminate*. Phytochemistry. 1998;47:339–43. https://doi.org/10.1016/S0031-9422(97)00593-1.

Sajwan KS, Ornes WH, Youngblood TV, et al. Uptake of soil applied cadmium, nickel and selenium by bush beans. Water Air Soil Pollut. 1996;91:209–17. https://doi.org/10.1007/BF00666258.

Schickler H, Caspi H. Response of antioxidant enzymes to nickel and cadmium stress in hyperaccumulator plants of the genus Alyssum. Plant Physiol. 1999;105:39–44. https://doi.org/10.1034/j.1399-3054.1999.105107.x.

Seregin IV, Kozhevnikova AD. Physiological role of nickel and its toxic effects on higher plants. Russ J Plant Physiol. 2006;53:257–77. https://doi.org/10.1134/S1021443706020178.

Sharma A, Dhiman A. Nickel and cadmium toxicity in plants. J Pharm Sci Innov. 2013;2:20–4. https://doi.org/10.7897/2277-4572.02213.

Sheoran IS, Singal HR, Singh R. Effect of cadmium and nickel on photosynthesis and the enzymes of the photosynthetic carbon reduction cycle in pigeon pea (*Cajanus cajan*). Photosynth Res. 1990;23:345–351. https://doi.org/10.1007/BF00034865.

Temp GA. Nickel in plants and its toxicity: Ustoichivost'k tyazhelym metallam dikorastushchikh vidov (Resistance of wild species to heavy metals). In: Alekseeva-opova NV editor. Leningrad: Lenuprizdat, 1991; pp. 139–146.

Turgeon R, Wolf S. Phloem transport: cellular pathways and molecular trafficking. Annu Rev Plant Biol. 2009;60:207–21. https://doi.org/10.1146/annurev.arplant.043008.092045.

Van Bel AJE. The phloem, a miracle of ingenuity. Plant Cell Environ. 2003;6:125–49.

Van der Pas L, Ingle RA. Towards an understanding of the molecular basis of nickel hyperaccumulation in plants. Plants. 2019;8:11. https://doi.org/10.3390/2Fplants8010011.

Vogel-Mikus K, Drobne D, Regvar M. Zn, Cd and Pb accumulation and arbuscular mycorrhizal colonization of pennycress *Thlaspi praecox* Wulf. (Brassicaceae) from the vicinity of a lead mine and smelter in Slovenia. Environ Pollut. 2005;133:233–42. https://doi.org/10.1016/j.envpol.2004.06.021.

Walker CD, Graham RD, Madison JT, et al. Effects of Ni deficiency on some nitrogen metabolites in cowpeas (Vigna unguiculata L. Walp.). Plant Physiol. 1985;79:474–479. https://doi.org/10.1104/pp.79.2.474.

White PJ. Long-distance transport in the xylem and phoem. In: Marschner P, editor. Mineral nutrition of higher plants. Berlin: Elsevier; 2012. p. 49–70.

Wood BW, Reilly CC, Nyczepir AP. Field deficiency of nickel in trees: symptoms and causes. Acta Hortic. 2006;721:83–98. https://doi.org/10.17660/ActaHortic.2006.721.10.

Yusuf M, Fariduddin Q, Hayat S, Ahmad A. Nickel: an overview of uptake, essentiality and toxicity in plants. Bull Environ Contam Toxicol. 2011;86:1–17. https://doi.org/10.1007/s00128-010-0171-1.

Chapter 18
Potentially Toxic Metals

Keywords Heavy metals · Arsenic · Cadmium · Chromium · Mercury · Lead

Heavy metals in soil-plant system are of great concern for soil quality and food production, in quality and quantity, for global food security. Arsenic (As), cadmium (Cd), chromium (Cr), mercury (Hg), and lead (Pb) are the heavy metals with the highest potential toxicity for crop production. In agricultural areas, heavy metal concentration in soil can be high due to application of organic and inorganic wastes from urban and industrial activities, fertilizers, soil conditioners, pesticides, waste-water, and atmospheric deposition. Current days demand smart crop management to improve crop yields. The uptake into the roots, the loading into the xylem, the acropetal transport to the leaves via sap flow, the redistribution via phloem, the accumulation in the shoot, and metabolic, physiologic, and morphological injury expression are fundamental for the crop quality and yield. Therefore, this chapter is focused on the mechanisms of: (i) uptake, (ii) transport, (iii) redistribution, (iv) metabolism, (v) accumulation (compartmentalization), and (vi) symptomatology of these potentially toxic metals.

Cassio Hamilton Abreu-Junior: Email – cahabreu@cena.usp.br; University of São Paulo, Piracicaba, Brazil
Paulo Henrique Silveira Cardoso: Email – paulohscardoso@usp.br; University of São Paulo, Piracicaba, Brazil
Flávio José Rodrigues Cruz: Email – fjrcpp@outlook.com; Rural Federal University of Pernambuco, Recife, Brazil
Thiago Assis Rodrigues Nogueira: Email – tar.nogueira@unesp.br; São Paulo State University, Jaboticabal, Brazil
Arun Dilipkumar Jani: Email – arun.jani@usda.gov; United States Department of Agriculture, Oregon, USA

© Springer Nature Switzerland AG 2021
R. de Mello Prado, *Mineral nutrition of tropical plants*,
https://doi.org/10.1007/978-3-030-71262-4_18

18.1 Introduction

Heavy metal is a term assigned to plant essential (Cu, Zn, Mo, and Ni) and nonessential (As, Cr, Cd, Hg, Pb, and Se) chemical elements with an atomic number above 20 or a specific mass greater than 5 g cm^{-3} (Malavolta 2006), associated with the contamination of natural resources. However, semi-metallic (or metalloids) and nonmetal elements, such as arsenic (As) and selenium (Se), are included in this term. Other definitions are also reported in the literature, such as potentially toxic elements (Abreu-Junior et al. 2019; Galhardi et al. 2020a), trace metals (Galhardi et al. 2020b), trace elements (Kabata-Pendias and Mukherjee 2007; Vilela et al. 2020), among other variations. Since heavy metal is the most widespread term found in scientific publications (Page and Feller 2015; Sheetal et al. 2016; Xiao et al. 2017; Cao et al. 2019), it is used in this chapter.

The origin of heavy metals in soil can be lithogenic (due to the parent material of the soil – rocks), pedogenic (due to a lithogenic source but with changes due to the process of soil formation – weathering) and anthropic (due to deposition activities / use of sources containing these elements) (Kabata-Pendias and Mukherjee 2007). In agricultural areas, heavy metal concentration in soil can be high due to application of organic and inorganic wastes from urban and industrial activities, fertilizers, soil conditioners, pesticides, wastewater, and atmospheric deposition (Shi et al. 2018; Peng et al. 2019).

When the availability of heavy metals in soil reaches toxic levels, crop yield can decline considerably (Rellán-Álvarez et al. 2006; Sahu et al. 2012; Lavres et al. 2019). Additionally, elevated heavy metal concentrations in the edible components of plants can negatively impact the health of humans and animals (Rabêlo et al. 2020; Ye et al. 2020).

18.2 Uptake, Transport and Redistribution of Potentially Toxic Metals

Uptake

Arsenic can be absorbed by roots as arsenate (As^{5+}) and arsenite (As^{2+}). By maintaining a chemical analogy with phosphate, As^{5+} can be absorbed via a high-affinity phosphate transport system, while absorption of As^{2+} occurs through intrinsic membrane proteins, similar to nodulin -26 (Bienert et al. 2008; Isayenkov and Maathuis 2008; Shukla et al. 2015). Under anoxic soil conditions, As^{2+} is absorbed via a silicic acid carrier (OsNIP2;1/Lsi1) (Verbruggen et al. 2009).

Mechanisms involved in Pb uptake by roots are poorly understood. It is believed that Pb^{2+} can probably be absorbed through ion channels or by passive mechanisms (Zimdahl 1976; Hughes et al. 1980; Marchetti 2013).

The absorption of Hg probably occurs through carriers or channels of iron (Fe), copper (Cu), and zinc (Zn), as these transport systems have broad affinities for ion metals (Patra and Sharma 2000; Clemens 2006; Esteban et al. 2008). In the soil, inorganic forms of Hg are the most available for root absorption, especially when linked to fulvic acid (Yu et al. 2004; Meng et al. 2012). However, organic forms are highly toxic to plants, because they are easily absorbed by plant roots and highly soluble in lipids (Clayden et al. 2013).

Cadmium is absorbed by roots as Cd^{2+} through the Zn transporter system of high and low affinity (Hacisalihoglu et al. 2001; Song et al. 2017).

Chromium is absorbed by roots as Cr^{3+} and Cr^{6+}. However, no specific absorption mechanism has been identified, although it is postulated that Cr uptake occurs through carriers of essential mineral elements (Shanker et al. 2005; Oliveira 2012; Singh et al. 2013).

Transport

In several plant species, only a small fraction of the As is transported to the shoot as it occurs with other toxic ions. On the other hand, in accumulating plant species (Brooks 1998), a considerable part of the As is transported to the aerial part. In most plant species, the As form that predominates in the xylem sap is As^{3+}. It varies around 60–100% of the total As (Zhao et al. 2009). Transport of As^{5+} is limited because it is reduced to As^{3+}, complexed by phytoquelatins and sequestered in vacuoles of root cells (Zhao et al. 2009).

In xylem, Pb transport occurs in complexed form with amino acids or organic acids (Roelfsema and Hedrich 2005; Vadas and Ahner 2009; Maestri et al. 2010), but it also can be transported in inorganic form (Pourrut et al. 2011). However, most of Pb is accumulated in the roots (Zimdahl 1976; Hughes et al. 1980; Marchetti 2013).

In higher plants, Hg can be transported in both inorganic and chelated forms. The Hg concentration in the aerial part of seedlings of *Zea mays* L., when supplied in inorganic or chelated form (e.g., sodium thiosulfate or iodide), is similar, indicating that these plants can transport this element in both forms. However, Hg concentration in the root tends to be higher compared to the aerial part (Li et al. 2020a, 2020b). Sodium or ammonium thiosulfate complexing agents are also involved in the transport of Hg from the root to the aerial part in leguminous plants (Moreno et al. 2005).

Transport of Cd from the root to the aerial components is dependent on membrane carriers located in root cells, which are in charge of loading and unloading Cd into the xylem. In root cells, Nramp5 (Natural resistance-associated macrophage protein 5) and heavy metal transporting ATPase are probably the most important carriers for the Cd loaded into the xylem. The HMA4 carrier is also involved in unloading Cd, present in the xylem, into the leaf cells. In the stem or branches, LCT1 (Low-affinity cation transporter) is the carrier that modulates the Cd unloading from the xylem. Cadmium can be transported into the xylem in cationic form (Cd^{2+}) or linked with some ligand (e.g., phytoquelatins) (Song et al. 2017).

The predominant form of Cr in the root cells is Cr^{3+}, due to the reduction of the Cr^{6+} form to Cr^{3+}, whose transport to the aerial part is limited (Zayed et al. 1998). This is because Cr^{3+} is immobilized in the vacuoles of the root cells, which results in greater accumulation of Cr in the plant root compared to the aerial part (Shanker et al. 2005). Besides, in aquatic accumulating plants, metal ions (e.g., both Cr^{3+} and Cr^{6+}) can be actively absorbed into the root symplasm via plasmalemma, and adsorbed on the cell walls via passive diffusion or moved acropetally on the cell roots (Mishra and Tripathi 2009). Conversely, there is considerable transport of Cr to the aerial part when plants are fertilized with EDTA chelating agent, which prevents reduction of Cr on the root cells and in xylem vessels (Myttenaere and Mousny 1974; Athalye et al. 1995; Cary et al. 1977), thus suggesting that complexed and Cr^{6+} forms are transported via xylem.

Redistribution

After the processes of absorption, transport via xylem, and accumulation of essential elements in plant organs, their redistribution occurs during the vegetative and reproductive growth stages of the plants. This physiological phenomenon occurs through symplastic transport via phloem as it occurs with sucrose. For inorganic elements, such as inorganic N, root is the major source organ and leaves, tubers, fruit/seeds(coat) are the major sink organs, but it depends on the source–drain interaction among the plant organs, depending on the plant status and organ development stage (Chang and Zhu 2017).

The redistribution of a metal ions is dependent on their mobility in the phloem and the strength of the source and drain of the plants (Van Bel 2003; Turgeon and Wolf 2009; Page and Feller 2015). In wheat plants, with leaves treated with radio-labeled metals, [109]Cd had low export from the second pair of leaves to other vegetative organs, due to low mobility in the phloem, and there was no mobilization of Cd from leaves to grain during the filling stage (Riesen and Feller 2005). However, there are accumulations of As in rice grain and Cd in wheat grain due to direct flow via root-xylem-phloem route (Harris and Taylor 2013; Suriyagoda et al. 2018).

Unlike Cd, As is more easily remobilized from vegetative tissues to developing grain. In rice plants treated with different As forms, remobilization of inorganic forms (arsenate and arsenite) from leaves to grain was lower compared to organic forms (dimethylarsinic and monomethylsonic acids) (Carey et al. 2011). This is because the mobility of organic forms of As is greater compared to inorganic forms in phloem. About 54%, 56%, 100%, and 89% of the As transported in the phloem is found in the forms As^{3+}, As^{5+}, monomethylsonic acid [MMA (V)], and dimethylsinic acid [DMA (V)], respectively (Kumarathilaka et al. 2018). These qualitative and quantitative differences regarding the transport of As in the phloem are due to genotypic differences (Kumarathilaka et al. 2018).

There is experimental evidence regarding the accumulation of Pb, Hg, and Cr in grain of various cereal species (Tkachuk and Kuzina 1972; Sharma et al. 1995;

Shakerian et al. 2012; Xie et al. 2018). However, mechanisms of remobilization of these heavy metals from vegetative to reproductive organs or from old to new leaves are not completely known given the toxicity of these heavy metals, even under very low concentrations.

18.3 Effects on Plant Metabolism

After being absorbed and transported to the aerial part of plants, heavy metals trigger physiological disorders that reduce plant growth. This effect is related to the type of interaction that heavy metals establish with plant metabolism. In general, oxidative stress, metabolic changes, and ionic imbalance can occur (Bhaduri and Fulekar 2012; DalCorso et al. 2013; Dubey et al. 2018).

When As is supplied to plants in As^{5+} form, more than 90% of its form is reduced to As^{3+} in cell (vacuoles) of roots and shoots. This As reduction mechanism constitutes a detoxification process (Xu et al. 2007), but As is still toxic to plants. One of the toxic mechanisms of As in plants is the replacement of inorganic phosphorus (Pi) by As^{5+} in biochemical reactions (Long and Ray 1973; Gresser 1981). In biochemical reactions related to central plant metabolism (glycolysis and oxidative phosphorylation), biosynthetic metabolism (phospholipid metabolism), cell signaling (phosphorylation and dephosphorylation) and nucleic acid metabolism (DNA and RNA), there may be disorders resulting from the toxic action of As^{5+} (Finnegan and Chen 2012). As^{3+} has great affinity for thiol group (-SH), being able to link to three -SH groups (Kitchin and Wallace 2006). For example, As^{3+} can bind to three different glutathione (GHS) molecules because it has a -SH group. As^{3+} can even bind to proteins rich in amino acid residues containing thiol group, interfering in the biological activity of proteins, resulting in physiological and biochemical disorders (Armendariz et al. 2016; Shahid et al. 2019).

The primary harmful effect of Pb in plant species is the reduction of root growth due to the paralysis of cell division in the apical root meristem (Eun et al. 2000). Under toxic Pb levels, there is loss of viability of root cells as observed in rice plants (*Oryza sativa*) (Huang and Huang 2008), with distension and lesions in the cell walls of the root system resulting from activation of enzymes that degrade the cellular wall (Kaur et al. 2013). Lead toxicity promotes mineral disorder in plants, with a decrease in the Ca concentration of the leaf due to the inhibition of Ca transporters and/or substitution of this element in the Ca-binding sites in biological structures (Habermann et al. 1983; White and Broadley 2003; Sharma and Dubey 2005; Wojas et al. 2007), which reduces plant growth (Sharma and Dubey 2005; Reddy et al. 2005; Zhou et al. 2018).

The mechanism of toxicity of Hg in plants consists of inducing the permeability of cell membranes, high affinity with sulfhydryl and phosphate groups, substitution of nutrients and alteration of the biological function of proteins (Patra and Sharma 2000; Patra et al. 2004). Mercury is accumulated in large amounts in the root system (Marrugo-Negrete et al. 2016; Li et al. 2020a, b; Xu et al. 2020). When Hg concentrations are excessive in plant tissue, it promotes considerable physiological changes with negative effects on the crop cycle (Mondal et al. 2015; Marrugo-Negrete et al. 2016; Hu et al. 2020).

Cadmium toxicity in plants is associated with its ability to bind with thiol, carboxylic, and histidyl groups of structural proteins and enzymes, changing their biological functions. In addition, as it presents chemical similarities with divalent cations, Cd can alter the functional activity of proteins by replacing ions at their activity sites, promoting nutritional disorders in vegetables (Huybrechts et al. 2019). These mechanisms of interaction between Cd and proteins induce changes in photosynthetic and oxidative metabolism (Benavides et al. 2005; López-Climent et al. 2011; Sun et al. 2015; Qin et al. 2020; Genchi et al. 2020).

18.4   Heavy Metal Concentration and Accumulation

Heavy metal concentration and accumulation in plant organs are directly related to the capacity of absorption, transport, and redistribution of these elements (Page and Feller 2015). Mostly, heavy metal concentration in plants is greater in the roots than in the shoots (Table 18.1). This can vary according to soil type (Zeng et al. 2011; Cao et al. 2019), the element form, and concentration (Azevedo et al. 2018; Fan et al. 2020; Li et al. 2020b), gene expressions (Li et al. 2020a; Zhang et al. 2020), and cultivars (Lavres et al. 2019; Ashraf et al. 2020), among other biotic and abiotic factors, conferring heavy metal tolerance or phytoremediation potential to the plants.

One of the first defense mechanisms of the plant to the excess heavy metal in the soil is the accumulation of these elements in the roots, especially in the cell wall. This was observed with Cd to *Urochloa decumbens* and *Panicum maximum* (Rabêlo et al. 2020), with Cr^{3+} to *Oryza sativa* L. (Fan et al. 2020), with Cr to *Leersia hexandra* (Liu et al. 2009), and Pb in *Conyza canadenses* (Li et al. 2016). Therein, the percentage of these elements was between 60% and 100% linked to the cell wall structures of the roots. This is because the positive charge of metals is easily complexed at the negative sites of the cell wall, such as carboxyl, hydroxyl, amine, and phosphate groups (Liu et al. 2010; Li et al. 2016).

The heavy metal accumulation on the cell wall is a strategy against toxicity due to its cell structure with minor metabolic activity in relation to the organelles and cytosol (Taiz et al. 2017). However, elements of chemical speciation can also influence heavy metal distribution on the roots. Fan et al. (2020) observed greater Cr^{6+} accumulation in the cytosol (44–53%) than in the cell wall (28–42%) of the roots, but the opposite was observed for Cr^{3+}. This can be one of the characteristics that confer greater Cr^{6+} toxicity to plants, requiring them to activate their antioxidant system to reduce stress caused by Cr^{6+}.

When transported to the shoots, heavy metal concentrations in the stem are intermediate between values for roots and leaves (Ashraf et al. 2020; Torres et al. 2020; Ye et al. 2020). The heavy metal concentration in the leaves can decrease from the oldest to the youngest leaf, due to the mobility of these elements in the conducting vessels. For example, Hg concentration in *Lycopersicon esculentum* ranged from 22

Table 18.1 Heavy metals concentration in parts of cultivated plants in soil or hydroponics

Element	Plant species	Rate	Plant part	Concentration mg kg^{-1}	Reference
As	Glycine max L.	25 μmol*	Root	13–20	Oller et al. (2020)
			Leaf	193–262	
	Fragaria × ananassa	1000 μL^{-1} As^{3+} *	Root	4.10	Torres et al. (2020)
			Stem	1.27	
			Fruit	0.30	
		1000 μL^{-1} As^{5+} *	Root	0.44	
			Stem	0.43	
			Fruit	0.20	
Cd	Zea mays L.	6 e 30 μmol *	Root	450–2585	Rellán-Álvarez et al. (2006)
			Leaf	12–48	
	Solanum tuberosum L.	0.44–1 mg kg^{-1} **	Tuber	0.10–0.94	Ye et al. (2020)
			Stem	1.20–12.2	
			Leaf	1.80–20.0	
	Brassica napus	0.24–0.83 mg kg^{-1} **	Root	0.17–3.17	Cao et al. (2019)
			Leaf	0.09–3.18	
			Seed	< 0.20	
Cr	Leersia hexandra	5–60 mg L^{-1} *	Root	3500–14,472	Liu et al. (2009)
			Leaf	2000–5275	
	Oryza sativa L.	12 mg L^{-1} Cr^{3+} *	Root	1400	Fan et al. (2020)
			Leaf	140	
		16 mg L^{-1} Cr^{6+} *	Root	531	
			Leaf	118	
Hg	Lycopersicon esculentum	10–50 μmol *	Root	625–1420	Cho and Park (2000)
			Leaf	17–52	
	Triticum aestivum	2.5–25 μmol *	Root	11–721	Sahu et al. (2012)
			Leaf	2.3–41	
	Zea mays L.	6 e 30 μmol *	Root	1805–18,054	Rellán-Álvarez et al. (2006)
			Leaf	16–58	
	Pisum sativum	0.1–100 μmol *	Root	12–2614	Azevedo et al. (2018)
			Leaf	3–142	
Pb	Brassica napus	13–153 mg kg^{-1} **	Root	0.26–35	Cao et al. (2019)
			Leaf	0.01–10	
			Pod	0.07–3.3	
	Oryza sativa L.	400–1200 mg kg^{-1} **	Root	1250–5000	Ashraf et al. (2020)
			Stem	100–1400	
			Leaf	50–450	
	Conyza canadensis	25–200 μmol *	Root	1400–5845	Li et al. (2016)
			Leaf	13–34	

* Cultivated in nutrient solution. ** Cultivated in soil

to 58; from 20 to 34, and from 12 to 21 mg kg^{-1} in the first, second, and third leaf, respectively (Cho and Park 2000). Contrary to what is observed in roots, heavy metal accumulation in leaves occurs preferably in the vacuole (Liu et al. 2009; Fan et al. 2020), which is responsible for storing substances, maintaining cell pH and osmotic control (Taiz et al. 2017).

Heavy metal accumulation in grain and fruit is in the smallest proportion in relation to other plant organs, and the allocation is dependent on both the transport and the redistribution of these elements in conducting vessels of the plant. As accumulation decreased 14-fold in strawberry fruit in relation to the roots (Torres et al. 2020); Pb in rice grains correspond from 0.1% to 0.7% of the total accumulated in the plant (Ashraf and Tang 2017); and Cr accumulation in maize and cowpea grains ranged from 0.6% to 7% and from 1.4% to 7%, respectively (Sousa et al. 2018).

Other factors of great importance in the accumulation and distribution of heavy metal are genotype and gene expressions of plants. Different cultivars of the same species may have different defense mechanisms against stress by heavy metals, increasing or decreasing their tolerance and, or, sensibility to these elements (Cao et al. 2019; Lavres et al. 2019; Ashraf et al. 2020; Ye et al. 2020).

The expression of certain genes, naturally or through genetic modifications, alters the plant's ability to absorb, transport, redistribute, and allocate heavy metal. The genetic modification of plants under heavy metal stress has been accomplished to increase plant tolerance, conferring yield and quality production maintenance or given the plant the ability to extract these elements from the soil, a technique called phytoremediation. For instance, the overexpression of metallochaperone *OsHIPP29* reduced the Cd accumulation in rice, while blocking this protein increased Cd accumulation (Zhang et al. 2020). Similarly, the expression of *merA* and *merB* genes reduced the Hg concentration in leaves, grain, and fruit of tobacco, rice, and tomato, respectively, cultivated in a contaminated substrate producing products safe for consumption (Li et al. 2020a). Heavy metal accumulator plants, called phytoremediation plants, has specialized transporters for the absorption and transport of metals in the aerial part, an important characteristic of phytoremediation plants (Pilon-Smits 2005). The expression of the MerC gene provided an increase in Hg absorption and transport in *Arabidopsis thaliana* (Kiyono et al. 2013).

18.5   Toxicity Symptoms

The appearance of visual symptoms due to toxicity with heavy metals in plants has not been well characterized and may present characteristics of nutrient deficiency, due to the nutritional imbalance that heavy metals may cause in plants (Shanker et al. 2005; Silva et al. 2014; Lavres et al. 2019). Symptoms caused by heavy metals in plants are changes in water balance and assimilation of nutrients, inhibition of photosynthesis and growth by leaf and root atrophy, chlorosis, necrosis, biomass reduction, senescence, and plant death (Smith et al. 2010; Carvalho et al. 2013; Singh et al. 2013, 2016; Xun et al. 2017).

The visual symptoms caused by high Cd concentration in plants are shriveled and curled leaves with brown margins, chlorosis, petioles and reddish veins, brown and short roots, withering and reduced plant growth. In addition, toxic levels of Cd in leaves range from 5 to 10 mg kg^{-1} (Lux et al. 2011).

Some studies have already managed to identify the harmful effects of some heavy metals on plants grown in nutrient solution. Oliveira (2009) found visual symptoms of Cd toxicity in bean (*Phaseolus vulgaris* L. cv. IPR Colibri) plants with concentration of 0.45 μmol L^{-1} after 5 days of metal exposure. The symptoms included reddish spots on the stem, and chlorosis, shriveling, and curling of the leaf blade. At concentrations of 0.90 and 4.5 μmol L^{-1}, it was also observed that the veins showed a reddish color (Fig. 18.1 a, b, c, d, e). Nogueira (2012) evaluated bean plants (cv. Pérola) in pots filled with 3 kg of an Oxisol and amended with 6 mg kg^{-1} of Cd applied as CdCl$_2$. This author also reported visual symptoms of Cd toxicity through curling of the leaf blade (Fig. 18.1 f).

In a study with rice (*Oryza sativa* L. cv. IAC 202) and soybean (*Glycine max* L. cv. BRS 133) grown in soils contaminated with heavy metals (Cd, Cu, Fe, Mn, Pb, and Zn), Silva et al. (2014) observed dark green veins and yellowing in the internerval region of rice leaves becoming totally chlorotic and whitish with time and severity of the toxicity, for later loss of mass and plant death after 10 days of germination. In soybean, chlorosis, dark brown spots on new leaves, and reduction of leaf biomass were observed. However, the plants completed the cycle. In both cases, the authors reported that these symptoms are associated with Fe deficiency caused by competition with heavy metals at the sites of absorption and transport in the plant membrane.

Fig. 18.1 Visual symptoms of Cd toxicity on young bean leaves (*Phaseolus vulgaris* L., cv. IPR Colibri and cv. Pérola), as a function of the concentration: without Cd application (a), 0.05 (b), 0.45 (c), 0.90 (d), 4.5 (e) μmol L^{-1} and 6 (f) mg kg $^{-1}$ of Cd. Photos: Oliveira (2009) and Thiago Assis Rodrigues Nogueira

Fig. 18.2 Lettuce (**a**) and soybean (**b**) plants showing shoot reduction in response to application (right pots) or not (left pots) of Cd. Photos: Thiago Assis Rodrigues Nogueira and Leônidas Carrijo Azevedo Melo

Fig. 18.3 Rice cultivars BRSMG Talento (**a**) and BRSMG Caravera (**b**) grown in soils with Cd (0.0, 0.65, 1.3, 3.9, 7.8, and 11.7 mg kg^{-1}) applied as CdCl$_2$. Photos: Thiago Assis Rodrigues Nogueira

Conversely, as the symptoms of Cd toxicity are caused by a set of factors induced by metabolic disorders and changes in nutrient functions, it is common to find only a decrease in plant growth – e.g., lettuce (*Lactuca sativa* L. cv. Elisa da Sakata) and soybean plants that received 6 and 10.4 mg kg^{-1} of Cd, respectively (Fig. 18.2). There is also a tolerance to toxic levels of metals within the same plant species (Fig. 18.3). Such aspects are worrisome because many crops can grow in environments contaminated with heavy metals, accumulate these elements in their tissues, and cause damage to human health when consumed.

References

Abreu-Junior CH, Lima Brossi MJ, Monteiro RT, et al. Effects of sewage sludge application on unfertile tropical soils evaluated by multiple approaches: a field experiment in a commercial *Eucalyptus* plantation. Sci Total Environ. 2019;655:1457–67. https://doi.org/10.1016/j.scitotenv.2018.11.334.

Armendariz AL, Talano MA, Travaglia C, et al. Arsenic toxicity in soybean seedlings and their attenuation mechanisms. Plant Physiol Biochem. 2016;98:119–27. https://doi.org/10.1016/j.plaphy.2015.11.021.

Ashraf U, Tang X. Yield and quality responses, plant metabolism and metal distribution pattern in aromatic rice under lead (Pb) toxicity. Chemosphere. 2017;176:141–55. https://doi.org/10.1016/j.chemosphere.2017.02.103.

Ashraf U, Mahmood MHR, Hussain S, et al. Lead (Pb) distribution and accumulation in different plant parts and its associations with grain Pb contents in fragrant rice. Chemosphere. 2020;248:126003. https://doi.org/10.1016/j.chemosphere.2020.126003.

Athalye VV, Ramachandran V, D'Souza TJ. Influence of chelating agents on plant uptake of ^{51}Cr, ^{210}Pb and ^{210}Po. Environ Pollut. 1995;89:47–53. https://doi.org/10.1016/0269-7491(94)00047-H.

Azevedo R, Rodriguez E, Mendes RJ, et al. Inorganic Hg toxicity in plants: a comparison of different genotoxic parameters. Plant Physiol Biochem. 2018;125:247–54. https://doi.org/10.1016/j.plaphy.2018.02.015.

Benavides MP, Gallego SM, Tomaro ML. Cadmium toxicity in plants. Braz J Plant Physiol. 2005;7:21–34. https://doi.org/10.1590/S1677-04202005000100003.

Bhaduri AM, Fulekar MH. Antioxidant enzyme responses of plants to heavy metal stress. Rev Environ Sci Biotechnol. 2012;11:55–69. https://doi.org/10.1007/s11157-011-9251-x.

Bienert GP, Thorsen M, Schüssler MD, et al. A subgroup of plant aquaporins facilitate the bidirectional diffusion of As(OH)$_3$ and Sb(OH)$_3$ across membranes. BMC Biol. 2008;6:26. https://doi.org/10.1186/1741-7007-6-26.

Brooks RR. Plants that hyperaccumulate heavy metals. Cambridge: University Press; 1998.

Cao X, Wang X, Tong W, et al. Distribution, availability and translocation of heavy metals in soil-oilseed rape (*Brassica napus* L.) system related to soil properties. Environ Pollut. 2019;252:733–41. https://doi.org/10.1016/j.envpol.2019.05.147.

Carey AM, Norton GJ, Deacon C, et al. Phloem transport of arsenic species from flag leaf to grain during grain filling. New Phytol. 2011;192:87–98. https://doi.org/10.1111/j.1469-8137.2011.03789.x.

Carvalho MTV, Amaral DC, Guilherme LRG, et al. *Gomphrena claussenii*, the first South-American metallophyte species with indicator-like Zn and Cd accumulation and extreme metal tolerance. Front Plant Sci. 2013;4:180. https://doi.org/10.3389/fpls.2013.00180.

Cary EE, Allaway WH, Olson OE. Control of chromium concentrations in food plants. 1. Absorption and translocation of chromium by plants. J Agric Food Chem. 1977;25:300–4. https://doi.org/10.1021/jf60210a048.

Chang TG, Zhu XG. Source–sink interaction: a century old concept under the light of modern molecular systems biology. J Exp Bot. 2017;68:4417–31. https://doi.org/10.1093/jxb/erx002.

Cho UH, Park JO. Mercury-induced oxidative stress in tomato seedlings. Plant Sci. 2000;156:1–9. https://doi.org/10.1016/S0168-9452(00)00227-2.

Clayden MG, Kidd KA, Wyn B, et al. Mercury biomagnification through food webs is affected by physical and chemical characteristics of lakes. Environ Sci Technol. 2013;47:12047–53. https://doi.org/10.1021/es4022975.

Clemens S. Toxic metal accumulation, responses to exposure and mechanisms of tolerance in plants. Biochimie. 2006;88:1707–19. https://doi.org/10.1016/j.biochi.2006.07.003.

DalCorso G, Manara A, Furini A. An overview of heavy metal challenge in plants: from roots to shoots. Metallomics. 2013;5:1117–32. https://doi.org/10.1039/c3mt00038a.

Dubey S, Shri M, Gupta A, et al. Toxicity and detoxification of heavy metals during plant growth and metabolism. Environ Chem Lett. 2018;16:1169–92. https://doi.org/10.1007/s10311-018-0741-8.

Esteban E, Moreno E, Peñalosa J, et al. Short and long-term uptake of Hg in white lupin plants: kinetics and stress indicators. Environ Exp Bot. 2008;62:316–22. https://doi.org/10.1016/j.envexpbot.2007.10.006.

Eun SO, Youn HS, Lee Y. Lead disturbs microtubule organization in the root meristem of *Zea mays*. Physiol Plant. 2000;110:357–65. https://doi.org/10.1111/j.1399-3054.2000.1100310.x.

Fan WJ, Feng YX, Li YH, et al. Unraveling genes promoting ROS metabolism in subcellular organelles of *Oryza sativa* in response to trivalent and hexavalent chromium. Sci Total Environ. 2020;744:140951. https://doi.org/10.1016/j.scitotenv.2020.140951.

Finnegan PM, Chen W. Arsenic toxicity: the effects on plant metabolism. Front Physiol. 2012;3:182. https://doi.org/10.3389/fphys.2012.00182.

Galhardi JA, Mello JWV, Wilkinson KJ. Bioaccumulation of potentially toxic elements from the soils surrounding a legacy uranium mine in Brazil. Chemosphere. 2020a;261:127679. https://doi.org/10.1016/j.chemosphere.2020.127679.

Galhardi JA, Leles BP, Mello JWV, et al. Bioavailability of trace metals and rare earth elements (REE) from the tropical soils of a coal mining area. Sci Total Environ. 2020b;717:134484. https://doi.org/10.1016/j.scitotenv.2019.134484.

Genchi G, Sinicropi MS, Lauria G, et al. The effects of cadmium toxicity. Int J Environ Res Public Health. 2020;17:3782. https://doi.org/10.3390/ijerph17113782.

Gresser MJ. ADP-arsenate. Formation by submitochondrial particles under phosphorylating conditions. J Biol Chem. 1981;256:5981–3.

Habermann E, Crowell K, Janicki P. Lead and other metals can substitute for Ca^{2+} in calmodulin. Arch Toxicol. 1983;54:61–70. https://doi.org/10.1007/BF00277816.

Hacisalihoglu G, Hart JJ, Kochian LV. High and low-affinity zinc transport systems and their possible role in zinc efficiency in bread wheat. Plant Physiol. 2001;125:456–63. https://doi.org/10.1104/pp.125.1.456.

Harris NS, Taylor GJ. Cadmium uptake and partitioning in durum wheat during grain filling. BMC Plant Biol. 2013;13:103. https://doi.org/10.1186/1471-2229-13-103.

Hu Y, Wang Y, Liang Y, et al. Silicon alleviates mercury toxicity in garlic plants. J Plant Nutr. 2020;43:2508–17. https://doi.org/10.1080/01904167.2020.1783302.

Huang TL, Huang HJ. ROS and CDPK-like kinase-mediated activation of MAP kinase in rice roots exposed to lead. Chemosphere. 2008;71:1377–85. https://doi.org/10.1016/j.chemosphere.2007.11.031.

Hughes MK, Lepp NW, Phipps DA. Aerial heavy metal pollution and terrestrial ecosystems. Adv Ecol Res. 1980;11:217–327. https://doi.org/10.1016/S0065-2504(08)60268-8.

Huybrechts M, Cuypers A, Deckers J, et al. Cadmium and plant development: an agony from seed to seed. Int J Mol Sci. 2019;20:3971. https://doi.org/10.3390/ijms20163971.

Isayenkov SV, Maathuis FJM. The *Arabidopsis thaliana* aquaglyceroporin AtNIP7;1 is a pathway for arsenite uptake. FEBS Lett. 2008;582:1625–8. https://doi.org/10.1016/j.febslet.2008.04.022.

Kabata-Pendias A, Mukherjee AB. Trace elements from soil to human. Berlin: Springer; 2007.

Kaur G, Singh HP, Batish DR, et al. Lead (Pb)-induced biochemical and ultrastructural changes in wheat (*Triticum aestivum*) roots. Protoplasma. 2013;1:53–62. https://doi.org/10.1007/s00709-011-0372-4.

Kitchin KT, Wallace K. Arsenite binding to synthetic peptides: the effect of increasing length between two cysteines. J Biochem Mol Toxicol. 2006;20:35–8. https://doi.org/10.1002/jbt.20112.

Kiyono M, Oka Y, Sone Y, et al. Bacterial heavy metal transporter MerC increases mercury accumulation in *Arabidopsis thaliana*. Biochem Eng J. 2013;71:19–24. https://doi.org/10.1016/j.bej.2012.11.007.

Kumarathilaka P, Seneweera S, Meharg A, et al. Arsenic accumulation in rice (*Oryza sativa* L.) is influenced by environment and genetic factors. Sci Total Environ. 2018;642:485–96. https://doi.org/10.1016/j.scitotenv.2018.06.030.

Lavres J, Silveira Rabêlo FH, Capaldi FR, et al. Investigation into the relationship among Cd bioaccumulation, nutrient composition, ultrastructural changes and antioxidative metabolism in lettuce genotypes under Cd stress. Ecotoxicol Environ Saf. 2019;170:578–89. https://doi.org/10.1016/j.ecoenv.2018.12.033.

Li Y, Zhou C, Huang M, et al. Lead tolerance mechanism in *Conyza canadensis*: subcellular distribution, ultrastructure, antioxidative defense system, and phytochelatins. J Plant Res. 2016;129:251–62. https://doi.org/10.1007/s10265-015-0776-x.

Li R, Wu H, Ding J, et al. Transgenic *merA* and *merB* expression reduces mercury contamination in vegetables and grains grown in mercury-contaminated soil. Plant Cell Rep. 2020a;39:1369–80. https://doi.org/10.1007/s00299-020-02570-8.

Li Y, Zhu N, Liang X, et al. Silica nanoparticles alleviate mercury toxicity via immobilization and inactivation of Hg(II) in soybean (*Glycine max*). Environ Sci Nano. 2020b;7:1807–17. https://doi.org/10.1039/D0EN00091D.

Liu J, Duan CQ, Zhang XH, et al. Subcellular distribution of chromium in accumulating plant *Leersia hexandra* Swartz. Plant Soil. 2009;322:187–95. https://doi.org/10.1007/s11104-009-9907-2.

Liu X, Peng K, Wang A, et al. Cadmium accumulation and distribution in populations of *Phytolacca americana* L. and the role of transpiration. Chemosphere. 2010;78:1136–41. https://doi.org/10.1016/j.chemosphere.2009.12.030.

Long JW, Ray WJJ. Kinetics and thermodynamics of the formation of glucose arsenate. Reaction of glucose arsenate with phosphoglucomutase Biochemistry. 1973;12:3932–7. https://doi.org/10.1021/bi00744a023.

López-Climent MF, Arbona V, Pérez-Clemente RM, et al. Effects of cadmium on gas exchange and phytohormone contents in citrus. Biol Plantarum. 2011;55:187–90. https://doi.org/10.1007/s10535-011-0028-4.

Lux A, Martinka M, Vaculík M, et al. Root responses to cadmium in the rhizosphere: a review. J Exp Bot. 2011;62:21–37. https://doi.org/10.1093/jxb/erq281.

Maestri E, Marmiroli M, Visioli G, et al. Metal tolerance and hyperaccumulation: costs and trade-offs between traits and environment. Environ Exp Bot. 2010;68:1–13. https://doi.org/10.1016/j.envexpbot.2009.10.011.

Malavolta E. Manual de nutrição mineral de plantas. São Paulo: Editora Agronômica Ceres; 2006.

Marchetti C. Role of calcium channels in heavy metal toxicity. Int Sch Res Notices. 2013;2013:184360. https://doi.org/10.1155/2013/184360.

Marrugo-Negrete J, Durango-Hernández J, Pinedo-Hernández J, et al. Mercury uptake and effects on growth in *Jatropha curcas*. J Environ Sci. 2016;48:120–5. https://doi.org/10.1016/j.jes.2015.10.036.

Meng B, Feng X, Qiu G. Inorganic mercury accumulation in rice (*Oryza sativa* L.). Environ Toxicol Chem. 2012;31:2093–8. https://doi.org/10.1002/etc.1913.

Mishra VK, Tripathi BD. Accumulation of chromium and zinc from aqueous solutions using water hyacinth (*Eichhornia crassipes*). J Hazard Mater. 2009;164:1059–63. https://doi.org/10.1016/j.jhazmat.2008.09.020.

Mondal NK, Das C, Datta JK. Effect of mercury on seedling growth, nodulation and ultrastructural deformation of *Vigna radiata* (L) Wilczek. Environ Monit Assess. 2015;187:241. https://doi.org/10.1007/s10661-015-4484-8.

Moreno FN, Anderson CWN, Stewart RB, et al. Induced plant uptake and transport of mercury in the presence of sulphur-containing ligands and humic acid. New Phytol. 2005;166:445–54. https://doi.org/10.1111/j.1469-8137.2005.01361.x.

Myttenaere C, Mousny JM. The distribution of chromium-51 in lowland rice in relation to the chemical form and the amount of stable chromium in the nutrient solution. Plant Soil. 1974;41:65–72. https://doi.org/10.1007/BF00017944.

Nogueira TAR. Disponibilidade de Cd em Latossolos e sua transferência e toxicidade para as culturas de alface, arroz e feijão. Tese Doutorado. Piracicaba: Universidade de São Paulo. 2012; https://doi.org/10.11606/T.64.2012.tde-22062012-091546.

Oliveira LA Silício em plantas de feijão e arroz: absorção, transporte, redistribuição e tolerância ao cádmio. Tese Doutorado. Piracicaba: Universidade de São Paulo. 2009; https://doi.org/10.11606/T.64.2009.tde-03122009-094223

Oliveira H. Chromium as an environmental pollutant: insights on induced plant toxicity. J Bot. 2012; 1–8 https://doi.org/10.1155/2012/375843

Oller ALW, Regis S, Armendariz AL, et al. Improving soybean growth under arsenic stress by inoculation with native arsenic-resistant bacteria. Plant Physiol Biochem. 2020;155:85–92. https://doi.org/10.1016/j.plaphy.2020.07.015.

Page V, Feller U. Heavy metals in crop plants: transport and redistribution processes on the whole plant level. Agronomy. 2015;5:447–63. https://doi.org/10.3390/agronomy5030447.

Patra M, Sharma A. Mercury toxicity in plants. Bot Rev. 2000;66:379–422. https://doi.org/10.1007/BF02868923.

Patra M, Bhowmik N, Bandopadhyay B, et al. Comparison of mercury, lead and arsenic with respect to genotoxic effects on plant systems and the development of genetic tolerance. Environ Exp Bot. 2004;52:199–223. https://doi.org/10.1016/j.envexpbot.2004.02.009.

Peng H, Chen Y, Weng L, et al. Comparisons of heavy metal input inventory in agricultural soils in North and South China: a review. Sci Total Environ. 2019;660:776–86. https://doi.org/10.1016/j.scitotenv.2019.01.066.

Pilon-Smits E. Phytoremediation. Annu Rev Plant Biol. 2005;56:15–39. https://doi.org/10.1146/annurev.arplant.56.032604.144214.

Pourrut B, Shahid M, Dumat C, et al. Lead uptake, toxicity, and detoxification in plants. In: Whitacre D, editor. Reviews of environmental contamination and toxicology volume 213. New York: Springer; 2011. p. 113–36.

Qin S, Liu H, Nie Z, et al. Toxicity of cadmium and its competition with mineral nutrients for uptake by plants: a review. Pedosphere. 2020;30:168–80. https://doi.org/10.1016/S1002-0160(20)60002-9.

Rabêlo FHS, Gaziola SA, Rossi ML, et al. Unraveling the mechanisms controlling Cd accumulation and Cd-tolerance in *Brachiaria decumbens* and *Panicum maximum* under summer and winter weather conditions. Physiol Plant. 2020; https://doi.org/10.1111/ppl.13160.

Reddy AM, Kumar SG, Jyonthsnakumari G, et al. Lead induced changes in antioxidant metabolism of horsegram (*Macrotyloma uniflorum* (Lam.) Verdc.) and bengalgram (*Cicer arietinum* L.). Chemosphere. 2005;60:97–104. https://doi.org/10.1016/j.chemosphere.2004.11.092.

Rellán-Álvarez R, Ortega-Villasante C, Álvarez-Fernández A, et al. Stress responses of *Zea mays* to cadmium and mercury. Plant Soil. 2006;279:41–50. https://doi.org/10.1007/s11104-005-3900-1.

Riesen O, Feller U. Redistribution of nickel, cobalt, manganese, zinc and cadmium via the phloem in young and maturing wheat. J Plant Nutr. 2005;28:421–30. https://doi.org/10.1081/PLN-200049153.

Roelfsema MRG, Hedrich R. In the light of stomatal opening: new insights into 'the Watergate'. New Phytol. 2005;167:665–91. https://doi.org/10.1111/j.1469-8137.2005.01460.x.

Sahu GK, Upadhyay S, Sahoo BB. Mercury induced phytotoxicity and oxidative stress in wheat (*Triticum aestivum* L.) plants. Physiol Mol Biol Plants. 2012;18:21–31. https://doi.org/10.1007/s12298-011-0090-6.

Shahid MA, Balal RM, Khan N, et al. Selenium impedes cadmium and arsenic toxicity in potato by modulating carbohydrate and nitrogen metabolism. Ecotoxicol Environ Saf. 2019;180:588–99. https://doi.org/10.1016/j.ecoenv.2019.05.037.

Shakerian A, Rahimi E, Ahmadi M. Cadmium and lead content in several brands of rice grains (*Oryza sativa*) in Central Iran. Toxicol Ind Health. 2012;28:955–60. https://doi.org/10.1177/0748233711430979.

Shanker AK, Cervantes C, Loza-Tavera H, et al. Chromium toxicity in plants. Environ Int. 2005;31:739–53. https://doi.org/10.1016/j.envint.2005.02.003.

Sharma P, Dubey RS. Lead toxicity in plants. Braz J Plant Physiol. 2005;17:35–52. https://doi.org/10.1590/S1677-04202005000100004.

Sharma DC, Chatterjee C, Sharma CP. Chromium accumulation by barley seedlings (*Hordeum vulgare* L.). J Exp Bot. 1995;25:241–51. https://doi.org/10.1007/BF00399719.

Sheetal KR, Singh SD, Anand A, et al. Heavy metal accumulation and effects on growth, biomass and physiological processes in mustard. Indian J Plant Physiol. 2016;21:219–23. https://doi.org/10.1007/s40502-016-0221-8.

Shi T, Ma J, Wu X. Inventories of heavy metal inputs and outputs to and from agricultural soils: a review. Ecotoxicol Environ Saf. 2018;164:118–24. https://doi.org/10.1016/j.ecoenv.2018.08.016.

Shukla T, Kumar S, Khare R, et al. Natural variations in expression of regulatory and detoxification related genes under limiting phosphate and arsenate stress in *Arabidopsis thaliana*. Front Plant Sci. 2015;6:898. https://doi.org/10.3389/fpls.2015.00898.

Silva MLDS, Vitti GC, Trevizam AR. Heavy metal toxicity in rice and soybean plants cultivated in contaminated soil. Rev Ceres. 2014;61:248–54. https://doi.org/10.1590/s0034-737x2014000200013.

Singh HP, Mahajan P, Kaur S, et al. Chromium toxicity and tolerance in plants. Environ Chem Lett. 2013;11:229–54. https://doi.org/10.1007/s10311-013-0407-5.

Singh S, Parihar P, Singh R, et al. Heavy metal tolerance in plants: role of transcriptomics, proteomics, metabolomics, and ionomics. Front Plant Sci. 2016;6:1–36. https://doi.org/10.3389/fpls.2015.01143.

Smith SE, Christophersen HM, Pope S, et al. Arsenic uptake and toxicity in plants: integrating mycorrhizal influences. Plant Soil. 2010;327:1–21. https://doi.org/10.1007/s11104-009-0089-8.

Song Y, Jin L, Wang X. Cadmium absorption and transportation pathways in plants. Int J Phytoremediation. 2017;19:133–41. https://doi.org/10.1080/15226514.2016.1207598.

Sousa RS, Nunes LAPL, Lima AB, et al. Chromium accumulation in maize and cowpea after successive applications of composted tannery sludge. Acta Sci Agron. 2018;40:1–7. https://doi.org/10.4025/actasciagron.v40i1.35361.

Sun S, Li M, Zuo J, et al. Cadmium effects on mineral accumulation, antioxidant defence system and gas exchange in cucumber. Zemdirbyste-Agriculture. 2015;102:193–200. https://doi.org/10.13080/z-a.2015.102.025.

Suriyagoda LDB, Ditteret K, Lambers H. Mechanism of arsenic uptake, translocation and plant resistance to accumulate arsenic in rice grains. Agric Ecosyst Environ. 2018;253:23–37. https://doi.org/10.1016/j.agee.2017.10.017.

Taiz L, Zeiger E, Moller IM, et al. Fisiologia e Desenvolvimento Vegetal. Porto Alegre: Artmed; 2017.

Tkachuk R, Kuzina FD. Mercury levels in wheat and other cereals, oilseed and biological samples. J Sci Food Agric. 1972;23:1183–95. https://doi.org/10.1002/jsfa.2740231006.

Torres AIG, Giráldez I, Martínez F, et al. Arsenic accumulation and speciation in strawberry plants exposed to inorganic arsenic enriched irrigation. Food Chem. 2020;315:126215. https://doi.org/10.1016/j.foodchem.2020.126215.

Turgeon R, Wolf S. Phloem transport: cellular pathways and molecular trafficking. Annu Rev Plant Biol. 2009;60:207–21. https://doi.org/10.1146/annurev.arplant.043008.092045.

Vadas TM, Ahner BA. Cysteine- and glutathione-mediated uptake of lead and cadmium into *Zea mays* and *Brassica napus* roots. Environ Pollut. 2009;157:2558–63. https://doi.org/10.1016/j.envpol.2009.02.036.

Van Bel AJE. The phloem, a miracle of ingenuity. Plant Cell Environ. 2003;26:125–49. https://doi.org/10.1046/j.1365-3040.2003.00963.x.

Verbruggen N, Hermans C, Schat H. Mechanisms to cope with arsenic or cadmium excess in plants. Curr Opin Plant Biol. 2009;12:364–72. https://doi.org/10.1016/j.pbi.2009.05.001.

Vilela EF, Guilherme LRG, Silva CA et al. Trace elements in soils developed from metamorphic ultrabasic rocks in Minas Gerais, Brazil. Geoderma Reg 21 2020; https://doi.org/10.1016/j.geodrs.2020.e00279

White PJ, Broadley MR. Calcium in plants. Ann Bot. 2003;92:487–511. https://doi.org/10.1093/aob/mcg164.

Wojas S, Ruszczyńska A, Bulska E. Ca^{2+}-dependent plant response to Pb^{2+} is regulated by *LCT1*. Environ Pollut. 2007;147:584–92. https://doi.org/10.1016/j.envpol.2006.10.012.

Xiao L, Guan D, Peart MR, et al. The respective effects of soil heavy metal fractions by sequential extraction procedure and soil properties on the accumulation of heavy metals in rice grains and brassicas. Environ Sci Pollut Res. 2017;24:2558–71. https://doi.org/10.1007/s11356-016-8028-8.

Xie L, Hao P, Cheng Y, et al. Effect of combined application of lead, cadmium, chromium and copper on grain, leaf and stem heavy metal contents at different growth stages in rice. Ecotoxicol Environ Saf. 2018;162:71–6. https://doi.org/10.1016/j.ecoenv.2018.06.072.

Xu XY, McGrath SP, Zhao FJ. Rapid reduction of arsenate in the medium mediated by plant roots. New Phytol. 2007;176:590–9. https://doi.org/10.1111/j.1469-8137.2007.02195.x.

Xu J, Zhang J, Lv Y, et al. Effect of soil mercury pollution on ginger (*Zingiber officinale* Roscoe): growth, product quality, health risks and silicon mitigation. Ecotoxicol Environ Saf. 2020;195:110472. https://doi.org/10.1016/j.ecoenv.2020.110472.

Xun Y, Feng L, Li Y, et al. Mercury accumulation plant *Cyrtomium macrophyllum* and its potential for phytoremediation of mercury polluted sites. Chemosphere. 2017;189:161–70. https://doi.org/10.1016/j.chemosphere.2017.09.055.

Ye Y, Dong W, Luo Y, et al. Cultivar diversity and organ differences of cadmium accumulation in potato (*Solanum tuberosum* L.) allow the potential for Cd-safe staple food production on contaminated soils. Sci Total Environ. 2020;711:134534. https://doi.org/10.1016/j.scitotenv.2019.134534.

Yu G, Wu H, Qing C, et al. Bioavailability of humic substance-bound mercury to lettuce and its relationship with soil properties. Communications in soil. Sci Plant Anal. 2004;35:1123–37. https://doi.org/10.1081/CSS-120030594.

Zayed A, Lytle CM, Qian J, et al. Chromium accumulation, translocation and chemical speciation in vegetable crops. Planta. 1998;206:293–9. https://doi.org/10.1007/s004250050403.

Zeng F, Ali S, Zhang H, et al. The influence of pH and organic matter content in paddy soil on heavy metal availability and their uptake by rice plants. Environ Pollut. 2011;159:84–91. https://doi.org/10.1016/j.envpol.2010.09.019.

Zhang BQ, Liu XS, Feng SJ, et al. Developing a cadmium resistant rice genotype with OsHIPP29 locus for limiting cadmium accumulation in the paddy crop. Chemosphere. 2020;247:125958. https://doi.org/10.1016/j.chemosphere.2020.125958.

Zhao FJ, Ma JF, Meharg AA. Arsenic uptake and metabolism in plants. New Phytol. 2009;181:777–94. https://doi.org/10.1111/j.1469-8137.2008.02716.x.

Zhou J, Zhang Z, Zhang Y, et al. Effects of lead stress on the growth, physiology, and cellular structure of privet seedlings. PlosOne. 2018;13:e0191139. https://doi.org/10.1371/journal.pone.0191139.

Zimdahl RL. Entry and movement in vegetation of lead derived from air and soil sources. J Air Pollut Control Assoc. 1976;26:655–60. https://doi.org/10.1080/00022470.1976.10470298.

Chapter 19
Visual and Leaf Diagnosis

Keywords Leaf sampling · Deficiency symptoms · Chemical analysis · Critical
level · Nutritional balance · Digital agriculture

Visual and leaf diagnosis is the most applied area of plant nutrition because it allows
assessing the nutritional status of crops to guide decision-making and to conduct a
correct fertilization practice, providing gains in crop productivity. This chapter contains (i) an introduction, (ii) basic aspects of visual diagnosis, (iii) leaf diagnosis
based on leaf analysis, (iv) leaf sampling criteria, (v) preparation of plant material
and chemical analysis, (vi) studies on leaf diagnosis in cultures, and (vii) interpretation of leaf analysis by DRIS/CND.

For any crop to be able to manifest its full genetic potential through the production of food, it is necessary to have at its disposal all the vital factors optimized
(climatic, genotype, edaphic, light, water, temperature, nutrients, etc.).

Regarding nutrition, it is necessary that the plant has at its disposal throughout its
life cycle nutrients in adequate quantities so that they can fulfill their functions in
plant metabolism.

Of the factors determining the profit of an agricultural enterprise, that is, product
price, cost of production and productivity, the latter is the main one, which in turn
is for the most part explained by the adequate nutritional status of the crop. It is
therefore necessary to know whether the plant or the crop is well nourished or not,
as the production of any crop depends on a proper nutritional status.

To optimize the plant's nutrition and prevent failures due to deficiencies or excess
of elements, a soil analysis should be considered a criterion for recommending cor-
rectives and fertilizers. The plant itself is also an object of diagnosis.

Thus, the assessment of the nutritional status of plants can be established by
performing chemical analyses of the soil and the plant (leaves), which are complementary techniques. Typically, the soil analysis indicates a potential availability of
an element to plants, whereas plant analysis reflects its current nutritional status,
which results from the integrated effect of all factors that affect nutrient availability
and plant development in a given environmental condition.

© Springer Nature Switzerland AG 2021
R. de Mello Prado, *Mineral nutrition of tropical plants*,
https://doi.org/10.1007/978-3-030-71262-4_19

In the case of the plant, there is leaf diagnosis. Two paths that can help in this sense, that is, one can resort to visual symptoms and chemical analyses of plant material.

Visual Diagnosis

To use the diagnosis by the visual method, it is necessary to ensure that the problem in the field is caused by the deficiency or the excess of a nutrient, since the incidence of pests and diseases, among others, can "mask" the fact that they generate symptoms similar to a nutritional deficiency. Thus, in cases of nutritional disorders, symptoms usually have the following characteristics:

Dispersion—the "nutritional problem" occurs homogeneously in the field, as in cases of diseases/pests; for example, they occur in isolated plants or in "tree clumps."

Symmetry—in a pair of leaves, nutritional disorder occurs in both leaves.

Gradient—in a branch or plant, the symptoms respect a gradient, presenting an aggravation of symptoms from old leaves toward new leaves or vice versa.

Thus, in visual diagnosis, the symptoms of deficiency/excess may vary according to crop. Normally, the deficiency symptoms occur in old leaves (for the mobile nutrients in the plant) or in new leaves or buds (for little mobile nutrients in the plant). Symptoms can also be observed in roots, featuring different types of symptoms (Fig. 19.1). Thus, as observed at the end of each chapter, the visual symptoms of nutritional deficiency can be grouped into six categories: (a) reduced growth, (b)

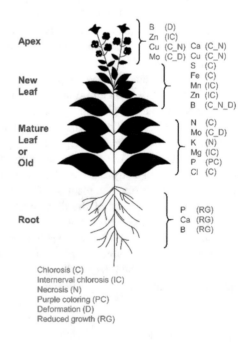

Fig. 19.1 Outline of a key symptom of nutritional disorder due to nutrient deficiency

uniform chlorosis or leaf spots, (c) internerval chlorosis, (d) necrosis, (e) purple coloring, and (f) deformations.

The visual diagnosis allows evaluating the symptoms of deficiency or excess of nutrients in a fast way. It is possible to make corrections in the fertilization program, with certain limitations. However, this method is criticized because of a number of limitations:

- In the field, the plant can suffer interference from agents (pests and pathogens) that may mask the accuracy of the detection of the problematic nutrient, as previously mentioned.
- In the field, the symptoms of deficiency may be different from those described in specialized publications, as in these works, authors report "severe" symptoms of nutritional disorder, and in the field such symptoms may be "mild."
- The deficiency symptom of a certain element may differ in different crops, that is, the knowledge of the deficiency symptom in a given species may not be valid for another. For example, Zn in fruit trees may experience symptoms such as small leaves and in maize it may be whitish new leaves.
- The same deficiency symptoms may occur for a different nutrient.
- A level of deficiency may occur that will reduce production without the plant developing any symptoms.
- Deficiency of two or more nutrients simultaneously makes it impossible to identify deficient nutrients.
- A condition of excess of a given nutrient can be mistaken for deficiency of another nutrient.
- A proper use of the visual diagnosis technique requires technicians with a significant expertise in the crop of their region.
- In addition, visual diagnosis does not quantify the level of deficiency or excess of the nutrient under study.

It should be added that only when the plant has an acute nutritional disorder does the manifestation of visual symptoms of characteristic deficiency or excess, which can be differentiated, clearly occur. However, at this point, a significant part of the production (about 40–50%) is already compromised, since a series of physiological damages have already been triggered and the visualization of the symptoms shows damage at the level of tissues, which, at this stage, is irreversible. Therefore, the use of visual diagnosis should not be the rule, but a complement to diagnosis.

Leaf Diagnosis

Thus, a method of diagnosis with a lower degree of limitation than visual diagnosis is leaf diagnosis through chemical analysis of leaves. It is a more commonly used method for monitoring the nutritional status of crops. The consideration of nutrient content in plants to indicate the nutritional status of plants was initially proposed in the 1930s by Lagatu and Maume (1934). Although the technique of chemical analysis of plants is relatively old, it is still little used today by Brazilian farmers.

The leaf diagnosis itself consists of the assessment of the nutritional status of a plant by collecting a sample of a plant tissue and comparing it with its preestablished pattern. This pattern consists of a plant that presents all nutrients in adequate proportions and capable of providing favorable conditions for the plant to express its maximum genetic potential for production.

The leaves are the most commonly used plant tissue for analysis, as this organ is the center of the plant's metabolism. Although other plant tissues can be used, such as part of the leaf (petiole) or even the seed, as Rafique et al. (2006) indicates, or fruits and juice (sugarcane), as the accumulation of nutrients in these tissues is not the same for all elements, especially those immobile or little mobile, it is unlikely that this organ will adequately reflect the nutritional status of all macro- and micronutrients of the crop.

The leaf chemical analysis can be interpreted by considering a single nutrient using the critical level method, the sufficiency range, or alternatively the relationship between nutrients of the method DRIS (Diagnosis and Recommendation Integrated System). There are several tools that can be used, preferably in an integrated manner, for the knowledge of the soil-plant system. They provide sufficient subsidies for interference, if applicable, in the adoption of fertilization practices, including making it more efficient.

Leaf diagnosis serves to identify the nutritional status of the plant by performing a chemical analysis of a plant tissue that is more sensitive in showing variations in nutrients in the center of the plant's physiological activities, in most cases the leaf. The use of leaf analysis as a diagnostic criterion is based on the premise that there is a relationship between the supply of nutrients and the levels of elements in leaves and that increases or decreases in concentrations are related to higher or lower yields. The nutrient content in the plant is an integral value of all factors that interact to affect it. For the purposes of interpreting the results of chemical analysis of plants, it is necessary to know the factors that affect the concentration of nutrients, the standardized sampling procedures, and the relevant relationships (basic premises for the use of leaf diagnosis):

(a) Supply of the nutrient by the soil × production. This means that in a more fertile soil, the production must be greater than in a low fertility soil;
(b) Nutrient supply by the soil × leaf content. With the increase in the nutrient supply in the soil, the leaf content also increases;
(c) Leaf content × production. The increase in the nutrient in the leaf explains the increase in production.

Specifically, regarding the relation between leaf content and production, there are several phases or zones (Fig. 19.2) which deserve to be discussed.

Deficient Zone
At this phase, there are symptoms of visible deficiency. This occurs in soils (or substrates) very deficient in the element that receive (still insufficient) doses of the nutrient. In this case, the response in dry matter production by the plant is very high, not allowing an increase in the leaf content of the element; there may even be dilution of the nutrient. This effect of diluting the nutrient due to the formation of

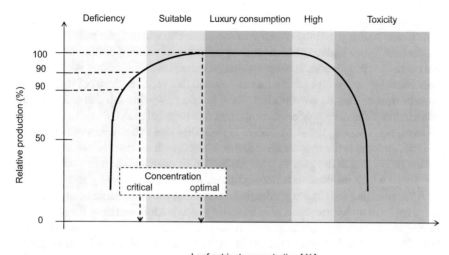

Fig. 19.2 Relationship between element contents and relative production. Relative production (%): [harvest in the presence of a x level of the element]/ harvest in the presence of the maximum level of the element] × 100

organic material is also known as the Steembjerg effect. The plant that presents a content of a given nutrient at this phase is interpreted as deficient.

Transition Zone

At this phase, although there are symptoms of nonvisible deficiency (hidden hunger), there is a direct relationship between leaf content and production. A leaf content that provides 80–95% of the maximum production is the critical level. Then, there is a content that reflects the maximum production (100%). This relationship is observed in soils (or substrates) with a mild nutrient deficiency, in which the response of growth and production to the application of the nutrient is low. In this case, proportional increases in leaf content and growth or production occur, that is, the greater absorption is offset by the formation of more organic material. Thus, a plant that presents a content of a given nutrient in this phase, corresponding to the production between the critical level and the maximum or optimum production, is interpreted as adequate.

Luxury Consumption Zone

At this phase, increasing the concentration of a nutrient does not result in an increase in production. This is observed in non-nutrient-deficient soils that receive doses of an element. In this case, the plant absorbs the nutrient, but this does not lead to growth. There is only an increase in its concentration (content) in the plant tissues. Thus, the plant that presents a content of a given nutrient at this phase, corresponding to the maximum or optimal production even before reaching a critical level of toxicity (which corresponds to a content that promotes a decrease from 5% to 20% of the maximum production) is interpreted as high.

Toxicity Range

This phase begins when the increase in the nutrient content significantly decreases production. When a given nutrient content causes a decrease of 5–20% or greater of the maximum production, it is interpreted as toxic. This relationship is observed in soils (or substrate) with excess of a nutrient and that receives doses of it. The plant absorbs it, increases its contents in tissues, but decreases growth due to its toxicity or an induced deficiency of another nutrient due to imbalance.

In research papers, the critical level of disability was initially proposed by Macy in 1936, and this is the most studied factor. Therefore, it corresponds to the concentration below which the growth rate (production or quality) is significantly reduced and above which production is not significant and not economically relevant.

After reaching maximum production, increasing the concentration of a nutrient in the leaf will no longer result in an increase in production, and thus, the plant starts to perform a "luxury consumption." It should be noted the luxury consumption leads to an accumulation of nutrients in the vacuoles of cells, which can be gradually released to meet the eventual nutritional needs of plants. After reaching the nutrient concentration range that characterizes luxury consumption, the addition of the nutrient in plant tissues may lead to a decrease in production, characterizing the toxicity zone.

Note that in leaf diagnosis, it is necessary for the plant to be at a time of maximum physiological activity, such as flowering or the beginning of fruiting. At that time, the leaf levels of nutrients have a high correlation with productivity. Due to this requirement in the chemical analysis of leaves at the height of the plant's development, leaf diagnosis is considered with little value for a contingent correction of nutrient deficiency in annual crops during a same crop production cycle. However, it generates safe information for the next harvest. However, in perennial crops such as coffee, citrus, etc., leaf diagnosis has a high potential for diagnosing the nutritional status of the plant and enabling correction in a same agricultural year with a satisfactory efficiency.

Leaf diagnosis has the advantage of using the plant itself as an extractor. Finally, leaf diagnosis has several applications:

(a) Assessment of nutrient requirements and exports in crops
(b) Identification of deficiencies that cause similar symptoms, making visual diagnosis difficult or impossible
(c) Assessment of nutritional status assisting management of fertilization programs

Leaf diagnosis is a direct method of assessing the nutritional status of crops, as it considers the contents of the nutrient present in the plant. However, there are indirect methods that evaluate the level of an organic compound or the activity of an enzyme, which considers that the nutrient is part of this organic compound or is an activator of such an enzyme, that is, a plant deficient in N must present a low amount of chlorophyll or a low nitrate reductase activity (NO_3^- induces the enzyme, as it is a substrate for it). In this sense, Malavolta et al. (1997) described biochemical tests for various nutrients that could be performed to assess the nutritional status of the plant. For example, N (reductase activity, glutamine synthetase, N amidic, asparagine), P (fructose-1,6-2P and photosynthesis; phosphatase activity), K (content of amides and pipecolic acid; content of putrescine), Mg (pipecolic acid), S (reaction with glutaraldehyde; free amino acids), Mn (peroxidases; chlorophyll a/b ratio), B (ATPase activity), Zn (ribonuclease; carbonic anhydrase; arginine content). In the case of P, other studies indicate that the Pi in vacuole cells can indicate the nutritional status of the plant (Bollons and Barraclough 1999).

19.1 Leaf Sampling Criteria

It is worth mentioning that leaf analysis results will only be valid if there is a standard for comparing them. There are variations interspecies and intraspecies, making it difficult to generalize patterns. Thus, a strict control of these criteria will guarantee the validity of the results of the chemical leaf analysis, its interpretation, and the correction of deficiencies in future fertilization. In addition, the fact that most errors occur in the sampling stage can compromise a fertilization program. Thus, problems arise from poor sampling and not from laboratory analytical problems or yet from the use of inadequate recommendation tables.

For a correct sampling of the diagnosis leaf itself, some criteria that are specific to each culture must be considered, such as:

- Leaf type
- Collection time
- Number of leaves per plot

As for leaf type, the newly ripen fully developed leaf is usually used, as it must present a greater sensitivity to reflect the real nutritional status of the plant, in addition to having suffered little effects from the redistribution of nutrients. This leaf is called diagnostic leaf or index leaf. Its standardization is important since an older leaf has a higher concentration of low mobile nutrients (Ca, S, and metallic micronutrients) and the newest leaf has a higher concentration of mobile nutrients (N, P, and K). In this sense, Chadha et al. (1980) observed in mango trees a greater stability or balance of macro- and micronutrients in leaves aged six to eight months (leaves with a mean age, that is, neither new nor old) (Table 19.1). However, only N increased in older leaves due to the application of nitrogen fertilizer at this time.

Research that seeks to identify the diagnosis leaf should isolate the most sensitive leaf to clearly discriminate the levels of nutrients in plant tissues that express deficient, adequate, and toxic levels. In addition, abrupt nutrient variations cannot occur over the sampling time.

Thus, the leaf to be collected must be the same from which the appropriate critical level/range was obtained, which makes up the standard of the tables for the respective crop.

The **sampling time** must be at defined physiological stages, as nutrients may vary along with plant age (Fageria et al. 2004). For example, when the rice plant reaches 20 days after emergence, the appropriate P content in the shoot is 4 g kg^{-1},

Table 19.1 Effects of leaf age on the nutrient concentration of mango trees

Leaf age	N	P	K	Ca	Mg	S	Zn	Cu	Mn	Fe
month	g kg⁻¹						mg kg⁻¹			
1	12.8	1.52	11.07	9.1	2.0	0.88	20	12	27	105
2	11.8	1.18	9.8	10.8	2.9	0.81	28	11	32	153
3	11.9	0.98	8.1	12.2	3.2	1.05	28	11	46	171
4	11.7	0.90	7.7	13.1	3.4	0.88	14	8	46	129
5	12.0	0.84	8.1	14.0	3.5	1.14	15	12	54	193
6	11.7	0.73	7.0	15.9	3.3	1.13	13	11	63	156
7	11.7	0.73	6.4	16.7	3.3	1.14	13	10	63	154
8	11.7	0.73	5.8	17.2	3.3	1.15	12	12	78	169
9	11.6	0.66	5.7	18.8	3.1	1.13	17	21	100	143
10	12.8	0.73	4.8	19.1	3.4	1.19	22	22	87	108
11	12.9	0.70	5.4	20.7	3.3	1.39	15	14	112	145
12	13.0	0.77	4.2	21.2	3.7	1.32	50	17	100	182

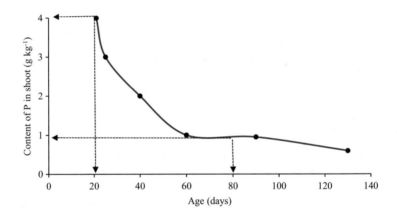

Fig. 19.3 Phosphorus contents in upland rice plants as a function of plant age.

and when the plant reaches the age of 80 days, the appropriate P content is 1 g kg⁻¹ of dry matter (Fig. 19.3). This means that the plant analysis data must be calibrated with the plant's age.

Normally, the sampling time must coincide with the greatest physiological activity of the plant (or its photosynthetic activity), often occurring at the time of the reproductive phase of plants, when the concentration of nutrients is greater.

The best fittings of equations between nitrogen fertilizer doses and leaf N content occur at the time of flowering compared to the fruiting time of guava (Table 19.2). This result indicates the best time for collecting guava leaves (Natale et al. 1994).

In any case, the leaf sampling period for a given crop must coincide with the leaf collection period considered by the researcher to establish the appropriate critical level/range or standards.

Table 19.2 Relationship of nitrogen doses and leaf content evaluated at the time of flowering and fruiting

Agricultural year	Sampling time	Fitting equation	R^2
89/90	Flowering	Linear	0.90
	Fruiting	Linear	0.56
90/91	Flowering	Quadratic	0.99
	Fruiting	Linear	0.92
91/92	Flowering	Quadratic	0.97
	Fruiting	Linear	0.78

Lin et al. (2006) found that the grain yield of rice correlated more with the leaf N collected 20 after flowering compared with that collected at the time of flowering.

Regarding the **number of leaves**, studies usually consider 25–100 leaves per sample. The sampling error normally decreases in a sampling that considers the largest number of leaves associated with the largest number of plants (Holland et al. 1967).

In the literature, there are indications of leaf sampling for various crops grown in tropical regions (Fig. 19.4). Raij et al. (1996) proposed these criteria. However, for some cultures, sampling standards supported by research have not yet been defined, and they could better indicate the nutritional status of the plant.

The leaf sampling criterion for star fruit was defined by Prado and Natale (2004). Criteria for sampling leaves of other crops, according to Raij et al. (1996):

Citrus—Collect the third leaf from the fruit that appeared in the spring (six months old) in branches with fruits (2–4 cm in diameter). Sample four leaves per plant (25 plants per plot)

Coffee—Collect the third pair from the apex of the fruit branch, at the medium height of the plant, in early summer, two leaves per plant in 50 plants per plot.

Passion fruit—Collect in the autumn the third or fourth leaf from the apex of unshaded branches or the leaf with floral axillary bud about to open. Sample 20 plants.

Mango—Collect leaves from the middle of the last vegetation flow of branches with flowers at the end. Sample four leaves per tree, 20 plants per plot.

Cotton—Collect the blade from the fifth leaf from the apex of the main stem during flowering. Sample 30 leaves per plot.

Bean—Collect the third leaf with petiole in the middle third of the plants, during flowering, from 30 plants.

Peanut—Collect the leaves of the apical tuft of the main branch during flowering. Sample 50 plants.

Sunflower—Collect the fifth to the sixth leaf below the capitulum during flowering. Sample 30 plants.

Banana—Remove the central 5–10 cm of the third leaf from the inflorescence, eliminating the central rib and peripheral halves. Sample 30 plants.

Wheat—Flag leaf collected at the beginning of flowering. Minimum 50 leaves.

Fig. 19.4 Leaf sampling criteria for chemical analysis of crops. Soybean—Collect the third leaf with petiole during flowering. 30 plants (Fig. 19.4a); Sugarcane—Collect the +1 leaf (highest leaf with a visible "TVD" collar), the central 20 cm, excluding the central rib, sample 30 plants during the phase of greatest vegetative development (Fig. 19.4b); Corn—Collect the central third of the leaf at the base of the ear when bolting (50% of the bolted plants), sample 30 leaves per ha (Fig. 19.4c); Guava—Collect the third pair of leaves with petiole from the end of the branch, 1.5 m from the soil, sample four pairs of leaves per tree in 25 plants per plot (Fig. 19.4d) (Natale et al. 1996)

Peach—Collect 26 fresh, fully expanded leaves from the median portion of the branches. Sample 25 plants per plot.

Pineapple—Collect, before floral induction, a freshly ripe "D" leaf (fourth leaf from the apex). Cut the leaves (1 cm wide), eliminating the basal portion without chlorophyll.

Avocado- collect in February/March newly expanded leaves aged five to seven months at the canopy average height. Sample 50 trees.

Eucalyptus—Collect freshly ripe leaves, usually the penultimate or antepenultimate leaf from the last 12 months, from the upper third of the canopy. At the end of winter, sample at least 20 trees per plot.

Acerola—Sample on the four sides of the plant, fully expanded young leaves from branches with fruits. Sample 50 plants.

Cashew*—Collect ripe leaves (fourth leaf) from newly grown plants in orchards in production. Sample four leaves per tree, 10 plants per plot.

Rice—Collect the flag leaf at the beginning of flowering. Minimum 50 leaves.

Rubber tree—On trees up to four years old, remove the two most developed leaves from the base of a terminal bouquet located outside the canopy and in full light. On trees over four years old, collect two more leaves developed in the last mature release on low branches in the canopy in shaded areas. Sample 25 plants in the summer.

Cocoa—Collect the second and the third green leaves from the apex of the branch at the average plant height, eight weeks after the main flowering. Sample four leaves per tree, 25 plants per plot.

Papaya—Collect 15 petioles of young leaves, fully expanded and mature (17th to 20th leaf from the apex), with a flower visible in the axillary bud.

Fig—Collect freshly ripe and fully expanded leaves from the median portion of the branches three months after budding. Samples of 25 plants per plot, totaling 100 leaves.

Grape—Collect the youngest newly ripe leaf counted from the apex of the vine's branches, removing a total of 100 leaves.

Potato—Collect the third leaf from the apical tuft on the 30th day. Sample 30 plants.

Watermelon/melon—Collect the fifth leaf from the tip, excluding the apical tuft from half to 2/3 of the plant cycle; 15 plants.

Strawberry—Collect the third or fourth newly developed leaf (without petiole) at the beginning of flowering; 30 plants.

Pod beans—Collect the fourth leaf from the tip from flowering to the beginning of pod formation; 30 plants.

Carrot—Collect the freshly ripe leaf, half to 2/3 of the growth; 20 plants.

Beet—Collect the newly developed leaf; 20 plants.

Pea—Collect the newly developed leaf at flowering; 50 leaflets.

Lettuce—Collect the freshly ripe leaf, half to 2/3 of the growth; 15 plants.

Cauliflower—Collect the newly developed leaf, formation of the head; 15 plants.

Tomato—Collect the leaf with petiole on the occasion of the first ripe fruit; 25 plants.

Pumpkin—Collect the ninth leaf from the tip at the beginning of fruiting; 15 plants.

Forage (grass)—Collect new shoots and green leaves during the active growth phase (November to February).

Forage (legumes)—Same as above, except for some legumes, such as perennial soybeans—Collect the tip of the branches from the apex to the third to the fourth developed leaves; pencilflower—collect the plant pointer (about 15 cm); white leadtree—collect new branches with a diameter of up to 5 mm; alfalfa—collect the upper third at the beginning of flowering.

Martinez et al. (1999) also reported criteria for sampling leaves for different cultures, in addition to those previously mentioned, such as azalea, bougainvillea, cashew, lily, cloves, chrysanthemum, hydrangea, castor, melon, pear, rose, and violet.

The procedure that must be followed in the field to collect the leaf sample is similar as that described for soil sampling. Thus, obtaining representative samples depends on sampling techniques capable of circumventing the heterogeneity that may occur in the plot. Therefore, here are some general guidelines for leaf sampling:

- Zig-zag walking.
- Walking in levels.
- Avoid plants close to roads or carriers.

In addition, leaf sampling should not be conducted under the following conditions:

- Plants with signs of pests and diseases (Table 19.3).
- Plots that received fertilization or pesticides less than 30 days before.
- Different varieties, as the nutritional status is influenced by the genetic factor, a fact widely reported in the literature for several cultures, as in rubber tree clones (Centurion et al. 2005).
- In the case of grafted perennial cultures, do not mix leaves of plants that had different canopies or rootstocks, as they influence nutritional status, such as yellow passion fruit canopy (Prado et al. 2005).
- In no case should leaves of different ages be mixed.
- In the case of perennial crops, leaves of productive branches and leaves of non-productive branches cannot be considered in a same sample.
- Dead or damaged tissue (mechanical).
- Avoid collecting leaves after the occurrence of high rainfalls, as there may be losses of some nutrients, such as N and K, a fact reported for sugarcane by Malavolta et al. (1997).

Finally, in the case of an isolated problem in a given crop. For example, to assess which nutrient could be causing a particular deficiency symptom in any plant, it is recommended to take leaf samples with very marked symptoms separated from other samples with less marked symptoms. There is still a need to collect leaves without symptoms, and in all cases, the sampled leaves must be of the same age and the same position in the plant.

Table 19.3 Composition of cations in cotton leaves collected from healthy plants or affected by "anthocyanosis"	Plant	
	Healthy	Sick
Element	g kg^{-1}	
K	18.3	15.9
Ca	37.0	31.1
Mg	6.3	4.9

19.2   Preparation of Plant Material and Chemical Analysis

The preparation of plant material and chemical analysis usually occur in a Plant Nutrition laboratory, which is an important stage in studies of leaf diagnosis. In this sense, Hanlon et al. (1995) reported that the interpretation of the chemical leaf analysis can be influenced of the physiological stage of the collected leaf, the analytical method, and the sample preparation procedures.

At the time of collecting the samples in the field, they must be identified according to the field by filling out a form with the following information:

Plant Sample Form
Sample no. ____. Producer Identification: _____

1. Identification

    Owner name:
    Property name:
    Address:
    Responsible for shipping:

2. Sample description

    Sampling date:
    Type of leaf sampled:
    Crop: Variety: Age:
    Date of last leaf spray:

3. Nutrients to be analyzed:

    ( ) macronutrients ( ) micronutrients (Fe, Mn, B, Zn, Cu) ( ) other ____

4. Desired recommendations:

After sampling the leaves in the field, some immediate procedures must be performed, such as (Malavolta 1992):

(a) If the sample can reach the laboratory within a maximum of two days after collection: place it in paper bags and send it to the laboratory.
(b) If the sample reaches the laboratory after two days after collection: wash it previously, in this sequence: "clean" running water, detergent solution (0.1%), and water; dry it in an oven set to a temperature close to 70 °C or in full sun (to stop leaf respiration); and put in a paper bag and send it to the laboratory. However, it is possible to conserve plant material for two to three days in a refrigerator (or Styrofoam with ice) without deterioration, thus not requiring drying in the field.

Then, upon reaching the laboratory, the leaf sample undergoes the following treatments:

(a) Registration: the sample receives a number that identifies it;

(b) Washing: in case of fresh leaves with "distilled" running water, detergent solution (0.1% v/v), hydrochloric acid solution (0.3% v/v), deionized water, and drying. The detergent solution aims to eliminate the earth particles, that is, especially Fe contamination, while the acid can remove metals previously applied in leaf fertilization (such as Zn) (Peryea 2005). The immersion time of leaves into the solution should not exceed 30 s to avoid losses due to diffusion of soluble elements of the leaf. A proper washing of leaves could even avoid final misinterpretations. A high content of elements in the analysis could indicate levels of toxicity. The element could remain on the leaf surface after an eventual spraying, and the analysis considered this element.

(c) Drying should be conducted as soon as possible to minimize biological and chemical changes. Eliminate excess water by draining, place the samples in paper bags, and dry them in an oven with forced air circulation at temperatures ranging from 65 to 70 °C (Bataglia et al. 1983) until reaching a constant mass, which should occur after 48–72 h. An alternative option is drying in a microwave oven with an operating time shorter than 30 min. This has promising results (Marcante et al. 2010; Teixeira et al. 2017) but requires further studies.

(d) Grinding: spraying in a mill to obtain fine and homogeneous material for analysis. The mill should be of stainless steel or plastic chambers to avoid contamination of the plant material with some micronutrients, such as Fe and Cu. Between successive grinds, the mill must be cleaned using a brush and 70% alcohol or compressed air.

(e) Storage: the ground leaves are placed in properly labeled paper bags, where they remain until the time of analysis.

Still in the laboratory, the sample to be analyzed must be submitted to different procedures for chemical analysis, such as (a) weighing, (b) obtaining the extract, and (c) determination of the element (Fig. 19.5).

To obtain the extract, it is necessary to perform digestion, which consists of the removal of the elements of organic compounds or elements adsorbed to such compounds (mineralization). There are different types of digestion available for this procedure (Bataglia et al. 1983; Malavolta et al. 1997). Thus, below we briefly present the analytical procedures for chemical analysis of plants according to Malavolta et al. (1997).

Dry digestion: material incineration (for B and Mo).

Wet digestion: using strong acids, which includes sulfuric digestion (for N) and nitro-perchloric digestion (for others, except for B, N, and Cl). Nitro-perchloric digestion is preferred because it operates at low temperatures without losses by volatilization of nutrients (PE – $HClO_4$= 208 °C; HNO_3= 85 °C). Cl is extracted by water.

To obtain the extract, it is necessary to perform digestion, which consists of the removal of the elements of organic compounds or elements adsorbed to such compounds (mineralization). There are different types of digestion available for this procedure (Bataglia et al. 1983; Malavolta et al. 1997). Thus, below we briefly present the analytical procedures for chemical analysis of plants according to Malavolta et al. (1997).

In the laboratory: chemical analysis

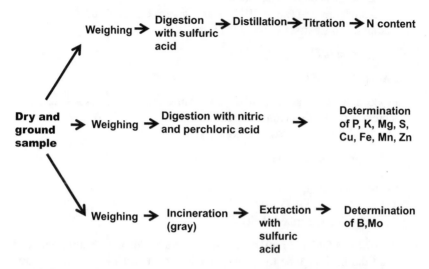

Fig. 19.5 Simplified scheme of the leaf analysis process in a plant nutrition laboratory.

Dry digestion: material incineration (for B and Mo).

Wet digestion: using strong acids, which includes sulfuric digestion (for N) and nitro-perchloric digestion for other nutrients (P, K, Ca, Mg, S, Cu, Fe, Mn, and Zn). Nitro-perchloric digestion is preferred because it operates at low temperatures without losses by volatilization of nutrients (PE – $HClO_4$= 208 °C; HNO_3= 85 °C). Cl is extracted by water.

Agitation digestion: using water (for Cl).

(a) Dry digestion (B, Mo):

- Weigh 0.2 g of dry matter (DM).
- Transfer the plant material to a crucible and incinerate it in an electric oven at 500–550 °C until a white ash is obtained (procedure lasts more or less three hours).
- Cool and add 10 mL of 0.1N HCl, dissolving the ash.
- Leave the residue at rest, thus obtaining the extract.

(b) Wet digestion (nitro-perchloric) (P, K, Ca, Mg, S, Cu, Fe, Mn, and Zn):

- Weigh 1 g of DM.
- Initially perform cold digestion: 5 mL HNO_3 g DM (leave to stand for one day).
- Add an additional 5 mL HNO_3 g DM.
- In the morning, leave the material in the digester block for the slow digestion of DM (there is a partial digestion of DM), and then let it cool.

- In the afternoon, add 2 mL g^{-1} DM) of perchloric acid and heat (in the afternoon) until a colorless extract appears.
- Avoid overheating (up to 210 °C), as it causes loss of P and S.
- Complete the extract solution to 50 mL.

(c) Wet digestion (sulfuric) (N):

- Weigh 0.1 g DM.

– Proceed to the digesting mixture (sulfuric acid + catalysts + salts); this increases the boiling temperature.
– In a 1000-mL beaker, add the mixture:

175 mL of distilled water
3.6 g of Na_2SeO_3
21.39 g of Na_2SO_4
4.0 g of $CuSO_4 \cdot 5H_2O$
200 mL of concentrated H_2SO_4.

– Add 0.1 g of DM and 7 mL of the mixture to a Kjeldahl flask or digestion tube.
– Take the flask or tube to the digester block, increasing the temperature by 40 °C every 30 min until reaching 350 °C, maintaining as such until the digestion is complete, which is characterized by obtaining a colorless or slightly greenish liquid (procedure lasts about three hours).

(d) Aqueous digestion by stirring (Cl):

- Weigh 100 mg of dry matter in a 50-mL conical flaskAdd 25 mL of distilled water, closing it with a rubber stopper
- Stir at approximately 100 rpm on a horizontal shaker for 10 min.
- Take a 10 mL aliquot of this extract and pipette it into a porcelain capsule, obtaining the extract

The nitro-perchloric digestion, in an open system, although it is a method widely used in Brazil, is not used in developed countries. In developed countries, digestion is performed in a microwave oven in a closed and fast system, unlike acid digestion, which is an open system and exposes chemicals to the environment, in addition to being a slower process.

After digestion, the element is determined, and the nutrients present in the solution extract are read. The determination methods are P by calorimetry; K and Ca by flame photometry or atomic absorption or titration; Mg by atomic absorption, calorimetry; S by gravimetry or turbidimetry; Cu by calorimetry or atomic absorption; Fe and Zn by calorimetry or atomic absorption; Cl by titration; and Mo by calorimetry.

Regarding Mo, there is practically no determination in routine laboratories given the lack of calibration for it, in addition to its leaf content being very low (<0.3 mg kg^{-1}), making it difficult to detect Mo by the device. However, using colorimetry (reaction between potassium iodide and hydrogen peroxide), the accuracy of Mo determination is adequate (Polidoro et al. 2006).

19.3   Other Chemical Analyses

Silicon

The method proposed by Elliott and Snyder (1991) is described here: 0.1 g of the sample (leaf tissue) is placed in a plastic tube. Then, 2 mL of hydrogen peroxide (30% or 50%) plus 3 mL of NaOH (1:1) are added. The tubes are shaken and then autoclaved for one hour at 123°°C and 1.5 atm of pressure. An aliquot of 2 mL of the digested material is taken, which is mixed with 2 mL of ammonium molybdate (1:5) to form the yellow silicon-molybdic acid complex. The ideal pH for the maximum complex formation is between 1 and 2. Therefore, when it is necessary to lower the pH of samples, HCl (50%) is added. To eliminate the interference of P and Fe, oxalic acid (75 g in 200 mL of distilled water) is used at the proportion of 2 mL per sample. The reading of Si in the extracts is made using a photocolorimeter at a wavelength of 410 ηm.

There is another Si analysis method based on the greenhouse-induced wet digestion proposed by Kraska and Breitenbeck (2010), which we adapted and describe below. Procedure: weigh 0.1 g of the dried vegetable tissue sample and place in a 50-mL polyethylene tube with a centrifugal screw cap previously washed with 0.1 M sodium hydroxide solution (NaOH) and with deionized water, and then dried.

To reduce foaming, add five drops of alcohol-octyl before adding hydrogen peroxide (H_2O_2) and NaOH. Dampen the samples with 2 ml of 30% H_2O_2, washing the walls of the tube with the sample (the quality of the peroxide is very important for extraction, because if it promotes the oxidation of carbon in the sample and if the peroxide is at a low concentration, only a partial digestion of samples may occur; therefore, the product packaging must be kept hermetically sealed).

Close the tube and place it in a forced air circulation oven at 95 °C. After 30 min, remove the tubes and add 4 mL of 50% NaOH to the hot samples. Shake the tubes slowly, close them, and return them to the oven heated to 95 °C. After four hours, remove the samples. This digestion time was determined for the analysis of leaf tissue in rice plants. However, the digestion time may vary depending on the type of plant material, making it necessary for the extract at the end of digestion to be a homogeneous solution. Example: For the sample of the leaf tissue of peanut plants, it took seven hours. During the digestion period in the oven, it is necessary to lightly stir the samples at one-hour intervals (stirring favors the digestion process). Dilute the samples in final volume with deionized water, adding solution up to 50 mL. The samples must remain at rest for 12 h. The determination of the silicon concentration is performed by spectrophotometer at 410 nm using the colorimetry method. Procedure: Collect a 1-mL aliquot of the digestion extract (in samples of non-Si accumulating plants, increase it to 2 mL); transfer the aliquot to plastic cups, and add a sufficient volume of deionized water to reach a final volume of 20 mL, add 1 mL of hydrochloric acid (50% v/v). The volume of acid may vary (1 or 2 mL), so that the pH value of the solution is close to 1.5 to 2.0; then add 2 mL of ammonium molybdate (100 g L^{-1}) and shake gently. A yellow color should appear in samples containing Si. The more yellow, the higher the concentration of Si in the solution. After five to ten minutes, add 2 mL of oxalic acid (75 g L^{-1}), gently shaking the solution to eliminate the interference of P and Fe; after two minutes, read on a spectrophotometer at 410 nm. The yellow color is not very stable, remaining so for only 15 min. Finally, only plastic containers should be used. The water to be used in the Si analysis is a critical issue: it is important that it is purified, without Si and, if possible, a pre-test should be conducted to certify it. It is necessary to distill and then deionize the water twice.

Nitrate

The nitrate content is an important index of food quality. Thus, its analysis is indicated for several crops, from vegetables to fodder. Mantovani et al. (2005) studied the extraction of nitrate with deionized water and its quantification using the procedures of the reducing column containing cadmium, distillation, salicylic acid, and reducing mixture containing zinc. The procedures of salicylic acid and reducing mixture containing zinc overestimate the levels of nitrate in the dry matter of lettuce, as they are highly subject to the presence of interferents and the effects of the extract color. The procedures of the reducing column containing cadmium and that of distillation are the most suitable for the quantification of nitrate in plant tissues. However, the simplicity and low cost of distillation compared to that of the reducing column indicate that distillation should be recommended.

Thus, we present the nitrate distillation procedure, originally proposed by Bremner and Keeney (1965).

Transfer samples of 0.2 g of dry matter and 20 ml of deionized water to plastic vials with a pressure cap and a capacity of 100 mL; subject them for one hour to agitation periods of 5 min, followed by 15 min of rest, in a water bath at around 60 °C. To clarifying the extracts, transfer the suspensions to 200-ml volumetric flasks after adding 5 ml of sodium tetraborate solution ($Na_2B_4O_7.10H_2O$) 50 g L^{-1}, 5 ml of potassium hexacyanoferrate (II) [$K_4Fe (CN)_6.3H_2O$] 150 g L^{-1}, and 5 ml of zinc sulfate ($ZnSO_4.7H_2O$) 300 g L^{-1}. After adding each solution, stir the contents, completing the volume with deionized water, and let it rest for 30 min. After extraction, filter the material with a qualitative filter paper and distill the extracts in a Kjeldahl microdistillator (Bremner and Keeney 1965). As an adaptation for plant tissue, use 5 mL of extract and 0.4 g of Devarda's alloy, so that all the nitrate in the sample is converted to ammonium in a single distillation. Then, the quantification of N of the distillate in the form of ammonium is performed by titration with standardized solution of H_2SO_4 0,00263 mol L^{-1}. Considering that the concentration of $N-NO_2^-$ present in the samples is negligible, the results of $N-NO_3^- + N-NO_2^-$ should be converted and expressed as $N-NO_3^-$.

Finally, after conducting the chemical analysis and using the glassware, it is necessary to clean them properly to avoid contamination that could influence (significantly) the next analytical results to be obtained. For this, consider a specific detergent for glassware, a special attention to solution dilution, and the time of soaking of the material. Submerge the material completely, not allowing it to remain outside the solution. Wash with a sponge or a brush suitable for each glassware, rinse for long under running water (three to five times) and then with deionized water and put it to dry in an oven (up to 30 °C).

19.4   Studies on Leaf Diagnosis in Crops

As previously mentioned, there are different tools to assess the nutritional status of crops, such as leaf diagnosis (critical level/appropriate range or DRIS), visual diagnosis, and soil chemical analysis.

Leaf Diagnosis (Critical Level or Appropriate Range)

In leaf diagnosis, to assess the nutritional status of plants, some procedures are required, such as leaf sampling, material preparation, chemical analysis in the laboratory, and obtaining analytical results (Fig. 19.6). These results may be used by the researcher to define critical levels and prepare tables of adequate levels (standards). If such tables already exist, the researcher makes only the comparisons and the due interpretations that can indicate whether the nutrients are in adequate levels, are deficient, or not in excess. Thus, there is a diagnosis of the crop nutritional status, which serves as a recommendation for fertilization or correction, with direct effects on the expression of productivity and the profitability of the agricultural exploration.

In the literature, there are tables with nutrient contents, which consist of the so-called normal crop patterns (high growth and production). Thus, the data in Table 19.4 show the leaf contents of macronutrients and micronutrients considered suitable for various crops in Brazil (Raij et al. 1996).

As stated, the results of leaf chemical analysis of a given sample can be compared with the standard values (tables). Three situations can then occur:

(a) The sample content is lower than the standard content. This indicates a possible deficiency.
(b) The sample content is equal to the content considered standard, indicating that there is neither deficiency nor toxicity.
(c) The sample content is higher than the content considered standard, indicating a possible element toxicity.

It is worth mentioning that adequate values of nutrients must be considered for the respective leaf sampling criterion, that is, the leaf that resulted in the sample must be the same as the standard one (which is "tabulated"), otherwise the interpretation of results is not reliable. In this sense, an adequate nutrient content depends on the diagnostic leaf, that is, taking as an example the corn culture, the values for

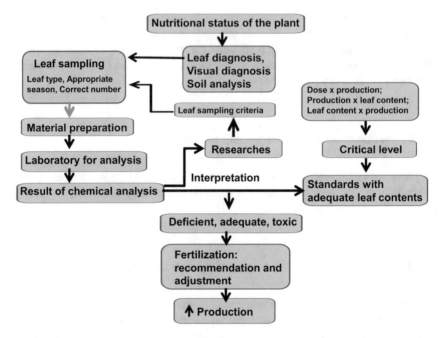

Fig. 19.6 Flowchart for assessing the nutritional status of plants and their development through the critical level or the appropriate range

the leaf below the ear are different from those of the opposite leaf and those below the ear (Table 19.5). The first methodology indicates a newer leaf and, therefore, a higher content of nutrients, especially mobile ones.

The adequate levels presented in the tables can be obtained by extrapolating plants grown in other countries and to Brazil, where errors may be embedded due to different edaphoclimatic conditions. A second way is through local experimentation conducted in Brazil. This last way of obtaining the adequate levels of nutrients is the most accurate to ensure the highest productivity using leaf diagnosis. For this, field experiments are installed. More than three doses of a given nutrient are commonly used in the presence of sufficient doses of the other elements. Thus, it is possible to obtain the previously mentioned relations, which are:

(a) Dose (supply of the nutrient by the soil) × production.
(b) Dose (soil nutrient supply) and/or soil nutrient content × leaf content (or production).
(c) **Leaf content × production**.

Thus, with the leaf content vs production ratio, it is possible to establish adequate levels of nutrients in plants. There are research works reporting these relationships for different nutrients and crops, such as N in mango, citrus, and coffee, where the critical content is 13 g kg^{-1} (or 1.3%), 27 g kg^{-1}, and 31 g kg^{-1}, respectively. In guava, the critical Ca level is 8.5 g kg^{-1} (Fig. 19.7). A coefficient of variation or an

Table 19.4 Values of adequate levels of macronutrients and micronutrients in leaves of some crops grown in tropical regions

Crop	N	P	K	Ca	Mg	S
	g kg^{-1}					
Rice	27–35	1.8–3.0	13–30	2.5–10	1.5–5.0	1.4–3.0
Corn	27–35	2.0–4.0	17–35	2.5–8.0	1.5–5.0	1.5–3.0
Wheat	20–34	2.1–3.3	15–30	2.5–10	1.5–4.0	1.5–3.0
Coffee	26–32	1.2–2.0	18–25	10–15	3.0–5.0	1.5–2.0
Cotton	35–43	2.5–4.0	15–25	20–35	3.0–8.0	4.0–8.0
Orange	23–27	1.2–1.6	10–15	35–45	2.5–4.0	2.0–3.0
Soybean	40–54	2.5–5.0	17–25	4–20	3.0–10	2.1–4.0
Sugarcane	18–25	1.5–3.0	10–16	2–8	1.0–3	1.5–3.0
	mg kg^{-1}					
	B	Cu	Fe	Mn	Mo	Zn
Rice	4–25	3–25	70–200	70–400	0.1–0.3	10–50
Corn	10–25	6–20	30–250	20–200	0.1–0.2	15–100
Wheat	5–20	5–25	10–300	25–150	0.3–0.5	20–70
Coffee	5–80	10–20	50–200	50–200	0.1–0.2	10–20
Cotton	30–50	5–25	40–250	25–300	–	25–200
Orange	36–100	4–10	50–120	35–300	0.1–1.0	25–100
Soybean	21–55	10–30	50–350	20–100	1.0–5.0	20–50
Sugarcane	10–30	6–15	40–250	25–250	0.05–0.2	10–50

Note: These adequate levels of nutrients are valid only for crops whose leaf sampling was conducted according to Raij et al. (1996)

Table 19.5 Adequate leaf content of macronutrients in corn culture according to the type of the diagnostic leaf

Nutrient	Sampling[a]	Sampling[b]
	g kg^{-1}	
N	27–35	27–33
P	2.0–4.0	2.5–3.0
K	17–35	21–30
Ca	2.5–8.0	2.0–5.0
Mg	1.5–5.0	2.1–4.0
S	1.5–3.0	2.0–3.0

[a]central third of the leaf at the base of the ear in male flowering (50% of bolted plants) (Raij et al. (1996); [b]limb of the opposite leaf and below the ear in female flowering (appearance of "hairs") (Malavolta 1992)

experimental error, common in field experiments, is associated to these relations of increased doses of a nutrient (or leaf content) and production assuming the adoption of adequate levels just above the critical level in a range of sufficiency (~100% of maximum production), so that future fertilization recommendations ensure the

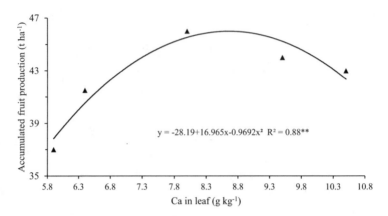

Fig. 19.7 Leaf Ca ratio and guava production

expected productivity. Thus, depending on the author, the interpretation of low, medium, adequate, and high leaf nutrient content is associated with relative production (<70%, 70–90%, 90–100%, and >100%, respectively).

Thus, for the interpretation of the results of the nutritional status of plants, the sufficiency range is the most appropriate when compared to the point value as a critical level. This is because, in addition to the variation coefficient obtained in obtaining the calibration curve, as previously mentioned, crops present genetic variations due to the diversity of variety and/or hybrids/clones. Therefore, the method of interpreting leaf nutrient by the appropriate range is less affected by variations in the environment and in the plant itself.

Fontes (2001) reports that in the crop diagnosis, one must take into account the critical level of the nutrient associated with a phytotechnical analysis (development) of the crop, since the concentration of the nutrient in the tissue is expressed in relative terms, that is, amount of nutrient/amount of dry matter.

Finally, another application of leaf diagnosis is to survey the nutritional status of a crop in a given region. With these data, it is possible to infer eventual problems in the management of crops, such as low Ca/Mg levels, which may indicate low limestone use, as well as low levels of P and K, which may indicate inefficient fertilization, among others.

As an example, there is a survey of the nutritional status of coffee crops in the forest zone (Minas Gerais). Most crops in that region have leaf levels below the critical level, such as Ca (74%), indicating a possible problem with liming and with P (91%), K (82%), and B (100%). Farmers may be using phosphate, potassium, and borate fertilization at levels below plant requirements (Table 19.6). Thus, with this information, institutions can guide producers in the region to adopt corrective agronomic practices, so that a satisfactory agricultural production be achieved.

Table 19.6 Nutritional survey of coffee trees in the forest area of Minas Gerais based on leaf analysis by critical level or threshold content

Nutrient	Critical level	Deficient crops
	g kg^{-1}	%
N	30.0	41
P	1.2	91
K	18.0	82
Ca	10.0	74
Mg	3.5	32
S	2.0	0
Cu	4.0	0
B	40.0	100

Leaf Diagnosis (DRIS—Diagnosis and Recommendation Integrated System)

As previously mentioned, the two interpretation criteria (critical level and appropriate range) discussed are based on the establishment of standards for productive plants and on the comparison of concentrations of nutrients in the samples with these standards. In both procedures, the absolute levels of nutrients are considered.

Following these criteria there is the relationship between the leaf nutrient content, obtained by the calibration curves, and plant productivity. Thus, the comparison of any sample with the standard should generate a valid indication only when there are the same edaphoclimatic conditions (temperature, water availability, other nutrients, soil reaction, among others) in which the calibration curve was fitted.

In field conditions and commercial fields, plots may not reproduce the same conditions of growth and development as those obtained by the calibration curve. In these circumstances, there are criticisms to the method of critical level or adequate range because of its limited capacity in predicting the crop nutritional status of a given nutrient, compromising the accuracy of future fertilization recommendations. DRIS, on the other hand, can be used in commercial plots, therefore close to the real cultivation conditions the producer faces.

The DRIS was conceived as a diagnostic process capable of overcoming the limitations of the conventional method (critical level or adequate range or sufficiency) and, mainly, minimizing the effects of dilution or concentration of nutrients in relation to variations in accumulation of dry matter in plant tissues.

The theoretical conception of DRIS is interesting agronomically, since it allows working the interactions because it considers relationships between two nutrients in a determined sample and a reference population. Thus, DRIS is considered a multivariate method, unlike the appropriate range (univariate).

There is another multivaried CND (Composition Nutritional Diagnosis) method that can be used to establish the plant's nutritional status. There is an unprecedented practical guide available for calculating the CND (Traspadini et al. 2018). An alternative to obtain the critical values for the sufficiency range method is the mathematical chance that applies probability distribution to data obtained from commercial crops (Wadt et al. 2013).

The conception of DRIS considers that in a crop or population of high production plants, there is a specific relationship between nutrients, for example N/K, N/P, P/K etc. Thus, there may be a situation in which a low crop production is related to a nutritional imbalance, and perhaps with a small addition of that nutrient, the problem is solved with a significant impact on production and profitability.

In fact, DRIS is a technique based on the comparison of indexes calculated considering relationships between nutrients. The system is based on the calculation of indexes for each nutrient considering its relationship with the others and comparing each relationship with the average relationships of a reference population (high production). For each nutrient indexes with negative values indicate deficiency and positive values indicate excess, while values close to zero correspond to a balanced nutrition. Thus, these results express numerically the influence of each nutrient on the nutritional balance of the plant.

The success of DRIS depends on the reliability of the data obtained for a reference population, which in turn depends on a high number of observations, which often is the "bottleneck" of DRIS. There is also fact that DRIS is not immune to adversities, which are common to other diagnostic methods. It is necessary that the application of DRIS be regional and not extrapolated to many producing regions, and that it maintains a satisfactory control of tissue sampling techniques for diagnosis.

The DRIS was developed for an interpretation that is less dependent on sampling variations with respect to the age and origin of the tissue, allowing an ordering of production limiting factors and emphasizing the importance of nutrient balance. Derivations of DRIS have been developed in the literature to reduce erroneous diagnoses in cases where there is a greater concentration or dilution and the nutrient relationships remain constant, such as DRIS-M (Hallmark et al. 1987).

Although the theoretical conception of physiological diagnosis, the precursor of DRIS is relatively old, as the first publication was that of Beaufils (1973), which widespread worldwide in the late 1980s (Sumner 1978), only at the end of the 1990s its potential began to be explored, mainly driven by the improvement of information technology in recent times.

Some procedures are necessary for the establishment of DRIS: from the collection of data that occurs in a similar way as that commonly adopted in the conventional method of diagnosis by the appropriate range to the performance of calculations to obtain the DRIS indexes, and finally to the nutritional balance of the crop.

Data Collection

(a) Production of the crop of interest.

    The productivity of each crop plot must be determined. It is important to obtain plots with different yields, ranging from low to high.

(b) Leaf sampling.

In these same plots, leaf sampling should be performed following the criteria normally adopted by the literature for the crop. Then, the nutrient content is determined using the method described by Bataglia et al. (1983).

Data Analysis

(a) Database

For this, the plots of the crop should be selected, which will be monitored during the agricultural year. A small number of fields may affect the success of DRIS; therefore, it is interesting to use as many fields as possible whose data is highly reliable.

All results of the analysis of leaves and productivity should be organized in a large database, discriminating the number and area of the field, age, cultivar, planting density, year of leaf sampling, and harvest. This large database is later analyzed according to population and the defined criteria. Initially, the population should be divided into two groups according to productivity: a high and a low population.

(b) Calculation of DRIS

Initially, to calculate the DRIS indexes, some procedures must be followed, such as calculate the number of possible ratios, choose the nutrient ratio, calculate the functions and the sum of the functions, and obtain the nutritional balance index.

(b_1) Number of possible relations

To obtain the number of possible relations, the following expression will be used: No. of relations $= n\,(n-1)$.

(b_2) Choice of the relationship to be studied

Initially, the mean, the variance, and the coefficient of variation of nutrients are calculated. The choice of the relation is necessary because in calculating the DRIS indexes, only one type of expression is used to relate each pair of nutrients.

There are two ways to choose the ratio of nutrients: the first is by the highest variance ratio ($S^2\,A/S^2\,B$), that is, taking the N/K ratio as an example, in which its variance ratio is greater than K/N, the first is selected (Letzsch 1985); and the second way, proposed by Nick (1998), is called "r value:" it is the calculation of the correlation coefficients (r) between the values of the plant response variable and the ratio between the pairs of nutrients, both in direct and in reverse ways. Thus, the relationship that results in the highest absolute value of the correlation coefficient is chosen.

(b_3) Calculation of functions

Several proposals with modifications of the original model proposed by Beaufils (1973) for the calculation of the nutrient functions of a sample and the calculation of the DRIS index have been proposed aiming to increase its accuracy in nutritional diagnosis. Thus, the Jones' proposal (1981) is presented here.

$f(Y/X) = [(Y/Xa) - (Y/Xn)] \times k/s$; where:

$f(Y/X)$ = Calculated function of the ratio between the nutrients Y and X

Y/Xa = Ratio of nutrients in the sample

Y/Xn = Standard nutrient ratio

s = Standard deviation of the Y/Xn ratio

k = Sensitivity constant

b4) Calculation of DRIS indexes

$Ix = [\Sum^m_{i=1} f(Y/Xi) - \Sum^n_{j-1} f(Xj/Y)] / (m+n)$; where

Ix = DRIS index for X

X = Nutrient for calculating the index

Y = Another nutrient

m = Function number whose factor X is in the denominator of the norm ratio

n = Function number whose factor X is in the numerator of the norm ratio

(b4) Nutritional balance index

The nutritional balance index (NBI) seeks to evaluate the mean of the absolute values of all nutrient indexes, both for M-DRIS and DRIS (Walworth and Summer 1987). This index can be useful in indicating the nutritional status of the plant. However, it does not indicate its causes. The higher the sum, the greater the indication that the plant is in a nutritional imbalance and, therefore, the lower its productivity.

$$NBIm = \left(|IY1| + |IY2| + \ldots + |IYn|\right) / n$$

The literature has a large number of scientific researches using DRIS for several crops, such as annual crops such as corn (Reis 2002) and soybeans (Beverly et al. 1986); semi-perennial crops, such as sugarcane (Reis Jr and Monnerat 2002); and perennials crops, such as coffee (Partelli et al. 2006); and fruit trees: cherry trees (Davee et al. 1986), apple trees (Parent and Granger 1989), vines (Chelvan et al. 1984), pecan (Beverly and Worley 1992), peach (Sanz 1999), mango (Raghupathi and Bhargava 1999), pineapple (Angeles et al. 1990), banana (Angeles et al. 1993; Villaseñor et al. 2020), papayas (Bowen 1992), lemon tree (Creste 1996), and orange trees (Beverly et al. 1984).

Some didactic examples can show the applicability of DRIS for the assessment of the nutritional status of plants, such as citrus. From three crops (samples), the NBI (nutritional balance index) equal to 1, 25, and 50 related to the production of ten, five, and two boxes per plant (Table 19.7).

Thus, an NBI close to or equal to 1 indicates that the nutrients are in balance in the plant, while a high NBI indicates a strong nutritional imbalance, also informing which nutrient is the most limiting whether due to deficiency (high negative number) or excess (high positive number). Thus, in the low-yielding plot (two boxes per plant), the nutrients Ca and P were the most limiting due to excess and Mg was the one that most negatively affected production due to deficiency.

Table 19.7 Assessment of the nutritional status of citrus using DRIS in plots with different yields

Parameter	N	P	K	Ca	Mg	S
Plot producing ten boxes per plant						
Leaf content in the sample, g kg^{-1}	29.0	1.3	19.0	45.0	5.7	2.0
DRIS Index	−2	−2	0	−1	2	3
NBI = 1						
Plot producing five boxes per plant						
Leaf content in the sample	28.6	1.3	13.5	40.4	4.0	3.0
DRIS Index	17	22	−19	16	−49	18
NBI = 25						
Plot producing two boxes per plant						
Leaf content in the sample	24.9	1.3	14.0	43.0	2.9	2.9
DRIS Index	17	42	−11	69	−136	18
NBI = 50						

The high NBI value indicates a nutritional imbalance, and consequently a low production. Many studies have indicated a significant correlation with crop production, such as coffee (Silva et al. 2003). This significant correlation expresses only a specific condition with no occurrences of non-nutritional events. In commercial crops, this correlation is not valid, as events such as diseases/pests/water deficit can occur before and after leaf sampling. This induces low IBN values, which may be associated with low productivity. Therefore, this correlation between production and NBI is not adequate.

Making a brief parallel to the interpretation of the chemical analysis of the citrus culture by the appropriate range and by the DRIS, both criteria were similar regarding the diagnosis of the nutritional status of the crop (Table 19.8). However, through the DRIS method it was possible to choose the three most unbalanced or the most limiting nutrients in ascending order: First place: B, second place: P, and third place: Mg (Fig. 19.8). In the field routine, it is interesting to always compare both diagnostic methods (DRIS and appropriate range), as there may be situations in which one method is more sensitive than the other to identify limitations in the sample. In coffee, Bataglia et al. (2004) reported that the appropriate range was a better criterion for interpreting the crop nutritional status than DRIS in high-yield crops, whereas in low-yield crops the opposite was true.

It is worth noting that the information provided by the DRIS indexes does not seem to be better than that provided by isolated contents. Therefore, objectively, the DRIS can assess nutritional status better, equally, or worse than the traditional critical levels or sufficiency range (Malavolta 2006).

It is worth noting that, due to the lack of universality of the rules, it is preferable to use specific rules instead of general rules (Silva et al. 2005). Therefore, it is more interesting to use DRIS standards at a regional level with similar edaphoclimatic characteristics, thus increasing the success of DRIS for a proper diagnosis.

According to Martinez et al. (1999), one of the difficulties of using DRIS for diagnosis is because absolute values of the calculated indexes may vary with the

Table 19.8 Nutritional diagnosis of results of chemical analysis of citrus leaves interpreted by appropriate range[a] and by DRIS

	N	P	K	Ca	Mg	S
	g kg^{-1}					
Leaf content	25.0	0.9	12.8	32.0	3.8	2.4
Appropriate range	23–27	1.2–1.6	10–15	35–45	2.5–4.0	2.0–3.0
Diagnosis	adequate	deficient	adequate	Deficient	adequate	adequate
DRIS Index						
Diagnosis	adequate	deficient	adequate	Adequate	Excess	adequate
	B	**Cu**	**Fe**	**Mn**	**Zn**	**Mo**
	mg kg^{-1}					
Leaf content	35.0	14.0	148.0	54.0	44.0	0.9
Appropriate range	36–100	4–10	50–120	35–300	25–100	0.1–1.0
Diagnosis	deficient	adequate	high	adequate	adequate	adequate
DRIS Index	−28	−2	11	1	−2	ND
Diagnosis	deficient	adequate	excess	adequate	adequate	–
Mean NBI	8.3					

[a]Appropriate range according to Quaggio et al. (1997)

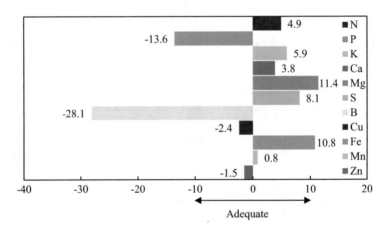

Fig. 19.8 DRIS indexes for citrus crop

calculation formula or the number of binary relationships involved, not allowing to evaluate, in each case, the potential for response to fertilization. In order to improve interpretation of the results of DRIS indexes, Federal University of Viçosa (Viçosa, Brazil) developed the method of the potential for response to fertilization (PRF). This method defines five classes of probability of response to fertilization and compares the index calculated for a given nutrient and the average nutritional balance index. Thus, there are the following classes of response to fertilization: class 1 (positive response), class 2 (positive or null response), class 3 (null response), class 4 (negative or null response), and class 5 (negative response).

However, this method does not consider a plant response that indicates whether the diagnosis is true or not. Therefore, it is important to assess the accuracy of DRIS or CND in commercial crops before their release to the producer. The procedures for assessing accuracy by assessing its quality in terms of errors and correct diagnosis were reported by Wadt et al. (2016). This accuracy test was conducted only for a few crops, such as sugarcane with CND (Silva et al. 2020) and DRIS (Silva et al. 2021) and eucalyptus seedlings (Morais et al. 2019).

The chemical analysis of leaves, in addition to being relevant for the interpretation of the nutritional status of plants due to the diagnosis of deficiency, toxicity or imbalance, as previously mentioned, is also useful for other purposes, according to Fontes (2001):

(a) Confirm the visual diagnosis of symptoms of deficiency/toxicity
(b) Identify "hidden hunger"
(c) Verify if the nutrient applied to the soil was indeed absorbed by the plant
(d) Characterize the concentration of nutrients in plants over the years
(e) Quantify the removal of nutrients by the harvested part
(f) Predict crop production (prognosis)
(g) Map soil fertility areas and estimate nutrient levels in diets available to animals

The chemical analysis of plants should always be considered as a complementary and not exclusive technique, as other criteria for diagnosis and as means of identifying nutritional problems, such as soil analysis and others (visual symptoms, use of indicator plants, and experimentation) (Souza and Carvalho 1985).

DRIS is an important sensitive and complementary tool to assess possible nutritional imbalances in crops. However, producers do not use it widely. This is possibly due to a little familiarity of technicians (of the laboratory or of the extension courses), a difficulty of obtaining regional or even national standards (there are no validation guarantees that these standards can be universally used), and the lack of information on field verification, calibration, and recommendation regarding the system. Thus, it is necessary to advance in terms of dissemination and of studies and expand testing networks with effective participation, integrating research, extension courses, and private initiative institutions (companies and producers' cooperatives).

Finally, as previously mentioned, the assessment of nutritional status can be conducted directly by leaf chemical analysis using the total content. This is thus an important and precise tool in diagnosis. However, the results obtained are not used for intervention in the same production cycle, especially in annual crops. This occurs at the advanced phase of the cycle, which is normally suitable for sample collection. Also, the time that elapses from leaf sampling and obtaining the results and diagnosis of the intervention may can take two to three weeks. However, in perennial crops, leaf diagnosis is the basis for a sustainable management of crop fertilization (Prado and Rozane 2020).

Thus, research is relevant to establish calibration curves in crops at different stages of the production cycle prior to those commonly recommended, associated with previous studies performing a proper diagnosis of leaf and alternative methods of rapid diagnosis, such as soluble nutrient content (leaf or sap) by means of specific ion meters using *kits* for N-NO$_3^-$ and K. There are also indirect methods, such as the chlorophyll meter (which measures the absorption of light by the chlorophyll), which evaluates the intensity of the green color and estimates the chlorophyll content. This last method is interesting in that the photosynthetic pigment, responsible for the green color, does not occur isolated in the chloroplast, but associated with proteins that form complex structures that contain most of the N in plants. Normally, plants subjected to N doses increase this nutrient content beyond their nutritional need, that is, this does not reflect in production (luxury consumption). However, this is not the case for chlorophyll. Thus, after reaching the maximum production, nitrogen fertilization does not influence the chlorophyll content, and therefore, there will be no response from the plant to the application of this nutrient. Therefore, there is a high correlation between the indirect measurement of chlorophyll and the N content in the plant and in production.

Thus, monitoring the N content in corn, based on the indirect measurement of chlorophyll (chlorophyll meter SPAD-502 and other devices), increases the efficiency of use of N compared to the not monitored management system (Rambo et al. 2007). Thus, the use of this technique has as its premise "fertilizing when necessary," which optimizes the management of fertilization.

It is appropriate to add that all factors that interfere with the intensity of the green color (cultivars, nutritional disorders of other elements, among others) should affect the reading results. Therefore, it requires that the producer has an excellent control over the area to be monitored, with only limitation of nitrogen.

In this way, meters estimate the chlorophyll content to obtain the critical level, which could be obtained at different crop growth stages. Another way of interpreting the results is through the Sufficiency Index (SI) (Schepers et al. 1992), which is obtained by the expression: SI = SPAD value of the sample/SPAD reference value. The interpretation by SI is interesting, because it can be done in the producer's own area and, therefore, minimize interferences that may occur using the conventional method by critical level. Through SI, the interpretation is specific to the cultivar, stage of development, edaphoclimatic conditions, and management practices in the area to be monitored.

The SI is obtained by the values of the problem-sample to be evaluated and by the values obtained in a small reference area. This reference is a microplot in the crop field, where, on purpose, abundant or even excessive amounts of N are applied and whose absolute value represents saturation, that is, 0.1. It is mandatory to obtain a reference value per year for each cultivar in each plot or in each production region with homogeneous agronomic conditions using excessive doses, about two to three times the recommended dose.

An SI of 0.95 is widely used as an adequate level, that is, only below this value is nitrogen fertilization recommended. However, Godoy et al. (2003) suggest that different SI be adopted during the corn cycle using the values of 0.98 (V4), 0.95 (V7), and 0.90 (close to bolting). Thus, it is possible for SI to vary according to the crop cycle, and it is interesting to standardize it for each stage of crop development.

The indexes expressed by the chlorophyll meters vary according to the manufacturer. That is why standardization is important.

One of the most explored aspects of digital agriculture is the combination of images from different types of sensors (visible range, multispectral, hyperspectral, chlorophyll fluorescence, etc.) and machine learning methods to optimize agricultural practices, such as determining the crop nutritional status quickly.

The phenological characteristics of crops are correlated with the vegetation index (VI) data from multispectral sensors embedded or not in drones. Thus, when mapping IV data of a plot, it is possible to make observations about the spatial variability of an important biological variable (height, leaf area, etc.). This biological information can indicate the need for a nutrient (most used for N) by the plant and, consequently, define the dose of this nutrient, allowing it to be applied at a variable rate in the field.

There is an attempt in precision agriculture to use this tool to optimize decision-making on fertilization. However, with the complexity of N interactions in the soil-plant-environment system, experimentation is needed to scientifically substantiate this practice, integrating sensing remote techniques and mathematical models using data of the main factors of production. This generates innovative platforms and assists in sustainable economic decisions.

There is still a long way to go regarding research, so that precision agriculture integrates dozens of factors of production, ensuring a high accuracy of agronomic decisions in the field.

References

Angeles DE, Summer ME, Barbour NW. Preliminary nitrogen, phosphorous, and potassium DRIS norms for pineapple. HortScience. 1990;25:652–5. https://doi.org/10.21273/HORTSCI.25.6.652.

Bataglia OC, Furlani AMC, Teixeira JPF, et al. Métodos de análise química de plantas. Campinas: Instituto Agronômico; 1983.

Beaufils ER. Diagnosis and recommendation integrated system (DRIS). A general scheme of experimentation and calibration based on principles developed from research in plant nutrition. University of Natal, Pietermaritzburg, 1973.

Beverly RB, Worley RE. Preliminary DRIS diagnostic norms for Pecan. HortScience. 1992;27:271–5. https://doi.org/10.21273/HORTSCI.27.3.271.

Beverly RB, Stark JC, Ojala JC, et al. Nutrition diagnosis of 'Valencia' oranges by DRIS. J Am Soc Hortic Sci. 1984;109:649–54.

Beverly RB, Sumner ME, Letzsch WS, et al. Foliar diagnosis of soybean by DRIS. Commun Soil Sci Planta. 1986;17:237–256. https://www.tandfonline.com/doi/abs/10.1080/00103628609367711.

Bollons HM, Barraclough PB. Assessing the phosphorus status of winter wheat crops: inorganic ortho- phosphate in whole shoots. J Agric Sci. 1999;133:285–95. https://doi.org/10.1017/S0021859699007066.

Bowen JE. Comparative DRIS and critical concentration interpretation of papaya tissue analysis data. Trop Agric. 1992;69:63–7.

Bremner JM, Keeney DR. Steam distillation methods for determination of ammonium, nitrate and nitrite. Anal Chim Acta. 1965;32:485–95. https://doi.org/10.1016/S0003-2670(00)88973-4.

Centurion MAPC, Centurion JF, Roque CG, et al. Effect of inter row management on chemical properties of the soil, nutritional status and development of rubber tree. Rev Árvore. 2005;29:185–93. https://doi.org/10.1590/S0100-67622005000200002.

Chadha KL, Samra JS, Thakur RS. Standardization of leaf-sampling technique for mineral composition of leaves of mango cultivar 'Chausa'. Sci Hortic. 1980;13:323–329. http://dx.doi.org/10.1016/0304-4238(80)90090-4.

Chelvan RC, Shikhamany SD, Chadha KL. Evaluation of low yielding vines of Thompson seedless for nutrient indices by DRIS analysis. Indian J Hortic. 1984;41:166–70.

Creste JE. Uso do DRIS na avaliação do estado nutricional do limoeiro-siciliano. Faculdade de Ciências Agronômicas: Tese; 1996.

Davee DE, Righetti T, Fallahi E, et al. An evaluation of the DRIS approach for identifying mineral limitations on yield in 'Napolean' sweet cherry. J Am Soc Hortic Sci. 1986;111:988–93.

Elliott CL, Snyder GH. Autoclave – induced digestion for the colorimetric determination of silicon in rice straw. J Agric Food Chem. 1991;39:1118–9. https://doi.org/10.1021/jf00006a024.

Fageria NK, Barbosa Filho MP, Stone LF, et al. Nutrição de fósforo na produção de arroz de terras altas. In: Yamada T, Abdalla SRS, editors. Fósforo na agricultura brasileira vol 1. Piracicaba: Potafos; 2004. p. 401–18.

Fontes PCR. Diagnóstico do estado nutricional das plantas. Viçosa: Universidade Federal de Viçosa; 2001. 122p.

Godoy LJG, Bôas RLV & Grassi Filho H. Adubação nitrogenada na cultura do milho baseada na medida do clorofilômetro e no índice de suficiência em nitrogênio (ISN). Agronomy, 2003;25:373–380.

Hallmark WB, Mooy CJ, Pesek J. Comparison of two DRIS methods for diagnosing nutrient deficiencies. J Fertil Issues. 1987;4:151–8.

Hanlon EA, Obreza TA, Alva AK. Tissue and soil analysis. In: Tucker DPH, Alv. 1995

Holland, DA, Little RC, Allen M, et al. Soil and leaf sampling in apple orchards. J Hortic Sci. 1967;42:403–417. https://doi.org/10.1080/00221589.1967.11514224.

Kraska JE, Breitenbeck GA. Simple, robust method for quantifying silicon in plant tissue. Commun Soil Sci Plant Anal. 2010;41:2075–85. https://doi.org/10.1080/00103624.2010.498537.

Lagatu H, Maume L. Le diagnostic foliare de la pomme de terre. Ann Écol Natl Agric. 1934;22:50–158.

Letzsch WS. Computer program for selection of norms for use in the Diagnosis and Recommendation Integrated System (DRIS). Commun Soil Sci Plant Anal. 1985;16:339–47. https://doi.org/10.1080/00103628509367609.

Lin X, Zhou W, Zhu D, et al. Nitrogen accumulation, remobilization and partitioning in rice (*Oryza sativa* L.) under an improved irrigation practice. Field Crops Res. 2006;96:448–54. https://doi.org/10.1016/j.fcr.2005.09.003.

Malavolta E. ABC da análise de solos e folhas: amostragem, interpretação e sugestões de adubação. São Paulo: Agronômica Ceres; 1992. 124p.

Malavolta E. Manual de nutrição mineral de plantas. São Paulo: Agronômica Ceres; 2006.

Malavolta E, Vitti GC, Oliveira SA. Avaliação do estado nutricional das plantas: princípios e aplicações. Piracicaba: Associação Brasileira de Potassa e do Fósforo; 1997. 319p.

Mantovani JR, Cruz MCP, Ferreira ME, et al. Comparison of procedures for nitrate determination in vegetable tissue. Pesq Agropec Bras. 2005;40:53–9. https://doi.org/10.1590/S0100-204X2005000100008.

Marcante NC, Prado RM, Camacho MA, et al. Determination of dry matter and macronutrient content in leaves of fruit trees using different drying methods. Cienc Rural. 2010;40:2398–2401. https://doi.org/10.1590/S0103-84782010001100025.

Martinez HEP, Carvalho JG, Souza RB. Diagnose foliar. In: Ribeiro AC, Guimarães PTG, Alvarez VH, editors. Recomendações para o uso de corretivos e fertilizantes em Minas Gerais. Viçosa: Comissão de Fertilidade do Solo do Estado de Minas Gerais; 1999. p. 144–68.

Martins APL, Reissmann CB. Plant material and laboratory routines in chemical-analytical procedures. Sci Agraria. 2007;8:1–17.

Morais TCB, Prado RM, Traspadini EIF, et al. Efficiency of the CL. DRIS and CND methods in assessing the nutritional status of eucalyptus spp. rooted cuttings. Forests. 2019;10:786–804. https://doi.org/10.3390/f10090786.

Natale W, Coutinho ELM, Boaretto AE, et al. Influence of the sampling period on the chemical composition of guava leaves (*Psidium guajava* L.). Revista de Agricultura. 1994;69:247–55.

Natale W, Coutinho ELM, Boaretto AE, et al. Goiabeira: calagem e adubação. Jaboticabal: Funep; 1996.

Natale W, Prado RM, Leal RM, et al. Effects of the zinc application on the development, nutritional status and dry matter production of passion fruit cuttings. Rev Bras Frutic. 2004;26:310–4. https://doi.org/10.1590/S0100-29452004000200031

Nick JA. DRIS para cafeeiros podados. Escola Superior de Agricultura Luiz de Queiroz: Dissertação; 1998.

Parent LE, Granger RL. Derivation of DRIS norms from a high-density apple orchard established in the Quebec Appalachian Mountains cherry. J Am Soc Hortic Sci. 1989;114:915–9.

Partelli FL, Vieira HD, Monnerat PH, et al. Comparison of two DRIS methods for diagnosing nutrients deficiencies in coffee trees. Pesq Agropec Bras. 2006;41:301–6. https://doi.org/10.1590/S0100-204X2006000200015.

Peryea FJ. Sample washing procedures influence mineral element concentrations in zinc sprayed apple leaves. Commun Soil Sci Plant Anal. 2005;36:2923–31. https://doi.org/10.1080/00103620500306098.

Polidoro JC, Medeiros AFA, Xavier RP, et al. Evaluation of techniques for determination of molybdenum in sugarcane leaves. Comm Soil Sci Plant Anal. 2006;37:77–91. https://doi.org/10.1080/00103620500408753.

Prado RM, Natale W. Leaf sampling in carambola trees. Fruits. 2004;52:281–9. https://doi.org/10.1051/fruits:2004027.

Prado RM, Vale DW, Romualdo LM. Phosphorus application to the nutritional status and dry matter production of passion fruit cuttings. Acta Scient Agron. 2005;27:493–498. https://doi.org/10.4025/actasciagron.v27i3.1461.

Prado RM, Rozane DE. Leaf analysis as diagnostic tool for balanced fertilization in tropical fruits. In: Srivastava AK, Hu C, editors. Fruit crops: diagnosis and management of nutrient constraints. Netherlands: Elsevier; 2020. p. 131–44.

Quaggio JÁ, Raij B Van, Piza Júnior CT. Fruit tree. In: Recommendations of Fertilization and Liming for the State of São Paulo. Raij B Van, Cantarella H, Quaggio JA et al (ed). Instituto Agronômico; 1997, pp.121–125.

Rafique E, Rashid A, Ryan J, et al. Zinc deficiency in rainfed wheat in Pakistan: magnitude, spatial variability, management, and plant analysis diagnostic norms. Comm Soil Sci Plant Anal. 2006;37:181–97. https://doi.org/10.1080/00103620500403176.

Raghupathi HB, Bhargava BS. Preliminary nutrient norms for 'Alphonso' mango using diagnosis and recommendation integrated systems. Indian J Agric Sci. 1999;60:648–50.

Raij BV, Cantarella H, Quaggio JA, et al. Recomendações de adubação e calagem para o Estado de São Paulo. Campinas: Instituto Agronômico & Fundação IAC; 1996.

Rambo L, Silva PRF, Strieder ML, et al. Monitoring plant and soil nitrogen status to predict nitrogen fertilization in corn. Pesq Agropec Bras. 2007;42:407–17. https://doi.org/10.1590/S0100-204X2007000300015.

Reis RA Jr, Monnerat PE. Sugarcane nutritional diagnosis with DRIS norms established in Brazil, South Africa, and the United States. J Plant Nutr. 2002;25:2831–51. https://doi.org/10.1081/PLN-120015542.

Sanz M. Evaluation of interpretation of DRIS system during growing season of the peach tree: Comparison with DOP method. Comm Soil Sci Plant Anal. 1999;30:1025–36. https://doi.org/10.1080/00103629909370265.

Schepers JS, Francis DD, Vigil M, et al. Comparison of corn leaf nitrogen concentration and chlorophyll meter reading. Comm Soil Sci Plant Anal. 1992;23:2173–87. https://doi.org/10.1080/00103629209368733.

Silva EBI; Nogueira FD, Guimarães PTG. Nutritional status of coffee tree evaluated by DRIS in response to potassium fertilization. Rev Bras Ciênc Solo. 2003;27:247–255. https://doi.org/10.1590/S0100-06832003000200005.

Silva GP, Prado RM, Wadt PGS, et al. Accuracy of nutritional diagnostics for phosphorus considering five standards by the method of diagnosing nutritional composition in sugarcane. J Plant Nutr. 2020;43:1485–97. https://doi.org/10.1080/01904167.2020.1730902.

Silva GP, Prado RM, Wadt PGS, et al. Accuracy measures for phosphorus in assessing the nutritional status of sugarcane using the comprehensive integrated diagnosis and recommendation system (DRIS). J Plant Nutr. 2021;44:9299. https://doi.org/10.1080/01904167.2020.1849299.

Souza DMG, Carvalho LJCB. Nutrição mineral de plantas. In: Goedert WJ, editor. Solos dos cerrados: tecnologias e estratégias de manejo. Planaltina: Embrapa; 1985. p. 75–98.

Sumner ME. A new approach for predicting nutrient needs for increased crop yields. Solutions. 1978;22:68–78.

Teixeira MP, Campos CNS, Prado RM, et al. Microwave drying of plant tissue for nutritional analysis of *Corymbia citriodora* (Hook.) and *Hevea brasiliensis* Muell. Arg. Agrociencia. 2017;51:555–60.

Traspadini EIF, Prado RM, Vaz GJ, et al. Guia prático para aplicação do método da diagnose da composição nutricional (CND): exemplo de uso na cultura da cana-de-açúcar. Campinas: EMBRAPA Informática Agropecuária; 2018. 30p.

Villaseñor D, Prado RM, Silva GP, et al. DRIS norms and limiting nutrients in banana cultivation in the South of Ecuador. J Plant Nutr. 2020;43:2785–2796. https://doi.org/10.1080/0190416 7.2020.1793183.

Wadt PGS, Anghinoni I, Guindani RHP, et al. Nutrition standards for flooded rice by the methods compositional nutrient diagnosis and mathematical chance. Rev Bras Ciênc Solo. 2013;37:145–56. https://doi.org/10.1590/S0100-06832013000100015.

Wadt PGS, Traspadini EIF, Martins RA, et al. Medidas de acurácia na qualificação dos diagnósticos nutricionais: teoria e prática. In: Prado RM. Jaboticabal: Cecílio Filho AB Nutrição e adubação de hortaliças. FCAV/UNESP/CAPES; 2016. p. 373–91.

Walworth JL, Summer ME. The diagnosis and recommendation integrated systems (DRIS). In: Stewart BA, editor. Advances in soil sciences. New York: Spring; 1987. p. 149–88.

Chapter 20
Interactions Between Nutrients

Keywords Antagonism · Synergism · K/N ratio · P/N ratio · K/Ca ratio · P/Zn ratio

Knowledge of the interaction between nutrients indicates the relationships between nutritional balance and the high productivity of crops. This chapter will address (i) studies on the most common interactions, (ii) relationships between nutrients in leaf analysis: N and K; N and S; K, Ca and Mg; Mg and Mn/Zn; S and Mo; N and P; and others, and (iii) final considerations

20.1 Studies on the Most Common Interactions

The interactions that take place between nutrients are of a very complex nature. Their effects reflect in the mineral composition of plants.

Although the nutrient absorption process is specific and selective, there is sometimes competition between them due to their similarity (ionic radius and charge), as both nutrients probably share the same transporter either through the specific ATPase, a coupled transport system, or a cotransport. Interaction is the influence or reciprocal action of one nutrient over the other in relation to the growth of plants. It is the differential response of a nutrient in combination with several levels (doses) of a second nutrient applied simultaneously (Olsen 1972). Gama (1977) stated that the antagonistic and synergistic effects between elements vary according to their proportion, species, cultivars, and plant development stage.

During the search for the maximum production of crops, an adequate content of the nutrients alone is important. However, the balance between nutrients in the soil-plant system also becomes a fundamental limiting factor. This is because the interactions between the elements in plant nutrition affect processes that occur in the soil, such as the contact between the nutrient and the root, and in the plant, such as the processes of absorption, transport, redistribution and metabolism, which may induce nutritional disorders either by deficiency or toxicity, with a consequent effect on crop production. Thus, chemical, physical, or biological changes in the soil and in the cultivated species may influence interactions.

© Springer Nature Switzerland AG 2021
R. de Mello Prado, *Mineral nutrition of tropical plants*,
https://doi.org/10.1007/978-3-030-71262-4_20

Thus, knowing such positive or negative interactions between nutrients (Table 20.1) allows for adjustments in fertilization.

Applying the limiting nutrient or using combined fertilizers: "N-P" and "N-K" increase fertilizer efficiency (use rate by the plant of the nutrient applied) (Malavolta et al. 1997).

Interactions between nutrients are usually classified into three types:

Antagonism The presence of one nutrient decreases the absorption of another, despite being at adequate or high concentrations in the soil. Decreased absorption can prevent toxicity, for example: Ca^{2+} prevents the excessive absorption of Cu^{2+}. This effect occurs when the concentration of one cation in the medium greatly increases and the absorption of another cation decreases, so that the plant seeks to keep the total positive charges constant.

Inhibition Consists in reducing the absorption of a nutrient; this is caused by the presence of another ion, which can be of two types: competitive and noncompetitive.

Competitive inhibition occurs when two elements compete for a same carrier site, decreasing the absorption of the nutrient with the lowest concentration. The transporter system cannot thus distinguish the two ions, for example: K^+ and Rb^+, SeO_4^{2-} and SO_4^{2-}, Zn^{2+}, and Ca^{2+}. Thus, this phenomenon can be corrected by increasing the concentration of the deficient nutrient in the soil solution.

Non-competitive inhibition occurs when binding occurs at different sites. There is an effective reduction in Vmax. Thus, the concentration of the nutrient does not affect its occurrence. Example: K^+ inhibits Mg^{2+} and Ca^{2+} absorption, $H_2PO_4^-$ inhibits Zn^{2+} absorption; and Mg^{2+} inhibits Zn^{2+} absorption.

Table 20.1 Interactions between the most common nutrients in cultivated plants

Nutrient applied	Effect on leaf content of												
	N	P	K	Ca	Mg	S	B	Cl	Cu	Fe	Mn	Mo	Zn
N	+		−	+		−	−						
P		+	−	+		−	−	−				+	−
K			+	−	−								
Ca			−	+	−					−			
Mg		+	−	−	+					−			−
S	−					+	−						
B							+						−
Cl						−		+					+
Cu									+	−	−	−	−
Fe									−	+	−		
Mn			−							−	+		−
Mo										−		+	
Zn		−											+

Synergism The presence of a given element increases the absorption of another nutrient: Example: Ca^{2+}, at not very high concentrations, increases the absorption of cations and anions due to its role in maintaining the integrity of plasmalemma, which affects the fertilization method.

The noninteractive effects can, for different reasons, cause dilution or concentration of nutrients by changing the chemical composition of plants. The dilution effect occurs in plants with moderate or severe deficiency, in which the application of the nutrient increases the dry matter and promotes nutrient dilution, not increasing its concentration in the leaf. The concentration effect occurs in extreme environmental situations, such as cold or water deficit. It slows growth and promotes the concentration of certain nutrients in the plant tissue.

20.2 Relationships Between Nutrients in Leaf Analysis

N × K Interaction

The N and K interaction obeys the law of the minimum, because when N is applied in a sufficient quantity to increase production, it becomes limited by the low levels of applied K (Dibb and Thompson 1985). Thus, the highest doses of nitrogen will only promote the highest production if accompanied by high doses of potassium (Fig. 20.1). A low N/K ratio (about 4) exerts the greatest positive effects on corn production (Fig. 20.2).

The N and K interaction is common (Faizy 1979; Bitzer 1982), as it promotes an increase in the use efficiency of N in the presence of K, consequently increasing production (Table 20.2).

The appropriate N and K ratio, in addition to an increased production, can result in other benefits:

1. Reduces lodging in corn (Usherwood 1982). The excess of N without application of K promotes plant lodging (Table 20.3). It is known that K may lead to the accumulation of a greater amount of carbohydrates in stems, benefiting the production of structural organic compounds in stems.
2. It increases the quality of grains, as there is a correlation between potassium and protein content in soybeans, beans, cotton, and corn (Blevins 1985). A possible explanation for the fact that crops with a high protein content require (and export) a large amount of K through grains is the involvement of K in the transport of N for protein synthesis.

Appropriate N and K ratios may vary depending on the culture cycle. In this sense, Adams and Massey (1984) observed that, from the beginning of tomato fruiting, this relationship changes dramatically. Aiming to maximize productivity and obtain a better fruit quality and a greater resistance to diseases, the authors suggest a K/N ratio of 1.2/1 at the vegetative stage and 2.5/1 at the reproductive stage.

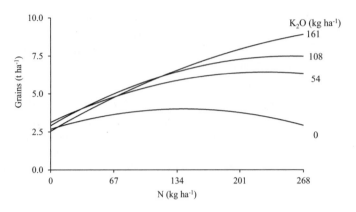

Fig. 20.1 Effects of nitrogen and potassium doses on corn production

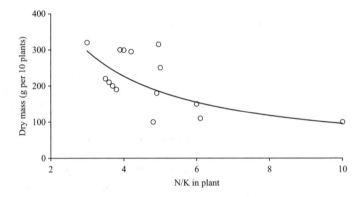

Fig. 20.2 Leaf N/K ratio and corn production

Farinelli et al. (2004) studied the response of rice to the application of N and K and observed a significant interaction. The best combination of doses was around 65 kg ha^{-1} of N and 20 kg ha^{-1} of K$_2$O, resulting in a higher productivity and crude protein per hectare.

Natale et al. (2006) observed that the highest dry matter production of passion fruit seedlings was associated with 44 g kg^{-1} of N and 18 g kg^{-1} of K (shoot), and 33 g kg^{-1} of N and 34 g kg^{-1} of K (roots).

Studies have indicated the positive effects of N especially in the ammoniacal form, which cause increases in the absorption of P. Two theses can explain this fact:

- P available in the soil: the absorption of N in the ammoniacal form decreases the pH of the rhizosphere, which in turn may increase the availability of P (Fig. 20.3). In addition, the absorption process itself is favored at low pH in the cells of the epidermis.
- P transport in the plant: there is an increase in the absorption and transport of P in the plant, since ammonium increases the dissociation rate of the phosphate-carrier complex in the xylem, increasing the concentrations of P in shoots.

Table 20.2 Effects of potassium doses on nitrogen nutrition efficiency in corn plants

	Production	N-efficiency	N-total absorbed
kg ha^{-1}	t ha^{-1}	kg grain/100 kg N	kg ha^{-1}
0	8.4		194
67	8.8		204
134	10.4		240

Table 20.3 Effects of potassium and nitrogen on lodging rate of the corn crop

K$_2$O (kg ha^{-1})	N (kg ha^{-1})		
	0	90	180
	Lodging rate (%)		
0	9	57	59
75	4	3	8
150	4 (3.7 t ha^{-1})	4 (7.7 t ha^{-1})	4 (8.1 t ha^{-1})

Studies have indicated that N has a synergistic effect on the P content in the leaf tissue and vice versa, and that there is a positive interactive effect on grain production: a product of N and P of 1.0 in the leaf tissue results in the highest production (Sumner and Farina 1986).

N and S Interaction

As already observed, N and S are two basic nutrients for protein synthesis. An inadequate supply of one of these nutrients causes an imbalance, resulting in damage to the harvested product and to production. This may occur with high doses of N with no application of S. Therefore, S is important to increase dry mass, and protein N is important to N reduction (non-protein) (Stewart and Porter 1969). The maximum production of corn dry matter was obtained with a N/S ratio close to 11 (Fig. 20.4).

K, Ca, Mg Interactions

The increase in K doses causes a decrease in Ca and Mg levels (Loué 1963), which, at extreme doses, decreases production. Leaf contents of K close to 1.8% provide an acceptable decrease in leaf Ca and Mg, resulting in a satisfactory production (Fig. 20.5). However, higher levels of K decrease the levels of leaf Ca and Mg, so that it should be avoided in the management of potassium fertilization. Meanwhile, the increase in Mg concentration in the solution does not affect K absorption (Fonseca and Meurer 1997). It is also widely known that Ca and Mg in the soil solution are antagonistic, that is, the excess of one impairs the absorption of the other (Moore et al. 1961).

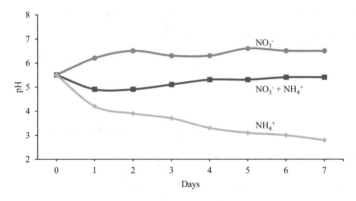

Fig. 20.3 Variations in the pH of the external solution when sorghum was supplied exclusively or combined with ammonium and nitrate (N-total = 300 mg/L)

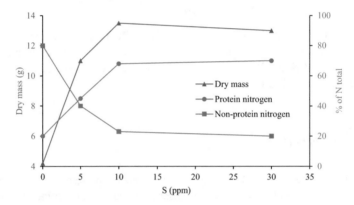

Fig. 20.4 Dry matter production and N distribution in shoots of corn plants 35 days after germination as a function of S

The preferential absorption of K is because it is a monovalent ion with a lower degree of hydration compared to divalent ones. The high Ca content in relation to K (low K/Ca ratio) can also be associated with a low production in eucalyptus (Fig. 20.6). The K/Ca leaf ratio close to 1.5 provides a production close to the maximum.

The harvest response by applying Mg is maximized when the K level is high or adequate (Malavolta et al. 1980).

It is common to associate Zn deficiency to high levels of P available in the soil or to fertilizers with a high phosphate content. A classic experiment in the literature with factorial N and Zn indicates that a low P/Zn ratio is associated with a low availability of P in the soil and leaves, which limits production. Intermediate relations occur when both nutrients are applied to the soil, resulting in a high grain production. Finally, a high P/Zn ratios are obtained with high doses of P, without the corresponding application of Zn, leading to Zn deficiency, which results in a decrease in production (Sumner and Farina 1986) (Fig. 20.7).

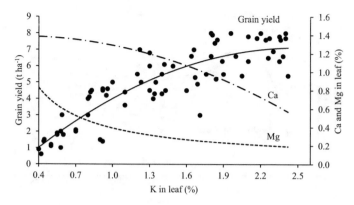

Fig. 20.5   Relationship between K, Ca, and Mg leaf contents in corn crops

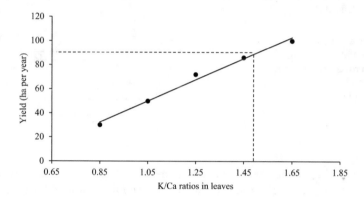

Fig. 20.6   Effects of the K/Ca ratio in eucalyptus leaves on productivity

The explanation for the Zn deficiency at high P/Zn ratios may be the precipitation reactions of P and Zn in the conductor vessels. They reduce the transport of Zn to shoots, or even cause a metabolic disorder by the imbalance between both nutrients.

There are some indications that potassium may favor the absorption of P and Zn, thus decreasing the intensity of the interaction P x Zn (Adriano et al. 1971).

Mg × Mn/Zn Interactions

The Mg interaction with Mn or Zn in plants is known (Moreira et al. 2003). In experiments with detached soybean roots using radioisotopes, the increase in Mg doses decreased the absorption of Mn and Zn by noncompetitive inhibition (Table 20.4).

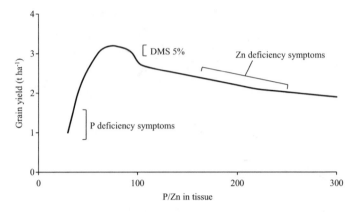

Fig. 20.7 P/Zn relationship in leaf tissue and corn grain production. The size of the circles indicates Zn deficiency

Table 20.4 Absorption of
^{65}Zn and ^{54}Mn by detached
soybean roots (means of four
cultivars) at different
doses of Mg

Mg	Zn	Mn
Mmol L^{-1}	% reduction in absorption	
0	–	
3.0	63	32
6.0	72	60

^{a65}Zn and ^{54}Mn = 2 µmol L^{-1}

Given the importance of the relationship between nutrients and crop production
(Malavolta 1996), there are indications of adequate relationships for some crops
such as coffee (Table 20.5).

S × Mo Interactions

The sulfate anion can decrease the absorption of the molybdate anion. This decrease
in the absorption of Mo may affect the plant's metabolism, causing deficiency of
N. Thus, fertilization with sulfur as gypsum should be used with care to not induce
nutritional disturbance in the plant by either deficiency of Mo or even of N.

N × P Interactions

The N x P relationship in cereals regarding a higher production was close to 7
(Duivenbooden et al. 1996). Specifically for rice, Mkamilo (2004) verified a nutritional balance with a N and P ratio equal to 5.6. We verified in several cultures that

Table 20.5 Relationships between leaf nutrients considered adequate for coffee

Relationship	Range	Relationship	Range
N/P	16–18	P/Cu	125–187
N/K	1.3–1.4	P/Zn	125–187
N/S	16–18	Ca/Mg	66–75
K/Ca	1.7–2.1	B/Zn	5.0–7.3
K/Mg	6.1–6.6	Cu/Zn	1
N/B	400–457	Fe/Mn	0.73–0.85
N/Cu	2000–3375		

the N and P ratio is important to explain the absorption of P, obtaining an adequate ratio of 5.6 and 8.7 for cereals and legumes, respectively.

A possible competition between the phosphate and nitrate anions is a factor of less nitrogen absorption under conditions of high concentrations of P in the substrate of cultivation of citrus rootstocks at the sowing phase (Fontanezzi 1989).

Other Interactions

Other interactions between nutrients can occur; however, they are little studied. The P x B interaction can occur because a low level of P may interfere with the metabolism of B, worsening the symptoms of both B deficiency and excess in *Brassica campestris* L. (Sinha et al. 2003).

The Cl^- may affect the $N-NO_3^-$ absorbed by plants either through carrier sites or through ion channel locations (Malavolta 2006).

20.3 Final Considerations

Soil science, referring to plant nutrition, is relatively new, and therefore, there are important challenges to the management of crop nutrition regarding achieving a maximum economic production with a rational use of nutrients and respecting the environment.

For this, two simultaneous actions are needed in the scope of extension activities and research. The first is to encourage the use of chemical analysis of leaves by farmers as a routine tool for managing nutrition and fertilizing crops. The second is to intensify basic and applied research in plant nutrition, such as (a) expanding the list of plant nutrients, (b) create knowledge on the mechanisms of absorption, transport, and redistribution of nutrients in plants, (c) foster interdisciplinary studies addressing the interface between soil microbiology (N-fixing microorganisms, phosphate solubilizers, among others) and phytopathology with plant nutrition, (d) use molecular biology to code genes responsible for the transport of solutes and nutrients through membranes, tolerance to excess of Al and heavy metals or the deficiency of N, P, and K in the selection of plants adapted to the soil, (e) study the nutrient use efficiency by plants, (f) explain the role of nutrition in obtaining high-quality agricultural products, (g) establish adequate levels of leaf nutrients and leaf sampling criteria in crops that have been little studied in Brazilian conditions, and (h) establish adequate nutrient management based on new agricultural trends, such as no-tillage, fertigation, protected cultivation, among others.

References

Adams P, Massey DM. Nutrient uptake by tomatoes from recirculating solutions. In: Abstracts of the 6rd international congress on soilless culture, International Society for Soilless, Wageningen Culture; 1984. p. 71.

Adriano DC, Paulsen GM, Murphy LS. Phosphorus-iron and phosphorus-zinc relationships in corn seedlings as affected by mineral nutrition. Agron J. 1971;63:56–9. https://doi.org/10.2134/agronj1971.00021962006300010013x.

Bitzer M. No-till corn highest yield with nitrogen and potassium. Better Crops. 1982;67(19)

Blevins DG. Role of potassium in protein metabolism in plants. In: Mund-Son RD, editor. Potassium in agriculture. ASA/CSSA/SSSA. Madison; 1985. p. 413–24.

Dibb DW, Thompson WR Jr. Interactions of potassium with other nutrients. In: Abstracts of the international symposium of potassium in agriculture. Atlanta: American Society of Agronomy; 1985. p. 515.

Duivenbooden NV, Wit CT, Van Keulen H. Nitrogen, phosphorus and potassium relations in five major cereals reviewed in respect to fertilizer recommendations using simulation modelling. Fertil Res. 1996;44:37–49.

Faizy SEDA. N-K interaction and and the net influx of ions across corn roots as affects by different NPK fertilizers. In: Colloquim of the international Potash Institute, 14. Sevilla. Proceedings; 1979.

Farinelli R, Penariol FG, Fornasieri Filho D, et al. Effects of nitrogen and potassium fertilization on agronomic characteristics of upland rice cultivated under no-tillage. Rev Bras Ciênc Solo. 2004;28:447–54. https://doi.org/10.1590/S0100-06832004000300006.

Fonseca JA, Meurer EJ. Inibição da absorção de magnésio pelo potássio em plântulas de milho em solução nutritiva. Rev Bras Ciênc Solo. 1997;21:47–50.

Fontanezzi GB da S. Efeito de micorriza vesicular-arbuscular e de superfosfato simples no crescimento e nutrição de porta-enxertos de citros. Dissertação, Universidade Federal de Lavras. 1989.

Gama MV. Efeitos do azoto e do potássio na composição mineral do trigo "Impeto" e do tomate "Roma". Agron Lusit. 1977;38:111–21.

Loué A. Contribuição para o estudo da nutrição catiônica do milho, principalmente a do potássio. Fertilité. 1963;20:1–57.

Malavolta E. Informações agronômicas sobre nutrientes para as culturas – Nutri-fatos. Piracicaba: Potafos; 1996.

Malavolta E. Manual de nutrição mineral de plantas. São Paulo: Agronômica Ceres; 2006.

Malavolta E, Vidal AA, Gheller AC, et al. Effects of the deficiencies of macronutrients in two soybean (Glycine max L. Merr.) varieties, Santa Rosa and UFV- grown in nutrient solution. An Esc Super Agric Luiz de Queiroz. 1980;37:473–84. https://doi.org/10.1590/S0071-12761980000100030.

Malavolta E, Vitti GC, Oliveira SA. Avaliação do estado nutricional das plantas: princípios e aplicações. Piracicaba: Associação Brasileira de Potassa e do Fósforo; 1997. 319p

Mkamilo GS. Maize–sesame intercropping in southeast tanzania: farmers' practices and perceptions, and intercrop performance. Tropical resource management Papers, No. 54. Wageningen University, Wageningen. (2004)

Moore DP, Overstreet R, Jacobson L. Uptake of magnesium & its interactions with calcium in excised barley roots. Plant Physiol. 1961;36:290–5. https://doi.org/10.1104/pp.36.3.290.

Moreira A, Malavolta E, Heinrichs R, et al. Magnesium influence on manganese and zinc uptake by excised roots of soybean. Pesq Agropec Bras. 2003;38:95–101. https://doi.org/10.1590/S0100-204X2003000100013.

Natale W, Prado RM, Almeida EV, et al. Nitrogenous and potassic fertilization on nutrition al status of yellow passion fruit seedlings. Acta Sci. 2006;28:187–92. https://doi.org/10.4025/actasciagron.v28i2.1036.

Olsen SR. Micronutrients interactions. In: Montverdt JJ, Gior-Dano PM, Lindsay WL, editors. Micronutrients in agriculture. Soil Science of America Monographs: Madison; 1972. p. 243–88.

Sinha P, Dube BK, Chatterjee C. Phosphorus stress alters boron metabolism of mustard. Commun Soil Sci Plant Anal. 2003;34:315–26. https://doi.org/10.1081/CSS-120017823.

Stewart BA, Porter LK. Nitrogen sulfur relationships in wheat (*Triticum aestivum* L.), corn (*Zea mays*), and beans (*Phaseolus vulgaris*). Agron J. 1969;61:267–71. https://doi.org/10.2134/agronj1969.00021962006100020027x.

Sumner ME, Farina MPW. Phosphorus interaction with other nutrients and lime in field cropping systems. Adv Soil Sci. 1986;5:201–36. https://doi.org/10.1007/978-1-4613-8660-5_5.

Usherwood NR. Interação do potássio com outros íons. In: Resumo do simpósio sobre potássio na agricultura brasileira. Piracicaba: Instituto da Potassa; 1982. p. 227.

Index

Printed in the United States
by Baker & Taylor Publisher Services